SMART CARD RESEARCH AND ADVANCED APPLICATIONS VI

IFIP – The International Federation for Information Processing

IFIP was founded in 1960 under the auspices of UNESCO, following the First World Computer Congress held in Paris the previous year. An umbrella organization for societies working in information processing, IFIP's aim is two-fold: to support information processing within its member countries and to encourage technology transfer to developing nations. As its mission statement clearly states,

> *IFIP's mission is to be the leading, truly international, apolitical organization which encourages and assists in the development, exploitation and application of information technology for the benefit of all people.*

IFIP is a non-profit making organization, run almost solely by 2500 volunteers. It operates through a number of technical committees, which organize events and publications. IFIP's events range from an international congress to local seminars, but the most important are:

- The IFIP World Computer Congress, held every second year;
- Open conferences;
- Working conferences.

The flagship event is the IFIP World Computer Congress, at which both invited and contributed papers are presented. Contributed papers are rigorously refereed and the rejection rate is high.

As with the Congress, participation in the open conferences is open to all and papers may be invited or submitted. Again, submitted papers are stringently refereed.

The working conferences are structured differently. They are usually run by a working group and attendance is small and by invitation only. Their purpose is to create an atmosphere conducive to innovation and development. Refereeing is less rigorous and papers are subjected to extensive group discussion.

Publications arising from IFIP events vary. The papers presented at the IFIP World Computer Congress and at open conferences are published as conference proceedings, while the results of the working conferences are often published as collections of selected and edited papers.

Any national society whose primary activity is in information may apply to become a full member of IFIP, although full membership is restricted to one society per country. Full members are entitled to vote at the annual General Assembly, National societies preferring a less committed involvement may apply for associate or corresponding membership. Associate members enjoy the same benefits as full members, but without voting rights. Corresponding members are not represented in IFIP bodies. Affiliated membership is open to non-national societies, and individual and honorary membership schemes are also offered.

SMART CARD RESEARCH AND ADVANCED APPLICATIONS VI

IFIP 18th World Computer Congress
TC8/WG8.8 & TC11/WG11.2 Sixth International Conference on
Smart Card Research and Advanced Applications (CARDIS)
22–27 August 2004
Toulouse, France

Edited by

Jean-Jacques Quisquater
UCL, Louvain-la-Neuve, Belgium

Pierre Paradinas
CNAM, Paris, France

Yves Deswarte
LAAS-CNRS, Toulouse, France

Anas Abou El Kalam
LAAS-CNRS, Toulouse, France

SPRINGER SCIENCE+BUSINESS MEDIA, LLC

Library of Congress Cataloging-in-Publication Data

A C.I.P. Catalogue record for this book is available from the Library of Congress.

Smart Card Research and Advanced Applications VI
Edited by Jean-Jacques Quisquater, Pierre Paradinas, Yves Deswarte, and
Anas Abou El Kalam
ISBN 978-1-4757-8010-9   ISBN 978-1-4020-8147-7 (eBook)
DOI 10.1007/978-1-4020-8147-7

Printed on acid-free paper.

Contents

Preface

This CARDIS is special in the list of all CARDIS conferences, nevertheless it follows the tradition where every 2 years the scientific smart card world from academic research organizations meet together with technologists from industries.

This year, CARDIS is celebrating its 10th anniversary, and the pioneers are less isolated now. More universities, more research and industrial centers have set up or launched research activities.

For this edition we received 45 papers and we selected 20 papers, with 20% from Asia or America and 60% from Europe. In contrast with the first editions where we received "marketing product presentation" – of course rejected – the quality across the years is more and more valuable, and CARDIS is renewed as "The conference" in smart card technology.

CARDIS'1994 was in Lille (France), CARDIS'1996 in Louvain-La-neuve (Belgium), CARDIS'1998 in Amsterdam (The Netherlands), CARDIS'2000 Bristol (UK), CARDIS'2002 San Jose (California) and now it returns to France and takes place within the IFIP 18th World Computer Congress. This may be also viewed as progress in the visibility of smart card technology and research.

The program committee was very accurate and fair during the review process, and the papers from the program committee members were reviewed more severely by more reviewers. The PC meeting hosted by CNAM (Paris) was very productive and at the end of the day the program main stream was set up.

We hope you will enjoy the conference as much as we liked to review and prepare the program of CARDIS'2004.

Jean-Jacques Quisquater (General Chair)
Pierre Paradinas (Program Chair)
Yves Deswarte (Local Chair)

Acknowledgements

Conference General Chair:

    Jean-Jacques Quisquater, UCL, Louvain-la-Neuve, Belgium

Local Organization Chair:

    Yves Deswarte, LAAS-CNRS, Toulouse, France

Program Committee Chair:

    Pierre Paradinas, CNAM, Paris, France

Program Committee:

    Boris Balacheff, Hewlett-Packard Labs, UK
    Edouard de Jong, Sun Microsystems, USA
    Yves Dewarte, LAAS-CNRS, France
    Josep Domingo-Ferrer, Universitat Rovira i Virgili, Spain
    Jean-Bernard Fischer, OCS, France
    Gilles Grimaud, RD2P, France
    Pieter Hartel, University of Twente, Netherlands
    Peter Honeymann, University of Michigan, USA
    Dirk Husemann, IBM Research, Switzerland
    Jean-Louis Lanet, INRIA-DirDRI, France
    Xavier Leroy, INRIA & Trusted Logic, France
    Mike Montgomery, Schlumberger, USA
    Erik Poll, Nijmegen University, Netherlands
    Joachim Posegga, SAP-Corporate Research, Germany
    Jean-Jacques Quisquater, UCL, Belgium
    Jean-Jacques Vandewalle, Gemplus Labs, France
    Serge Vaudenay, EPFL, Switzerland
    J. Verschuren, TNO-EIB, The Netherlands
    Tim Wilkinson, Hive Minded, USA

Additional Referees:

Gildas Avoine

Thomas Baignères

Claude Barral

Asker Bazen

Rob Bekkers

Arnaud Boscher

Richard Brinkman

Jordi Castella-Roca

Liqun Chen

Chong Cheun Ngen

Alessandro Coglio

Ricardo Corin

Bert den Boer

Eric Deschamps

Jean-François Dhem

Christophe Giraud

Dieter Gollman

Jochen Haller

Keith Harrison

Engelbert Hubbers

Dieter Hutter

Mounir Idrassi

Pierre Jansen

Marc Joye

Pascal Junod

Caroline Kudla

Laurent Lagosanto

Yee Wei Law

Gabriele Lenzini

Yi Lu

Antoni Martínez Ballesté

Jean Monnerat

Christophe Muller

Martijn Oostdijk

Gilles Piret

David Plaquin

Joseph R. Kiniry

Jim Rees

Philip Robinson

Peter Schmitt

Berry Schoenmakers

Francesc Sebé

Jan Seedorf

David Simplot-Ryl

François-Xavier Standaert

Susan Thompson

David Ware

Martijn Warnier

The original version of this book was revised.
An erratum to this book can be found at DOI 10.1007/978-1-4020-8147-7_21

ENFORCING HIGH-LEVEL SECURITY PROPERTIES FOR APPLETS

Mariela Pavlova[1], Gilles Barthe[1], Lilian Burdy[1], Marieke Huisman[1] and Jean-Louis Lanet[2]
[1]*INRIA Sophia Antipolis, France and* [2]*INRIA Dir DRI, France*

Abstract   Smart card applications often handle privacy-sensitive information, and therefore must obey certain security policies. Typically, such policies are described as high-level security properties, stating for example that no pin verification must take place within a transaction.

Behavioural interface specification languages, such as JML (Java Modeling Language), have been successfully used to validate functional properties of smart card applications. However, high-level security properties cannot directly be expressed in such languages. Therefore, this paper proposes a method to translate high-level security properties into JML annotations. The method synthesises appropriate annotations and weaves them throughout the application. In this way, security policies can be validated using existing tools for JML. The method is general and applies to a large class of security properties.

To validate the method, it has been applied to several realistic examples of smart card applications. This allowed us to find violations against the documented security policies for some of these applications.

Keywords: Smart devices, security, specification, verification

1.   Introduction

Nowadays, most efforts in smart card security focus on adequate countermeasures against hardware attacks. However, logical attacks, caused by *e.g.* illegal control flow or uncaught exceptions, form a new major threat for security and privacy. An example of such an attack is a malicious GSM applet that performs illegal calls to the method sendSMS.

To ensure user confidence, smart card application providers therefore have to guarantee the dependability of their software. This can be achieved by following certification procedures, such as "Common Criteria[1]", focusing on security aspects. But such procedures are relatively heavy, and they are also concerned with aspects unrelated to software security. Therefore, industry often prefers to do a more lightweight analysis or software audit.

Such an analysis typically consists in a manual deep code review, for which no tool support is available. Therefore, this is a costly procedure, and there is no formal guarantee of its results. The quality of this analysis can be improved by using program verification techniques. Therefore, industry is investigating how these techniques can be used to provide high quality software. For example, in the context of smart cards, program verification has been successfully used to verify functional properties of applications, discovering subtle programming errors that remain undetected by intensive testing [3, 5].

Unfortunately, the cost of employing program verification techniques remains an important obstacle for most industrials. Our experiences, which are confirmed by two recent road-maps for smart card research[2], show that the difficulty of learning a specification language whose internals may be obscure to programmers, and the large amount of work required to formally specify and verify applications constitute major obstacles to the use of program verification techniques in industry. Therefore, recent work on formal methods for Java and Java Card[3] tries to tackle these problems.

To reduce the difficulty of learning a specification language, the Java Modeling Language (JML)[4] [6] has been designed as an easily accessible specification language. It uses a Java-like syntax with some specification-specific keywords added. JML allows developers to specify the properties of their program in a generalisation of Hoare logic, tailored to Java. By now, it has been generally accepted as *the* behavioural interface specification language for Java (Card).

While JML is easily accessible to Java developers, actually writing the specifications of a smart card application is labour-intensive and error-prone, as it is easy to forget some annotations. There exist tools which assist in writing these annotations, *e.g.* Daikon [11] and Houdini [12] use heuristic methods to produce annotations for simple safety and functional invariants. However, these tools cannot be guided by the user—they do not require any user input—and in particular cannot be used to synthesise annotations from realistic security policies.

The main contribution of this paper is a method that, given a security policy, automatically annotates a Java (Card) application, in such a way that if the application respects the annotations then it also respects the security policy. The generation of annotations proceeds in two phases: synthesising and weaving.

1 Based on the security policy we *synthesise* core annotations, specifying the behaviour of the methods directly involved.

2 Next we propagate these annotations to all methods directly or indirectly invoking the methods that form the core of the security policy, thus *weaving* the security policy throughout the application.

The need for such a propagation phase stems from the fact that we are interested in doing static verification. We need tool support for the propagation, because a typical security property may involve methods from different classes, as illustrated below. The annotations that we generate all use JML static ghost variables: special specification-only variables, that can be modified via a special ghost-assignment annotation. Since we use only static ghost variables, the properties are independent of the particular class instances available.

The annotations we generate can be checked with existing verification tools *e.g.* JACK (Java Applet Correctness Kit) [7], Jive [17], Krakatoa [15], Loop [2] and ESC/Java [14]. We use JACK as it provides the best compromise between soundness, efficiency, scalability and usability.

To show the usefulness of our approach, we applied the algorithm to several realistic examples of smart card applications. When doing this, we actually found violations against the security policies documented for some of these applications.

This paper is organised as follows. Section 2 introduces several typical high-level security properties. Next, Section 3 presents the process to weave these properties throughout applications. Subsequently, Section 4 discusses the application of our method to realistic examples. Finally, Sections 5 and 6 present related work and draw conclusions.

2. High-level Security Properties for Applets

Over the last years, smart cards have evolved from proprietary into open systems, making it possible to have applications from different providers on a single card. To ensure that these applications cannot damage the other applications or the card itself, strict security policies—expressed as high-level security properties—must be obeyed. Such properties are high-level in the sense that they have impact on the whole application and are not restricted to single classes. Below we will present several examples. It is important to notice that we restrict our attention to source code-level security of applications.

The properties that we consider can be divided in several groups, related to different aspects of smart cards. First of all there are properties dealing with the so-called *applet life cycle*, describing the different phases that an applet can be in. Many actions can only be performed when an applet is in a certain phase. Second, there are properties dealing with the transaction mechanism, the Java Card solution for having atomic updates. Further there are properties restricting the kind of exceptions that can occur, and finally, we consider properties dealing with access control, limiting the possible interactions between different applications. For each group we present some example properties. For all these properties encodings into JML annotations exist.

We would like to emphasise that there exist many more relevant security properties for smart cards, for example specifying memory management, information flow and management of sensitive data. Identifying all relevant security properties for smart cards, and expressing them formally, is an important ongoing research issue.

Applet life cycle. A typical applet life cycle defines phases as *loading, installation, personalisation, selectable, blocked* and *dead* (see *e.g.* [16]). Each phase corresponds to a different moment in the applet's life. First an applet is loaded on the card, then it is properly installed and registered with the Java Card Runtime Environment. Next the card is personalised, *i.e.* all information about the card owner, permissions, keys *etc.* is stored. After this, the applet is selectable, which means that it can be repeatedly selected, executed, and deselected. However, if a serious error occurs, for example there have been too many attempts to verify a pin code, the card can get blocked or even become dead. From the latter state, no recovery is possible.

In many of these phases, restrictions apply on who can perform actions, or on which actions can be performed. These restrictions give rise to different security properties, to be obeyed by the applet.

> **Authenticated initialisation** Loading, installing and personalising the applet can only be done by an authenticated authority.
>
> **Authenticated unblocking** When the card is blocked, only an authenticated authority can execute commands and possibly unblock it.
>
> **Single personalisation** An applet can be personalised only once.

Atomicity. A smart card does not include a power supply, thus a brutal retrieval from the terminal could interrupt a computation and bring the system in an incoherent state. To avoid this, the Java Card specification prescribes the use of a transaction mechanism to control synchronised updates of sensitive data. A statement block surrounded by the methods `beginTransaction()` and `commitTransaction()` can be considered atomic. If something happens while executing the transaction (or if `abortTransaction()` is executed), the card will roll back its internal state to the state before the transaction was begun.

To ensure the proper functioning and prevent abuse of this mechanism, several security properties can be specified.

No nested transactions Only one level of transactions is allowed.

No exception in transaction All exceptions that may be thrown inside a transaction, should also be caught inside the transaction.

Bounded retries No pin verification may happen within a transaction.

The second property ensures that the commitTransaction will always be executed. If the exception is not caught, the commitTransaction would be ignored and the transaction would not be finished. The last property excludes pin verification within a transaction. If this would be allowed, one could abort the transaction every time a wrong pin code has been entered. As this rolls back the internal state to the state before the transaction was started, this would also reset the retry counter, thus allowing an unbounded number of retries. Even though the specification of the Java Card API prescribes that the retry counter for pin verification cannot be rolled back, in general one has to check this kind of properties.

Exceptions. Raising an exception at the top level can reveal information about the behaviour of the application and in principle it should be forbidden. However, sometimes it is necessary to pass on information about a problem that occurred. Therefore, the Java Card standard defines so-called ISO exceptions, where a pre-defined status word explains the problem encountered. These exceptions are the only exceptions that may be visible at top-level; all other exceptions should be caught within the application.

Only ISO exceptions at top-level No exception should be visible at top-level, except ISO exceptions.

Access control. Another feature of Java Card is an isolation mechanism between applications: the firewall. The firewall ensures that several applications can securely co-exist on the same card, while managing limited collaboration between them: classes and interfaces defined in the same package can freely access each other, while external classes can only be accessed via explicitly shared interfaces. Inter-application communication via shareable interfaces should only take place when the applet is selectable, in all other phases of the applet life cycle only authenticated authorities are allowed to access the applet.

Only selectable applications shareable An application is accessible via a shareable interface only if it is selectable.

3. Automatic Verification of Security Properties

As explained above, we are interested in the verification of high-level security properties that are not directly related to a single method or class, but that guarantee the overall well-functioning of an application. Writing appropriate JML annotations for such properties is tedious and error-prone, as they have to be spread all over the application. Therefore, we propose a way to construct such annotations automatically. First we synthesise core-annotations for methods directly involved in the property. For example, when specifying that no nested transactions are allowed, we annotate the methods `beginTransaction`, `commitTransaction` and `abortTransaction`. Subsequently, we propagate the necessary annotations to all methods (directly or indirectly) invoking these core-methods. The generated annotations are sufficient to respect the security properties, *i.e.* if the applet does not violate the annotations, it respects the corresponding high-level security property.

Whether the applet respects its annotations can be established with any of the existing tools for JML. We use JACK [7], which generates proof obligations for different provers, including the AtelierB prover[5] and Simplify[6]. Both are automatic verifiers for first-order logical formulae. Since for most security properties the annotations are relatively simple—but there are many—it is important that these verifications are done automatically, without any user interaction. The results in Section 4 show that for the generated annotations all correct proof obligations can indeed be automatically discharged.

Before presenting the overall architecture of our tool set and outlining the algorithm for propagation of annotations, we briefly present a few JML keywords, that are relevant for the examples presented here.

3.1   JML in a Nutshell

JML [6] uses a Java-like syntax to write predicates, extended with several specification-specific constructs, such as \forall, \exists *etc.* Method specifications consist of preconditions (**requires**), postconditions (**ensures**), and exceptional postconditions (**exsures**), *i.e.* the condition that has to hold upon abnormal termination of a method. We can also specify so-called **assignable** clauses, stating which variables may be modified by a method. Class invariants (keyword **invariant**) describe properties that have to be preserved by each method.

To write more abstract and implementation-independent specifications, JML provides several means of abstraction. One of these are so-called ghost-variables, which are visible only in specifications. Their declaration is preceded by the keyword **ghost**. A special assignment annotation **set** allows to update their value. Using invariants they can be related to concrete variables.

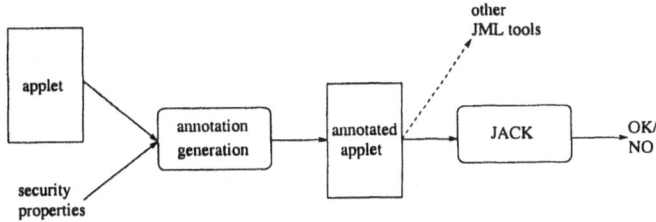

Figure 1. Tool set for verifying high-level security properties

A large class of security properties can be expressed using static ghost variables of primitive type; these are typically used to keep track of the control state of the application (including the ones presented in Section 2). Therefore, here we only study the propagation of annotations containing static ghost variables of primitive type. However, our propagation technique easily can be generalised to concrete (static) variables, as long as we do not have to handle aliasing.

To give an example JML specification, we show a fragment of the core-annotation for the **No nested transactions** property. A static ghost variable TRANS is declared that keeps track of whether there is a transaction in progress. It is initialised to 0, denoting that there is no transaction in progress.

```
/*@ static ghost int TRANS == 0; @*/
```

The method beginTransaction is annotated as follows.

```
/*@ requires TRANS == 0;
  @ assignable TRANS;
  @ ensures TRANS == 1; @*/
public static native void beginTransaction()
                        throws TransactionException;
```

Since the method is native, one cannot describe its body. However, if it had been non-native, an annotation //@ set TRANS = 1; would have been generated, to ensure that the method satisfies its specification.

3.2 Architecture

Figure 1 shows the general architecture of the tool set for verifying high-level security properties. Our annotation generator can be used as a front-end for any tool accepting JML-annotated Java (Card) applications. As input we have a security property and a Java Card applet. The output is a JML Abstract Syntax Tree (AST), using the format as defined for the standard JML parser. When pretty-printed, this AST corresponds to a JML-annotated Java file. From this annotated file, JACK generates appropriate proof obligations to check whether the applet respects the security property.

3.3 Automatic Generation of Annotations

Section 4 presents example core-annotations for some of the security properties presented in Section 2, here we focus on the weaving phase, *i.e.* how the core-annotations are propagated throughout the applet. We define functions mod, pre, post and excpost, propagating assignable clauses, preconditions, postconditions and exceptional postconditions, respectively. These functions have been defined and implemented for the full Java Card language, but to present our ideas, we only give the definitions for a representative subset of statements: statement composition, method calls, conditional and `try-catch` statements and special set-annotations. We assume the existence of domains MethName of method names, Stmt of Java Card statements, Expr of Java Card expressions, and Var of static ghost variables, and functions call and body, denoting a method call and body, respectively.

All functions are defined as mutual recursive functions on method names, statements and expressions. When a method call is encountered, the implementation will check whether annotations already have been generated for this method (either by synthesising or weaving). If not it will recursively generate appropriate annotations. Java Card applets typically do not contain (mutually) recursive method calls, therefore this does not cause any problems. Generating appropriate annotations for recursive methods would require more care (and in general it might not be possible to do without any user interaction).

Propagation of assignable clauses. First we define a function mod that propagates assignable clauses for static ghost variables.

DEFINITION 1 (**mod**) *We define functions* mod: MethName $\rightarrow \mathcal{P}(\mathsf{Var})$, mod: Stmt $\rightarrow \mathcal{P}(\mathsf{Var})$, *and* mod: Expr $\rightarrow \mathcal{P}(\mathsf{Var})$ *by rules like (where* m, n: MethName, s_1, s_2: Stmt, c: Expr *and* x: Var*):*

$$
\begin{aligned}
\mathsf{mod}(m) &= \mathsf{mod}(\mathsf{body}(m)) \\
\mathsf{mod}(s_1; s_2) &= \mathsf{mod}(s_1) \cup \mathsf{mod}(s_2) \\
\mathsf{mod}(\mathsf{call}(n)) &= \mathsf{mod}(n) \\
\mathsf{mod}(\texttt{if } (c)\ s_1\ \texttt{else}\ s_2) &= \mathsf{mod}(c) \cup \mathsf{mod}(s_1) \cup \mathsf{mod}(s_2) \\
\mathsf{mod}(\texttt{try}\ s_1\ \texttt{catch}\ (E)\ s_2) &= \mathsf{mod}(s_1) \cup \mathsf{mod}(s_2) \\
\mathsf{mod}(\texttt{set}\ x = c) &= \{x\}
\end{aligned}
$$

Propagation of preconditions. Next, we define a function pre for propagating preconditions. This function analyses a method body in a sequential way—from beginning to end—computing which preconditions of the methods called within the body have to be propagated. To understand the reasoning behind the definition, we will first look at an example. Suppose we are checking the **No nested transactions** property for an application, which contains a method m, whose only method calls are those shown, and which does not contain any set annotations.

```
void m() { ... // some internal computations
         JCSystem.beginTransaction();
         ... // computations within transaction
         JCSystem.commitTransaction(); }
```

Core-annotations are synthesised for beginTransaction and commit-Transaction. The annotations for beginTransaction are shown in Section 3.1 above, while commitTransaction requires TRANS == 1 and ensures TRANS == 0. As we assume that TRANS is not modified by the code that precedes the call to beginTransaction, the only way the precondition of this method can hold, is by requiring that it already holds at the moment m is called. Thus, the precondition of beginTransaction has to be propagated. In contrast, the precondition for commitTransaction (TRANS == 1) has to be established by the postcondition of begin-Transaction, because the variable TRANS is modified by this method. Thus, preconditions containing only unmodified variables should be propagated. Propagating pre- or postconditions can be considered as passing on a method contract. Method bodies can only pass on contracts for variables they do not modify; once they modify a variable it is their duty to ensure that the necessary conditions are satisfied.

We assume the existence of a domain **Pred** of predicates using static ghost variables only, and function fv, returning the set of free variables.

DEFINITION 2 (**pre**) *We define* pre: MethName $\to \mathcal{P}(\text{Pred})$, pre: Stmt $\to$ $\mathcal{P}(\text{Var}) \to \mathcal{P}(\text{Pred})$, *and* pre: Expr $\to \mathcal{P}(\text{Var}) \to \mathcal{P}(\text{Pred})$ *by rules like (where* m, n : MethName, s_1, s_2 : Stmt, c : Expr, $V: \mathcal{P}(\text{Var})$ *and* x : Var*):*

$$
\begin{aligned}
\text{pre}(m) &= \text{pre}(\text{body}(m), \emptyset) \\
\text{pre}(s_1; s_2, V) &= \text{pre}(s_1, V) \cup \text{pre}(s_2, V \cup \text{mod}(s_1)) \\
\text{pre}(\text{call}(n), V) &= \{p \mid p \in \text{pre}(n) \wedge (\text{fv}(p) \cap V) = \emptyset\} \\
\text{pre}(\text{if } (c) \, s_1 \text{ else } s_2, V) &= \text{pre}(c, V) \cup \text{pre}(s_1, V \cup \text{mod}(c)) \cup \\
& \quad\; \text{pre}(s_2, V \cup \text{mod}(c)) \\
\text{pre}(\text{try } s_1 \text{ catch } (E) \, s_2, V) &= \text{pre}(s_1, V) \cup \text{pre}(s_2, V \cup \text{mod}(s1)) \\
\text{pre}(\text{set } x = c) &= \{\,\}
\end{aligned}
$$

In the rules defining pre on Stmt and Expr, the second argument denotes the set of static ghost variables that have been modified so far. When calculating the precondition for a method, we calculate the precondition of its body, assuming that so far no variables have been modified. For a statement composition, we first propagate the preconditions for the first sub-statement, and then for the second sub-statement, but taking into account the variables modified by the first sub-statement. When propagating the preconditions for a method call, we propagate all preconditions of the called method that do not contain modified variables. Since we are restricting our annotations to expressions containing static ghost variables only, in the rule for the conditional statement we cannot take the outcome of the conditional expression into account. As a consequence, we sometimes generate too strong annotations, but in practice this does not cause problems. Moreover, it should be emphasised that this only can make us reject correct applets, but it will never make us accept incorrect ones. Similarly, for the `try-catch` statement, we always propagate the precondition for the `catch` clause, without checking whether it actually can get executed. Again, this will only make us reject correct applets, but it will never make us accept incorrect ones. Finally, a set annotation does not give rise to any propagated precondition.

Notice that by definition, we have the following property for the function pre (where s is either in Stmt or Expr, and V is a set of static ghost variables).

$$p \in \mathsf{pre}(s, V) \Leftrightarrow (p \in \mathsf{pre}(s, \emptyset) \land (\mathsf{fv}(p) \cap V) = \emptyset)$$

Propagation of postconditions. In a similar way, we define functions post and excpost, computing the set of postconditions and exceptional postconditions that have to be propagated for method names, statements and expressions. The main difference with the definition of pre is that these functions run through a method from the end to the beginning. Moreover, they have to take into account the different paths through the method. For each of these possible paths, we calculate the appropriate (exceptional) postcondition. The overall (exceptional) postcondition is then defined as the disjunction of the postconditions related to the different paths through the method.

Example. For the example discussed above, our functions compute the following annotations.

```
/*@ requires TRANS == 0;
  @ assignable TRANS;
  @ ensures TRANS == 0; @*/
```

```
void m() {
   ... // some internal computations
   JCSystem.beginTransaction();
   ... // computations within transaction
   JCSystem.commitTransaction(); }
```

This might seem trivial, but it is important to realise that similar annotations will be generated for all methods calling m, and transitively for all methods calling the methods calling m *etc.* Having an algorithm to generate such annotations enables to check automatically a large class of high-level security properties.

3.4 Annotation Generation and Predicate Transformer Calculi

A natural question that arises is whether there is a relation between our propagation functions and well-known program transformation calculi as the weakest precondition (wp) and strongest postcondition (sp).

The conceptual difference between our propagation functions and standard program transformation calculi is that, given method m our functions extract a method contract for m, while the program transformation calculi compute the proof obligations that, given all method contracts, allow to decide whether the implementation of m is correct.

A formal relation between pre and (a variant of) the wp-calculus can be established. Since we only consider ghost variables, we need to consider an abstract version of the wp: $\mathsf{wp}^\#$, which does not consider concrete variables. Most rules of this abstract wp-calculus are unchanged, but rules as for the conditional statement cannot consider the outcome of the conditional expression.

$$\mathsf{wp}^\#(\mathtt{if}(c)s_1 \ \mathtt{else} \ s_2, Q) = \mathsf{wp}^\#(c, \mathsf{wp}^\#(s_1, Q)) \wedge \mathsf{wp}^\#(c, \mathsf{wp}^\#(s_2, Q))$$

The abstract wp-calculus is sound, that is every program that can be proven correct with the abstract wp-calculus, can also be proven correct with the standard wp-calculus.

LEMMA 3 *For any statement s, and any predicates P and Q, containing static ghost variables only, we have:*

$$\forall P, Q \colon \mathsf{Pred}, s \colon \mathsf{Stmt}. (P \Rightarrow \mathsf{wp}^\#(s, Q)) \Rightarrow (P \Rightarrow \mathsf{wp}(s, Q))$$

Now we can prove a correspondence between pre and $\mathsf{wp}^\#$.

THEOREM 4 (CORRESPONDENCE) *For any statement s, its abstract weakest precondition is equivalent to the calculated precondition, in conjunc-*

tion with a universally quantified expression F.

$$\exists F \colon \mathsf{Pred}.\mathsf{wp}^{\#}(s, \lambda x.\mathsf{true}) = (\mathsf{pre}(s, \emptyset) \wedge \forall \mathsf{mod}(s).F)$$

This property formalises the conceptual difference described above: the function pre extracts the "external" part of the $\mathsf{wp}^{\#}$ (the method contract), while the quantified expression F corresponds to the "internal" proof obligations. The proofs of both properties proceed by structural induction. We believe similar equivalences can be proven for the function post and the sp-calculus. However, we are not aware of any adaptation of the sp-calculus to Java, therefore we did not study this.

4.   Results

For several realistic examples of Java Card applications, we checked whether they respect the security properties presented in Section 2, and actually found some violations. This section presents these results, focusing on the atomicity properties.

4.1   Core-annotations for Atomicity Properties

The core-annotations related to the atomicity properties specify the methods related to the transaction mechanism declared in class JCSystem of the Java Card API. As explained above, a static ghost variable TRANS is used to keep track of whether there is a transaction in progress. Section 3.1 presents the annotations for method beginTransaction; for commitTransaction and abortTransaction similar annotations are synthesised. After propagation, these annotations are sufficient to check for the absence of nested transactions.

To check for the absence of uncaught exceptions inside transactions, we use a special feature of JACK, namely pre- and postcondition annotations for statement blocks (as presented in [7]). Block annotations are similar to method specifications. The propagation algorithm is adapted, so that it not only generates annotations for methods, but also for designated blocks. As core-annotation, we add the following annotation for commitTransaction.

```
/*@ exsures (Exception) TRANS == 0; @*/
public static native void commitTransaction()
                              throws TransactionException;
```

This specifies that exceptions only can occur if no transaction is in progress. Propagating these annotations to statement blocks ending with a commit guarantees that if exceptions are thrown, they have to be caught within the transaction.

Finally, in order to check that only a bounded number of retries of pin-verification is possible, we annotate the method `check` (declared in the interface `Pin` in the standard Java Card API) with a precondition, requiring that no transaction is in progress.

```
/*@ requires TRANS == 0; @*/
public boolean check(byte[] pin, short offset, byte length);
```

4.2   Checking the Atomicity Properties

As mentioned above, we tested our method on realistic examples of industrial smart card applications, including the so-called Demoney case study, developed as a research prototype by Trusted Logic[7], and the PACAP case study[8], developed by Gemplus. Both examples have been explicitly developed as test cases for different formal techniques, illustrating the different issues involved when writing smart card applications. We used the core-annotations as presented above, and propagated these throughout the applications.

For both applications we found that they contained no nested transactions, and that they did not contain attempts to verify pin codes within transactions. All proof obligations generated *w.r.t.* these properties are trivial and can be discharged immediately. However, to emphasise once more the usefulness of having a tool for generating annotations, in the PACAP case study we encountered cases where a single transaction gave rise to twenty-three annotations in five different classes. When writing these annotations manually, it is very easy to forget some of them.

Finally, in the PACAP application we found transactions containing uncaught exceptions. Consider for example the following code fragment.

```
void appExchangeCurrency(...) {
  ...
  /*@ exsures (Exception) TRANS == 0; @*/
  { ...
  JCSystem.beginTransaction();
  try {balance.setValue(decimal2); ...}
  catch (DecimalException e) {
    ISOException.throwIt(PurseApplet.DECIMAL_OVERFLOW); }
  JCSystem.commitTransaction();
  } ... }
```

The method `setValue` that is called can actually throw a decimal exception, which would lead to throwing an ISO exception, and the transaction would not be committed. This clearly violates the security policy as described in Section 2. After propagating the core-annotations, and computing the appropriate proof obligations, this violation is found automatically, without any problems.

5.   Related Work

Our approach to enforce security policies relies on the combination of: an annotation assistant that generates JML annotations from high-level security properties, a proof obligation generator for annotated applets, using *e.g.* a weakest precondition calculus, and an automated or interactive theorem prover to discharge all generated proof obligations. Experience suggests that our approach provides accurate and automated analyses that may handle statically a wide range of security properties.

Proof-carrying code [18] provides another appealing solution to enforce security policies statically, but it does not directly address the problem of obtaining appropriate specifications for the code to be downloaded. In fact, our mechanism may be used in the context of proof-carrying code as a generator of verification conditions from high-level security properties.

Run-time monitoring provides a dynamic measure to enforce safety and security properties, and has been instrumented for Java through a variety of tools, see *e.g.* [1, 4, 20]. Security automata provide another means to specify security policies and to monitor program executions. Different forms of automata (edit automata, truncation automata, insertion automata, *etc.*) have been proposed, to prevent or react against violations of security policies, see *e.g.* [19, 22, 13, 10]. Inspired by aspect-oriented programming, Colcombet and Fradet [8] propose a technique to compose programs in a simple imperative language with optimised security automata. However, run-time monitoring is not an option for smart card applications, in particular because of the card's limited resources.

6.   Conclusions

We have developed a mechanism to synthesise JML annotations from high-level security properties. The mechanism has been implemented as a front-end for tools accepting JML-annotated Java programs; we use it in combination with JACK. The resulting tool set has been successfully applied to the area of smart cards, both to verify secure applications, and to discover programming errors in insecure ones. Our broad conclusion is that the tool set contributes to effectively carrying out formal security analyses, while also being reasonably accessible to security experts without intensive training in formal techniques.

Currently, we are developing solutions to hide the complexity of generating core annotations from the user. To this end, we plan to develop appropriate formalisms for expressing high-level security properties, and a compiler that translates properties expressed in these formalisms into appropriate JML core-annotations. Possible formalisms include security automata, for which appealing visual representations can be given, or more traditional logics, such as temporal logic. In the latter case, we believe that it will be necessary to rely on a form of security patterns reminiscent of the specification patterns developed by Dwyer *et al.* [9], and also to consider extensions of JML with temporal logic [21].

Further, we intend to apply our methods and tools in other contexts, and in particular for mobile phone applications. In particular, this will require extending our tools to other Java technologies that, unlike Java Card, feature recursion and multi-threading.

Notes

1. `http://commoncriteria.com/`
2. `http://www.ercim.org/reset` and `ftp://ftp.cordis.lu/pub/ist/docs/ka2/`
3. Java Card is a dialect of Java, tailored explicitly to smart card applications.
4. `http://www.jmlspecs.org`
5. `http://www.atelierb.societe.com/`
6. `http://research.compaq.com/SRC/esc/Simplify.html`
7. `http://www.trusted-logic.fr`
8. `http://www.gemplus.com/smart/r_d/publications/case-study`

References

[1] D. Bartetzko, C. Fischer, M. Möller, and H. Wehrheim. Jass – Java with Assertions. In K. Havelund and G. Roşu, editors, *ENTCS*, volume 55(2). Elsevier Publishing, 2001.

[2] J. van den Berg and B. Jacobs. The LOOP compiler for Java and JML. In T. Margaria and W. Yi, editors, *Tools and Algorithms for the Construction and Analysis of Systems (TACAS 2001)*, number 2031 in LNCS, pages 299–312. Springer, 2001.

[3] P. Bieber, J. Cazin, V. Wiels, G. Zanon, P. Girard, and J.-L. Lanet. Checking Secure Interactions of Smart Card Applets: Extended version. *Journal of Computer Security*, 10(4):369–398, 2002.

[4] G. Brat, K. Havelund, S. Park, and W. Visser. Java PathFinder - second generation of a Java model checker. In *Workshop on Advances in Verification*, 2000.

[5] C. Breunesse, N. Cataño, M. Huisman, and B. Jacobs. Formal Methods for Smart Cards: an experience report. Technical Report NIII-R0316, NIII, University of Nijmegen, 2003. To appear in *Science of Computer Programming*.

[6] L. Burdy, Y. Cheon, D. Cok, M. Ernst, J. Kiniry, G.T. Leavens, K.R.M. Leino, and E. Poll. An overview of JML tools and applications. In T. Arts and

W. Fokkink, editors, *Formal Methods for Industrial Critical Systems (FMICS 03)*, volume 80 of *ENTCS*. Elsevier, 2003.

[7] L. Burdy, A. Requet, and J.-L. Lanet. Java Applet Correctness: a Developer-Oriented Approach. In *Formal Methods (FME'03)*, number 2805 in LNCS, pages 422–439. Springer, 2003.

[8] T. Colcombet and P. Fradet. Enforcing trace properties by program transformation. In *Proceedings of POPL'00*, pages 54–66. ACM Press, 2000.

[9] M. Dwyer, G. Avrunin, and J. Corbett. Property Specification Patterns for Finite-state Verification. In *2nd Workshop on Formal Methods in Software Practice*, pages 7–15, 1998.

[10] U. Erlingsson. *The Inlined Reference Monitor Approach to Security Policy Enforcement*. PhD thesis, Department of Computer Science, Cornell University, 2003. Available as Technical Report 2003-1916.

[11] M.D. Ernst, J. Cockrell, W.G. Griswold, and D. Notkin. Dynamically discovering likely program invariants to support program evolution. *IEEE Transactions on Software Engineering*, 27(2):1–25, 2001.

[12] C. Flanagan and K.R.M. Leino. Houdini, an annotation assistant for ESC/Java. In J.N. Oliveira and P. Zave, editors, *Formal Methods Europe 2001 (FME'01): Formal Methods for Increasing Software Productivity*, number 2021 in LNCS, pages 500–517. Springer, 2001.

[13] K. Hamlen, G. Morrisett, and F.B. Schneider. Computability classes for enforcement mechanisms. Technical Report 2003-1908, Department of Computer Science, Cornell University, 2003.

[14] K.R.M. Leino, G. Nelson, and J.B. Saxe. ESC/Java user's manual. Technical Report SRC 2000-002, Compaq System Research Center, 2000.

[15] C. Marché, C. Paulin-Mohring, and X. Urbain. The Krakatoa tool for JML/Java program certification. *Journal of Logic and Algebraic Programming*, 58(1-2):89–106, 2004.

[16] R. Marlet and D. Le Métayer. Security properties and Java Card specificities to be studied in the SecSafe project, 2001. Number: SECSAFE-TL-006.

[17] J. Meyer and A. Poetzsch-Heffter. An architecture of interactive program provers. In S. Graf and M. Schwartzbach, editors, *Tools and Algorithms for the Construction and Analysis of Systems (TACAS 2000)*, number 1785 in LNCS, pages 63–77. Springer, 2000.

[18] G.C. Necula. Proof-Carrying Code. In *Proceedings of POPL'97*, pages 106–119. ACM Press, 1997.

[19] F.B. Schneider. Enforceable security policies. Technical Report TR99-1759, Cornell University, October 1999.

[20] L. Tan, J. Kim, and I. Lee. Testing and monitoring model-based generated program. In *Proceeding of RV'03*, volume 89 of *ENTCS*. Elsevier, 2003.

[21] K. Trentelman and M. Huisman. Extending JML Specifications with Temporal Logic. In H. Kirchner and C. Ringeissen, editors, *Algebraic Methodology And Software Technology (AMAST'02)*, number 2422 in LNCS, pages 334–348. Springer, 2002.

[22] D. Walker. A Type System for Expressive Security Policies. In *Proceedings of POPL'00*, pages 254–267. ACM Press, 2000.

ON-THE-FLY METADATA STRIPPING FOR EMBEDDED JAVA OPERATING SYSTEMS

Christophe Rippert
INRIA Futurs, IRCICA/LIFL, USTL — Lille 1 *

Christophe.Rippert@lifl.fr

Damien Deville
INRIA Futurs, IRCICA/LIFL, USTL — Lille 1 *

Damien.Deville@lifl.fr

Abstract   Considering the typical amount of memory available on a smart card, it is essential to minimize the size of the runtime environment to leave as much memory as possible to applications. This paper shows that on-the-fly constant pool packing can result in a significant reduction of the memory footprint of an embedded Java runtime environment. We first present JITS, an architecture dedicated to building fully-customized Java runtime environments for smart cards. We then detail the optimizations we have implemented in the class loading mechanism of JITS to reduce the size of the loaded class constant pool. By suppressing constant pool entries as they become unnecessary during the class loading process, we manage to compact constant pools of loaded classes to less than 8% of their initial size. We then present the results of our mechanism in term of constant pool and class size reductions, and conclude by suggesting some more aggressive optimizations.

Keywords: Java class loading, constant pool packing, embedded virtual machine

Introduction

Embedding Java applications on resource-limited devices is a major challenge in a highly heterogeneous world where computing power is found in all kind of unusual devices. The portability of Java is an invaluable asset for the programmer who needs to deploy applications on

---

*This work is partially supported by grants from the CPER Nord-Pas-de-Calais TACT LOMC C21, the French Ministry of Education and Research (ACI Sécurité Informatique SPOPS), and Gemplus Research Labs.

these heterogeneous platforms. However, embedded Java virtual machines are typically very restricted because of the limitations of the underlying hardware. For instance, the Java Card virtual machine [Chen, 2000] does not support multi-threading or garbage collection due to the typical computing power and memory space available on smart cards [Rippert and Hagimont, 2001]. Memory is an especially scarce resource in most embedded systems due to technical constraints and prohibitive costs which prevent the miniaturization of large memory banks. For instance, a smart card typically includes 1–4 KB of RAM used as working space, 32–64 KB of persistent writable memory (usually EEPROM) used to preserve data when the card is not connected to a power source (and sometimes also a working space if the RAM is not large enough), and 256 KB of ROM which usually contains the kernel of the runtime environment. Thus, reducing the size of the virtual machine and its runtime memory consumption are critical objectives if complex applications are to be executed on the system.

Reducing the memory space consumed by classes obviously means trying to obtain smaller code and smaller data. Previous work has shown that bytecode compression can be used to reduce the memory space used by the code [Bizzotto and Grimaud, 2002]. However, the compressed code size usually cannot be reduced to more than $\frac{2}{3}$ of the initial code size, which does not result in a significant reduction of the overall class size considering that most classes include much more data than code. So it seems interesting to try and compress the constant pool of each class, which stores most of the data used by the class (*e.g.* immediate values, external method names and prototypes, etc.). A careful analysis of the constant pool shows that many of its entries are only needed during the class loading process and can be removed before execution. Moreover, some data is duplicated in different classes and could be factorized. Thus, we have devised a new class loading mechanism which compacts the constant pool on-the-fly by suppressing entries as soon as they are unnecessary, and implemented it in JITS, our architecture for building customized Java operating systems. A valuable asset of our mechanism is that it does not imply disabling important features of the virtual machine, such as dynamic type checking or garbage collection. JITS advocates a very different philosophy than the Java Card environment, since Java Card can be seen as a customization of the specification of Java, whereas JITS implements the standard Java specification while making it possible to customize the code of the environment to fit to the underlying hardware.

We first present the JITS platform we have developed to build customized Java virtual machines for embedded systems. We then detail the class loading scheme we have chosen in JITS and present the optimizations we have implemented to reduce the memory space needed by loaded classes. Some evaluations of the memory consumption of various loaded classes are then presented, and we conclude by detailing the future optimizations we plan to implement in JITS. A trace showing the evolutions of the constant pool of a classical embedded application is described in an appendix.

1.   JITS: Java In The Small

JITS is an architecture dedicated to building embedded Java operating systems. JITS is composed of a full-featured virtual machine (including garbage collection, multi-threading, etc.) and a complete Java 1 API. Developers can use the services and packages provided to build a tailored Java Runtime Environment fitting the needs of the application and exploiting the resources available in the best way. Developers can therefore choose which services they want to include, which contrasts with other embedded environments which usually provide a restricted Java runtime environment with little support for customization [Deville et al., 2003]. For example, a developer building a Java Card compliant environment does not need to include a TCP/IP stack and can replace it by a much smaller APDU automaton.

JITS also offers some tools dedicated to help building the embedded environment, as presented in Figure 1. These tools include a program dedicated to generate the binary image of the environment which will be embedded in the device (we call this binary image a "Rom" though it can be stored in other kinds of memory on the embedded device). This program, called the Romizer, first loads all classes selected to be part of the embedded API, and brings them to an initialized state using the loading scheme presented below. After loading the classes, the Romizer takes a snapshot of the objects created in memory and dumps it to a C file which will be compiled with the core of the virtual machine to build the binary image of the runtime environment. The Romizer is a program entirely written in Java and can be run on any virtual machine[1], which differs from standard romization schemes which usually impose a dedicated building environment [Sun Microsystems, 2000][Sun Microsystems, 2002]. Similarly, the JITS API can be used as any other Java API by programs executed on a standard virtual machine. JITS is programmed mostly in Java, so as to ease its porting to various hard-

---

[1]Though it needs the part of the JITS API in charge of loading classes, namely the classes `Class`, `ClassLoader`, `Field` and `Method`.

ware platforms and to reduce as much as possible its memory footprint. Native code is limited to parts which cannot be programmed in Java and implemented using strict ANSI C to ease the porting.

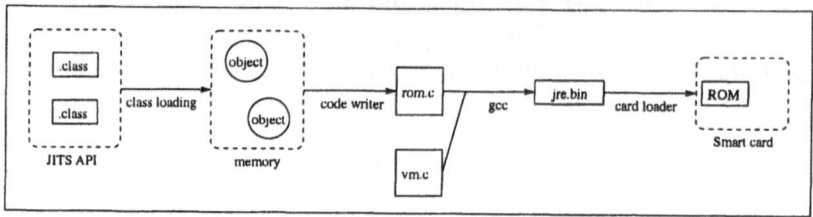

Figure 1. The romization process

It is important to note that JITS is not a replacement of Java Card since it is not dedicated only to smart cards. It is meant to build dedicated Java operating systems for various types of hardware. It can be used to build a Java Card compatible runtime environment, but the generated system will be implemented in a very different way than standard Java Card environments. A major difference concerns the class loading mechanism, since in Java Card class loading is done outside the card, when converting classes to .cap files [Schwabe and Susser, 2003], whereas in JITS the class loader is part of the runtime environment. This means that the same class loader is used during romization and when dynamically loading new classes. Thus, the class loader in JITS takes into account the limitations of the underlying hardware and is devised to minimize its memory consumption when optimizing the memory footprint of loaded classes, whereas the .cap file converter can use as much memory as needed since it is always executed off-card. This on-card reduction of the memory footprint of loaded class is to our knowledge very rarely supported by embedded Java runtime environments.

2.   Class loading in JITS

2.1   Principles

The class loading mechanism in JITS is different from the one implemented in a standard Java runtime environment [Lindholm and Yellin, 1999]. In Java, classes are loaded and linked only when they are actually used (*i.e.* when one of their methods is called or one of their fields is accessed). On the other hand, in Java Card, classes are compacted in .cap files containing closed packages, which means that all classes are pre-linked when inserted in the .cap file. In JITS we chose an intermediate scheme. JITS class loading mechanism supports the standard Java class loading scheme, but also permits to recursively load and link all classes referenced by the class currently being loaded[2]. This scheme is useful for an embedded Java runtime environment executing on a platform which might not be connected permanently to the network and which therefore needs to load all classes available when actually connected. Another difference with the standard Java class loading scheme is that JITS provides both the standard `defineClass` method which takes a `.class` file stored in a byte array as parameter, but also a `defineClass` method taking an `InputStream` as a parameter. This permits to create the internal representation of the class being loaded on-the-fly without having to load the whole `.class` file in memory, thus preserving memory.

Most classes loaded by JITS go through the four states presented in Figure 2. Primitive types and arrays are exceptions to this scheme, since they are directly created by the virtual machine without having to load any class file. This class loading scheme is used both for classes loaded during romization and for classes dynamically loaded when the virtual machine runs on the embedded system, except for the initialization of static fields as explained below. For classes loaded during romization, our mechanism permits to reduce the footprint of the binary image which will be loaded in ROM. For classes loaded during the execution of the environment, the loading scheme we propose permits to reduce the space consumed in EEPROM where dynamically loaded classes are stored. So we are able to preserve memory both when the environment is created and during its execution.

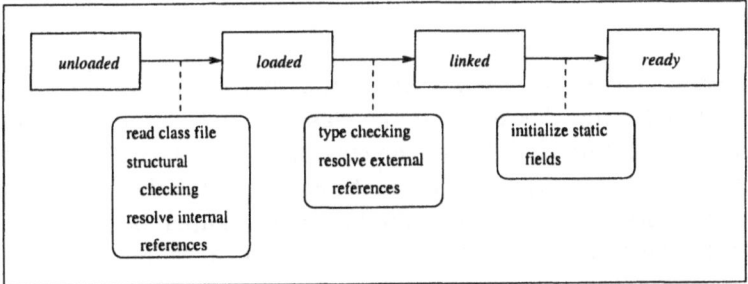

Figure 2.   The four states of class loading

---

[2]Recursive linking is also supported in Java Card through the export file mechanism, though this linking is not done automatically as in standard Java.

2.2   State *unloaded*

A class is *unloaded* when its `Class` object is first created by a class loader (*i.e.* by using a `new Class()` instruction). This `Class` object is basically an empty container which is filled as the class is loaded from its `.class` file. Figure 3 details the structure of a `.class` file as specified in Section 4.1 of [Lindholm and Yellin, 1999]. A class in state *unloaded* is typically a class which is referenced by another class but which has not yet been loaded by a class loader. This differs from the applet loading scheme in Java Card where a `.cap` file is made of all classes needed by the application. In JITS, classes are loaded one by one and are only required to be loadable when they are actually used.

```
ClassFile {
  id_info;        // magic and version number
  constant_pool;  // stores all constants used by the class
  base_info;      // access flags, class name and superclass
  interface_list; // interfaces implemented by the class
  field_list;     // fields of the class
  method_list;    // methods of this class (including their bytecode)
  attribute_list; // attributes (e.g. debug info, etc.)
}
```

Figure 3.   `.class` file structure

2.3   State *loaded*

A class is loaded when the `loadClass` method of a class loader is called. After having checked that the class has not already been loaded and having found its class file in the classpath, the class loader calls the `load` method of class `Class`. This method first reads the basic information of the class (*i.e.* its version number, name, superclass, etc.) before loading its constant pool. When loading a class, JITS ignores attributes not useful during execution of the program (*e.g.* line number table, source file, etc.). This can save a significant memory space, especially if the class file contains lots of debugging information.

The constant pool of classes is loaded from the `.class` file in two tables, named `atable` and `vtable`. The `atable` is an array of `Object` which is used at first to store `Utf8` constants (represented as `String` objects), whereas the `vtable` is an array of `int` in which immediate values are encoded. The `atable` is used later on to store other kinds of objects, such as `Class`, `Method` or `Field` objects.

The constant pool is then prelinked, which consists in resolving the accesses to the structures which represent metadata in the constant pool (*i.e.* the Constant_info structures defined in Section 4.4 of [Lindholm and Yellin, 1999]). For instance, a class is represented in the .class file by a structure (called Constant_Class_info) containing an index pointing to an array of characters (a Constant_Utf8_info entry) representing its name. In JITS, a class constant is represented by a corresponding Class object stored in the atable. Thus, the Constant_Utf8_info entry does not need to be mapped in memory. The Class object is created if it does not already exist (which means that the referenced class has not yet been loaded or referenced). If the class has already been loaded, we use the Class object created when the referenced class was in state *unloaded.* If it has already been referenced, we use the Class object created when the class was first referenced. Thus, we preserve memory since the Class object is needed to load the referenced class anyway and we do not create any intermediate object to represent it. We apply the same transformation to metadata representing strings and name-and-type constants. Figure 4 details the transformation applied to Constant_NameAndType_info entries. These structures are used to describe fields and methods. A name-and-type is composed of the name of the entity (field or method) and a string representing its type (using the convention presented in Section 4.3 of [Lindholm and Yellin, 1999]).

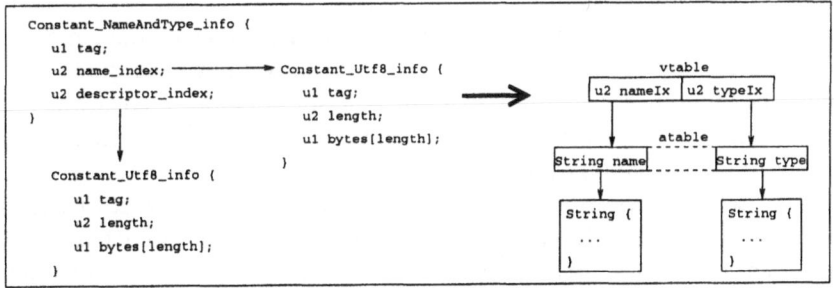

Figure 4.   Constant pool prelinking (phase one)

A second pass of the prelinker transforms the metadata representing fields, methods, and interface methods (*e.g.* Constant_Fieldref_info, Constant_Methodref_info and Constant_InterfaceMethodref_info) into an int stored in the vtable. This int is composed of the 16-bit index of the corresponding Class object and the 16-bit index of the Field or Method object representing the constant. These objects are added to the `atable`. Once more, we are able to preserve some memory by discarding unused `Constant_Utf8_info` entries. Figure 5 details the transformation applied to `Constant_Fieldref_info` entries.

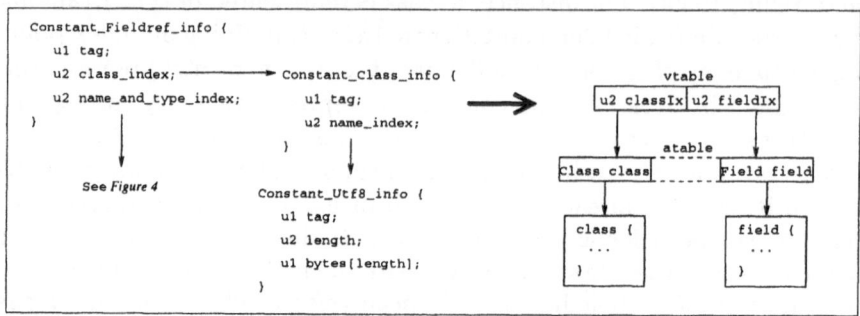

Figure 5.   Constant pool prelinking (phase two)

After loading the constant pool, the load method reads the interfaces implemented by the class, then its fields and its methods. The static fields of the class are stored in two tables, `aStaticZone` which contains reference fields, and `vStaticZone` for immediate values. Reading the methods consists of loading the bytecode, reading the exception table, loading stack maps if they are included in the class file, and finally building the class virtual method table. When loading the bytecode of a method, some instructions are replaced by an optimized version which will be interpreted faster at runtime and can also save some memory space. These optimized instructions are usually known as *quick* byte-codes. For instance, the `anewarray` instruction includes a constant pool index pointing to the type of the elements of the array. This instruction is replaced by `anewarray_quick`, which takes as a parameter an index pointing to an entry in the `atable` containing a `Class` object of the array component type. Thus, we can suppress the `Constant_Class_info` and `Constant_Utf8_info` entries representing the type of the elements of the array.

Another interesting example of instruction replacement concerns the `ldc`, `ldc_w` and `ldc2_w` instructions which are used to load constants from the constant pool onto the operand stack. When loading the bytecode, these instructions are replaced by their *quick* counterparts which directly access the immediate value stored in the `vtable` without needing the `Constant_Integer_info`, `Constant_Float_info`, `Constant_Long_info` and `Constant_Double_info` structures which can be discarded. Thus, a `ldc` instruction is replaced by a `ldc_quick_a` instruction if the constant is a reference, a `ldc_quick_i` if the constant is an `int`, and a

ldc_quick_f if the constant is a `float`. It would be possible to use the
same instruction for both `int` and `float` constants since they are both
32-bit immediate values, but that would compromise the type-checker
which needs to be able to differentiate `int` and `float`. By replacing `ldc`
instructions by a type-specific opcode, we can preserve necessary type
information without keeping complete constant pool entries, and so preserve both memory space and functionalities of the virtual machine.

2.4   State *linked*

Classes reach the *linked* state after being linked to each others. The
linking process starts by recursively loading all the classes referenced by
the constant pool of the class being linked. Then every method of the
class is prelinked, which consists of type-checking its bytecode, compacting `invokevirtual` instructions, and marking the constant pool
entries used by the method code.   During prelinking of a method,
`invokevirtual` instructions are compacted if the index of the method
in the constant pool and the number of arguments of the method are
both less than 256. Compacting these instructions simply consists of
replacing the index of the method in the constant pool, which is encoded in 16 bits in the instruction, by the number of arguments of the
method and its index in the virtual method table of the class declaring
it. Thus, at runtime the interpreter can call the method directly without accessing the constant pool, which speeds up the calling process.
It also saves memory space, since the constant pool entry representing
the called method can be deleted. During method prelinking, constant
pool entries which are used by the bytecode are marked so that unused
entries can be detected during the compaction of the constant pool.

Static fields referenced in the `vtable` are then converted to references
pointing to the `vStaticZone` and `aStaticZone`. Static fields are treated
differently than virtual fields since their value can be accessed directly
since we know the class to which they belong (whereas finding a virtual
field requires looking up the inheritance tree to find the first class defining
that field). The index pointing to the `Field` object representing the
field is replaced by a 16-bit immediate value containing the 13-bit offset
of the field in the corresponding static zone and the 3-bit type of the
field (which is necessary in order to know which static zone contains
the field and how many bytes should be read). Thus, constant pool
entries representing static fields can be suppressed. Figure 6 presents
the compacting of static fields.

The constant pool is then packed and resized, thereby losing all unused
entries. Finally, each method is linked, which basically means modifying

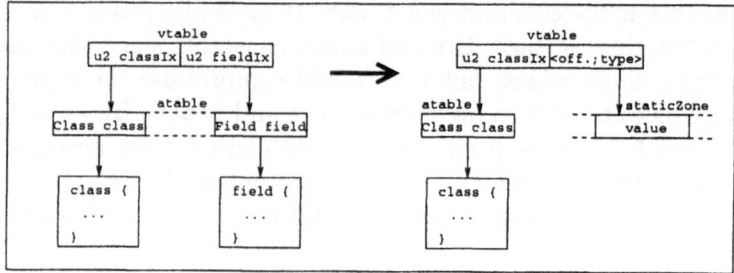

Figure 6.   Linking of static fields

the bytecode by replacing indices to the original constant pool entries
by indexes to the corresponding compacted constant pool entries.

2.5   State *ready*

A class reaches the final state *ready* after initializing its static fields
to their initial values. If the class is loaded during romization, this is
done by using the underlying virtual machine class loader to load the
class and then copying the values set by the static initializer to the
JITS instance of the class. This rather heavy mechanism is necessary
since the <clinit> method of a class cannot be called directly from a
Java program executing on a standard virtual machine. On the other
hand, if the class is being dynamically loaded by a running JITS virtual
machine, <clinit> methods are called directly by the virtual machine
as specified in [Lindholm and Yellin, 1999]. A final optimization can
be done here, as the <clinit> method of each class can be removed
after it has been used to initialize the static fields of the class. This is
done simply by removing the Method object representing the <clinit>
method from the linked list of the methods included in each class, and
letting the garbage collector free the corresponding memory space.

3.   Benchmarks

We monitored the memory footprint of the JITS API when loaded using the scheme presented above. The API currently contains most classes
from the base package java.lang, and some classes from java.awt,
java.io and java.net, including a full TCP/IP stack.

We first counted the number of constant pool entries discarded while
loading the classes. Results are presented in Figure 7, with state *unloaded* refering to the number of entries in the .class files.

Class state	unloaded	loaded	linked
Number of entries	8,416	3,067	1,426
% of initial number	100%	36.44%	16.94%

Figure 7.   Number of constant pool entries for the whole JITS API

These results show that most of the reduction of the number of constant pool entries is done while loading the class, i.e. when resolving accesses to the constant pool and removing unnecessary indirections. We still manage to divide by two the number of entries while linking, i.e. by compacting `invokevirtual` instructions and packing static fields (which implies suppressing unreferenced metadata for methods and fields).

We then monitored the memory footprint of the constant pool in bytes. We tried and suppress as many strings as possible since they are the most space-consuming data in the constant pool. Unfortunately, some of them (*e.g.* field names, method descriptors, etc.) are needed by the `java.lang.reflect` package, so we need to keep them if we want to support introspection. Figure 8 presents the size of the constant pool with and without those strings to illustrate the cost of supporting introspection.

Class state	unloaded	with introspection		without introspection	
		loaded	linked	loaded	linked
Size in bytes	152,154	48,203	40,455	19,435	11,687
% of initial size	100%	32.68%	26.59%	12.77%	7.68%

Figure 8.   Size of the constant pool for the whole JITS API

The size of the constant pool can be reduced to less than 8% of its original size if introspection is not supported. This is due to the fact that direct references to `Constant_String_info` represent only a small part of all the `Constant_Utf8_info` constant pool entries, so most of them can be eliminated during loading. If those strings are not removed, we manage to pack the constant pool to nearly one fourth of its original size, while preserving a complete support for introspection.

Since most of our optimizations concern compacting the constant pool (apart from disregarding unused attributes, which are seldom included in `.class` files except when debugging), we can use the results presented in Figure 8 to compute the size reduction for entire classes of the JITS API.

The total size of all `.class` files is 271,117 bytes[3], which includes 152,154 bytes for constant pools. Since we manage to reduce the size of constant pools to 40,455 bytes with support for introspection and to 11,687 without introspection, we obtain a memory footprint for the API which is only 58.8% of the total size of the `.class` files (48.19% without support for introspection) in state *linked*. Suppressing the `<clinit>` method of each class allows to save 54812 bytes for the whole API (including 50KB from `java.awt`), which means that the final memory footprint of JITS API in state *ready* is only 38.58% of the total size of the `.class` files with support for introspection and 27.97% without it. Figure 9 sums up the reduction of the total size of classes with and without support for introspection.

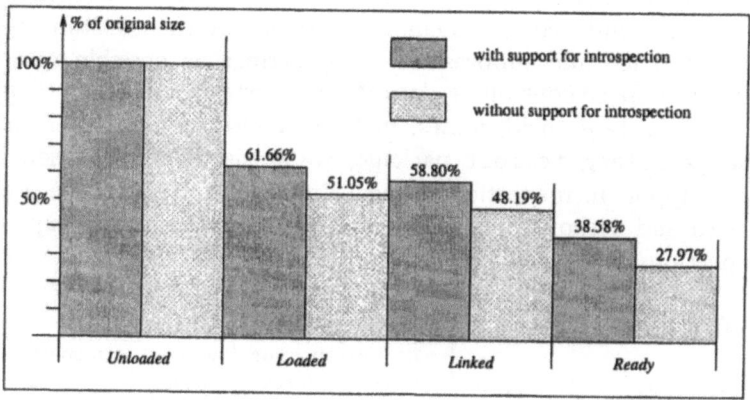

Figure 9. Reduction of the total size of classes

These results are similar to those obtained using the *JEFF* class format [The J-Consortium, 2002], which reduces the size of the `.class` files by merging constant pools of different classes. We make a similar optimization when factorizing entities used to represent data. For instance, most classes in a Java API includes an entry in their constant pool representing the class `String` (since most classes implement the `toString` method). In JITS, we replace each entry by a reference to one `Class` object representing the class `String`. Thus we are able to benefit from optimizations similar to those done in the *JEFF* class format,

---

[3]The total size of the whole API is over 260KB. However, this includes packages like `java.net` and `java.awt` which will probably not be included on a very constrained platform such as a smart card.

while loading standard .class files and staying compliant with Sun's specification.

4.   Future work

An optimization similar to the one applied to static fields can be done for private virtual fields and methods. In JITS, objects are implemented as a C structure containing a pointer to the related class and the virtual fields of the object. When the getfield and putfield bytecodes are interpreted, the virtual machine accesses the required field by adding the offset stored in the bytecode to the base address of the object. Thus, it is possible to suppress all constant pools entries referencing private fields since all accesses to these fields are made in the class declaring them and so the getfield and putfield instructions can be modified to contain the proper offset. Similarly, entries describing private virtual methods can be removed from the constant pool. This optimization cannot be applied to protected, package-accessible or public fields or methods, since they could be accessed by a method of a class loaded dynamically after romization. In that case, the constant pool entry representing the target field or method would be necessary to link the new method.

However, if we define an additional state, called *package-closed*, we can apply this optimization to all non-public fields. The state *package-closed* is reached by classes in a package when no new class can be added to that package. Locking a package this way can be useful for instance to prevent an application from modifying a fundamental package such as java.lang. If a class is *package-closed*, all constant pools entries corresponding to its non-public fields can be suppressed since all accesses can be linked before romization of the package.

Similarly, it is possible to define a state *closed* to be able to extend this optimization even to public fields. A class reaches the state *closed* if we can assure that no dynamically loaded class will need to be linked to this class. In practice, this state is most useful for embedded virtual machines romized with all the applications and that do not need to dynamically load new classes. These last two optimizations implies disabling some features of the virtual machine (namely restricting or even forbidding dynamic class loading) so they will be made optional when implemented in JITS.

Preliminary results show that these optimizations would allow a reduction of the constant pool to below the 7.68% lower limit presented in Figure 8. In state *closed*, all Constant_Utf8_info representing name or type metadata would become useless, as well as all metadata representing fields or methods in the constant pool. Thus, we can assume that

7.68% is the upper limit of the results we can expect when state *closed* is reached by a class, noting of course that closing a class prevents loading of any new class referencing it.

Another optimization especially interesting for smart cards would consist in minimizing the reorganizations of the constant pool of dynamically loading classes. These classes are stored in EEPROM, a type of memory much slower than RAM and which life expectancy diminishes with each write. So it would increase the performance of the dynamic loading process and prolong the lifetime of the smart card if we could devise a loading mechanism which compacts the constant pool of a class with only the minimal amount of entry reorganizations.

Conclusion

This paper shows that it is possible to greatly reduce the memory footprint of the class loading mechanism by applying on-the-fly packing of the constant pool of loaded classes. This allows saving memory space on the embedded system without sacrificing functionality of the virtual machine, since for instance we can still type-check the bytecode of the class while suppressing type information from the constant pool. Coupled with the flexibility of JITS which permits to choose precisely which components need to be included in the runtime environment (as a garbage collector would not be relevant to a Java Card compliant platform for instance), this makes it possible to generate a Java virtual machine fully tailored for the target device, thus exploiting the limited resources in the best possible way.

Acknowledgments

We would like to thank our shepherd, Erik Poll, for his helpful advice on enhancing this paper.

Appendix

We present in Figure A.1 a piece of the classical example of the Purse application to illustrate the reduction of the constant pool during romization. This partial example shows the drastic reduction of the constant pool which can be achieved by discarding unnecessary entries.

Figure A.2 presents meaningful information from the constant pool (*i.e.* vtable and atable) of class Purse after it has been loaded. Method prototypes and type descriptions are still included in the atable since they are necessary to link the class. Besides the initial value for the field Purse.id, the vtable only contains references to entries of type class which are stored in the atable.

Figure A.3 shows the constant pool after linking the class, which permitted discarding most method prototypes and type descriptions. The vtable still includes the

```
public class Purse extends Object {
    private static int id;
    private Float sum;
    static { id = 0x12345678; }
    public Purse(float b) {
        this.sum = new Float(b);
    }
    final public void credit(float n) {
        this.sum = new Float(this.sum.floatValue() + n);
    }
}
```

Figure A.1.   The original Java source of the Purse class

```
vtable:
  [0]: 0                      [3]: 18
  [1]: 17                     [4]: 19
  [2]: 0x12345678
atable:
  [0]: null
  [1]: id                     [14]: fr/lifl/rd2p/jits/test/Purse
  [2]: java/lang/Float        [15]: java/lang/Object
  [3]: sum                    [16]: floatValue
  [4]: ()F                    [17]: class fr/lifl/rd2p/jits/lang/Float
  [5]: <init>                 [18]: class fr/lifl/rd2p/jits/test/Purse
  [6]: (F)V                   [19]: class fr/lifl/rd2p/jits/test/Object
  [7]: I                      [20]: JMethod <init> ()V access=0x1001
  [8]: ()V                    [21]: JField private fr/lifl/rd2p/jits/-
  [9]: Ljava/lang/Float;            lang/Float sum
  [10]: credit                [22]: JMethod <init> (F)V access=0x5001
  [11]: <clinit>              [23]: JField private static int id
  [12]: getSum               [24]: (F)Ljava/lang/Float;
  [13]: JMethod floatValue ()F access=0x1001
```

Figure A.2.   The constant pool in state *loaded*

```
vtable:
  [0]: 0x12345678
atable:
  [0]: class fr/lifl/rd2p/jits/lang/Float
  [1]: class fr/lifl/rd2p/jits/test/Purse
```

Figure A.3.   The constant pool in state *linked*

value used to initialize the static field Purse.id since this has not yet be done by the

Romizer. Similarly, the `atable` contains the fully qualified name of the class `Purse` since it is needed by the static initializer to find the `Purse.id` field.

```
vtable:
 empty
atable:
 [0]: class fr/lifl/rd2p/jits/lang/Float
```

Figure A.4.   The constant pool in state *ready*

Figure A.4 shows the final state of the constant pool, after the static initializer of class `Purse` has been executed and discarded. The constant pool entries which were associated with it have been removed too, resulting in a constant pool with only 1 entry left. This entry describing class `Float` could probably be suppressed in most Java runtime environments since class `Float` is part of the package `java.lang` which is typically completely romized. However, we chose to keep it in case the programmer decides not to include support for floating point arithmetics in the base system and then load it dynamically during execution.

References

[Bizzotto and Grimaud, 2002] Bizzotto, G. and Grimaud, G. (2002). Practical Java Card Bytecode Compression. In *Proceedings of RENPAR14 / ASF / SYMPA*.

[Chen, 2000] Chen, Z. (2000). *Java Card Technology for Smart Cards: Architecture and Programmer's Guide*. Addison Wesley.

[Deville et al., 2003] Deville, D., Galland, A., Grimaud, G., and Jean, S. (2003). Smart Card operating systems: Past, Present and Future. In *The 5th NORDU/USENIX Conference*.

[Lindholm and Yellin, 1999] Lindholm, T. and Yellin, F. (1999). *The Java Virtual Machine Specification, Second Edition*. Addison Wesley.

[Rippert and Hagimont, 2001] Rippert, C. and Hagimont, D. (2001). An evaluation of the Java Card environment. In *Proceedings of the Advanced Topic Workshop "Middleware for Mobile Computing"*.

[Schwabe and Susser, 2003] Schwabe, J. E. and Susser, J. B. (2003). *Token-Based Linking*. US Patent Application number US 2003/0028686 A1. http://www.uspto.gov/.

[Sun Microsystems, 2000] Sun Microsystems (2000). *J2ME Building Blocks for Mobile Devices*. http://java.sun.com/products/kvm/wp/KVMwp.pdf.

[Sun Microsystems, 2002] Sun Microsystems (2002). *The CLDC Hotspot Implementation Virtual Machine*. http://java.sun.com/products/cldc/wp/CLDC_HI_WhitePaper.pdf.

[The J-Consortium, 2002] The J-Consortium (2002). *JEFF Draft Specification*. http://www.j-consortium.org/jeffwg/index.shtml.

PRIVACY ISSUES IN RFID BANKNOTE PROTECTION SCHEMES

Gildas Avoine

Swiss Federal Institute of Technology (EPFL)
Security and Cryptography Laboratory (LASEC)
CH-1015 Lausanne Switzerland

gildas.avoine@epfl.ch

Abstract   Radio Frequency Identification systems are in the limelight for a few years and become pervasive in our daily lives. These smart devices are nowadays embedded in the consumer items and may come soon into our banknotes. At *Financial Cryptography 2003*, Juels and Pappu proposed a practical cryptographic banknote protection scheme based on both Optical and Radio Frequency Identification systems. We demonstrate however that it severely compromises the privacy of the banknotes' bearers. We describe some threats and show that, due to the misuse of the secure integration method of Fujisaki and Okamoto, an attacker can access and modify the data stored in the smart device without optical access to the banknote. We prove also that despite what the authors claimed, an attacker can track the banknotes by using the access-key as a marker, circumventing the randomized encryption scheme that aims at thwarting such attacks.

Keywords: RFID, Privacy, Banknote Protection.

1   Introduction

The main goal of Radio Frequency Identification (RFID) systems is to identify objects in an environment by embedding tags onto these objects. A tag is a tiny device capable of transmitting, over a short distance, a unique serial number and other additional data. For instance, goods in stores can be tagged in order to prevent shoplifting, or to speed up the goods registration process by using wireless scanning instead of human or optical scanning. The security issues of such systems are therefore two-fold. On one hand, it must be impossible to thwart the system by modifying the tag's data or even creating fake tags; on the other hand, tags should not compromise the bearers' privacy.

It is vital to ensure the security of RFID systems, since many organizations have already turned to use such devices for many large-scale applications. In particular, the European Central Bank (ECB) decided to use some RFIDs to protect Euro banknotes [13]. Although Euro banknotes already include physical security features, ECB believes that RFIDs will add further protection: electronic tags will give governments and law enforcement agency the means to track banknotes in illegal transactions. We do not know yet if such chips will be embedded into all Euro banknotes, or just those of a high denomination. Japanese government also plans to embed RFIDs into new 10'000 Yen notes [9]. Though these examples may be just rumors until now, we can consider that such devices will be used for such applications soon. Up to now, RFIDs are already used in less sensitive applications. For instance, the tire manufacturer Michelin decided to implant RFID tags inside the rubber sidewall of its tires. These tags contain the tire's unique ID and maybe some other data such as origin, maximum inflation pressure, size, etc. The data stored in these tags are readable by a receiver positioned up to 30 inches away from the tire. These tags could pinpoint, for example, tires belonging to a defective batch. The purpose currently is to identify and track tires, but it could be adapted to allow tags to communicate directly with the vehicle's dashboard to indicate if the tires are properly inflated, overheated, overloaded, or if the tire tread is seriously worn [7]. Michelin Tires' RFID system is currently being tested in some taxis and rental cars, but it could be extended to all Michelin Tires after 2005.

When technology advances, research on the privacy and security aspects of such devices lags far behind. Security flaws could however result in large-scale consequences from a sociological and economic point of view, by flooding the market with fake-tagged items. By proposing the first cryptanalysis of a scheme specially designed for RFID systems, we show that, up to now, such systems can not be used in practical applications without endangering the users' privacy.

In this paper, we first describe the main characteristics of the RFIDs and present in Section 3 the Juels – Pappu banknote protection scheme [8], which uses RFIDs. We then describe in Section 4 some potential threats to this scheme and show that, due to the misuse of the secure integration method of Fujisaki and Okamoto, an attacker can access and modify the data stored in the smart device without optical access to the banknote. We also prove that despite the claims of the authors, an attacker can track the banknotes by using the access-key as a marker, circumventing the randomized encryption scheme that aims to thwart such attacks.

2 Radio Frequency Identification Systems

In this section, we describe the technical aspects of RFID systems, which consist of three elements:

- The RFID tag (*transponder*) that carries the identifying data;

- The RFID readers (*transceivers*) that read and write the tags' data.

- The back-end database, connected to the readers, that records information related to the tags data.

While tags are low-capability devices, as explained below, readers and back-end database have powerful capability of storage, computation and communication. Therefore readers and back-end database can communicate through a secure channel.

2.1 Tags

In order to use RFID tags in large-scale applications, the per-unit cost of such devices should be very low. Currently, the cost of such devices is a few tens US cents [12] and would further drop to 5 US cents [10]. On the other hand, practical applications require that the tag size be as tiny as a few millimeters square. Cost and size requirements mean that power consumption, processing time, storage and logical gates are extremely limited. From a technical point of view, a tag consists of an antenna and a microchip capable of storing data and performing logical operations such as bit-string comparisons of keys. One can distinguish *active* tags which have a battery to run the microchip's circuit and to broadcast a signal to the reader, from *passive* tags which have no battery. Active tags are suitable to track high cost items over long ranges but they are too expensive to be embedded into banknotes. Only passive tags are suitable for this application. Tags memory is small and can store only a few hundreds bits in the best case [4]. Tags contain a few thousands gates, which is below the threshold of embeding cryptographic algorithms. For instance AES typically requires about 20,000 gates and SHA-1 requires about 15,000 gates. From a security point of view, tags can not be considered as tamper-resistant: all physical threats, such as laser etching, ion-probes, clock glitching, etc. [12] are applicable to recover the data stored in the tag. Therefore the tag cannot securely store secret keys in the long term, for instance.

2.2 Communication

As will be explained in Section 2.3, the reader can transmit various commands to the tag. In order to do this, the reader broadcasts Radio

Frequency radiation as long as necessary to bring sufficient power to the tag. The tag modulates the incoming radiation with its stored data. We actually consider two channels: the *forward channel* from the reader to the tag which can operate over long distance and the *backward channel* from the tag to the reader which can operate over a shorter distance. Both channels can be eavesdropped by an attacker since it is obviously impossible to use cryptographic features.

2.3   Interface

We give below the common commands that are available on a tag:

- `read`: allows every reader to obtain the data stored in the tag memory.

- `write`: allows every reader to write data in the tag memory.

Some other commands can be available on a tag:

- `sleep`: this command is keyed so that the reader has to send a key in order to put the tag into the sleep state. Then the tag does not respond to the reader's queries until it receives the `wake` command with the legitimate key.

- `wake`: after this command, the tag starts afresh to respond to the reader. It is a keyed command associated with the `sleep` command.

- `kill`: this command destroys the tag definitively.

Moreover, Juels and Pappu [8] suppose that the following commands are available:

- `keyed-read`

- `keyed-write`

These commands are similar to the `read`/`write` commands except that they are keyed.

3   Juels – Pappu banknote protection scheme

3.1   Interested parties

Juels and Pappu proposed a banknote protection scheme whose goal is to resist banknotes counterfeiting and track traffic flows by some law enforcement agency, nevertheless guaranteeing the privacy of the banknote handlers. First we describe all the interested parties who are involved in the scheme.

- *Central bank.* The central bank aims at creating banknotes and at avoiding banknote forgery. It is therefore in its interest to have unique banknote serial numbers and to protect the secret key, which is used to sign the banknotes.

- *Law enforcement agency.* The goal of this agency is to arrest forgers. In order to achieve this, it needs to track banknotes and detect fake ones easily, even in areas of dense traffic, such as airports.

- *Merchants.* Merchants handle large quantities of banknotes. It is conceivable that they will try to compromise their clients' privacy. Merchants may comply with the law enforcement agency by reporting irregularities in banknote data.

- *Consumers.* Banknotes bearers want to protect their privacy. They want therefore to limit banknotes tracking even if it means not respecting existing laws.

3.2   Concept and requirements

Up to now, banknote security solely relies on optical features, which can be checked either by human-scanning or machine-scanning. In [8] security relies on both optical and electronic features. Banknotes thus have two data sources:

- *Optical*: data can be encoded in a human-readable form and/or in a machine-readable form such as a two-dimensional bar code. It contains banknote serial number as well as denomination, origin, etc.

- *Electronic*: data can be read by wireless communication. Data are signed by the central bank and encrypted with the law enforcement agency public key and a random number.

Electronic data are stored in a RFID tag, which consists here of two cells whose access is key-protected. The access-key can be (re-)generated from the banknote optical data. One of the two cells, denoted γ, is universally readable but keyed-writable. The other cell, denoted δ, is both keyed-readable and keyed-writable. In [8], the proposed scheme consists in writing in γ the serial number of the banknote signed by the central bank and encrypted with the law enforcement agency public key. If this encrypted value was static, then an attacker could still track the banknote using this value as a marker. To overcome this weakness, the signature on the serial number is re-encrypted by merchants as often as possible, using obviously a probabilistic encryption scheme. Since the signature is available from the optical data, encryption is performed from scratch and does not need to be homomorphic. After the re-encryption is performed, the new encrypted value is put into γ and the used random value r is put into δ. Since γ and δ are keyed-writable, one must have optical contact with the banknote to obtain the access-key and thereby to re-encrypt the banknote. We will detail this procedure in Section 3.3.

We give below the requirements that [8] should guarantee.

- *Consumer privacy.* Only the law enforcement agency is able to track the banknotes using the RFID interface. Even the central bank is not allowed to track banknotes.

- *Strong tracking.* Law enforcement agency are able to identify a banknote (by its serial number) even without optical contact.

- *Minimal infrastructure.* In order to be user-friendly, the system should not require that banknote bearers possess special equipment. For their part, retail banks and shops should only buy devices at reasonable cost. Furthermore, they should not be required to set up a permanent network connection.

- *Forgery resistance.* A forger has to have optical contact with a banknote in order to create a fake one with the same serial number. A forger should not be able to create a fake banknote with a new serial number and moreover he should not be able to change the banknote denomination.

- *Privilege separation.* The data stored in the tag should only be alterable given optical contact with banknotes.

- *Fraud detection.* If the data stored by the tag are wrong, then a merchant who has optical access to the banknote should be able to detect the forgery.

In order to illustrate these requirements, Juels and Pappu give two examples of privacy attacks that a banknote protection system should withstand. We give these two examples here because we will show, in Section 4, that their scheme is actually not resistant to these attacks.

EXAMPLE 1 *"Bar X wishes to sell information about its patrons to local Merchant Y. The bar requires patrons to have their drivers' licenses scanned before they are admitted (ostensibly to verify that they are of legal drinking age). At this time, their names, addresses, and dates of birth are recorded. At the same time, Bar X scans the serial numbers of the*

RFID tags of banknotes carried by its patrons, thereby establishing a link between identities and serial numbers. Merchant Y similarly records banknote serial numbers of customers from RFID tags. Bar X sells to Merchant Y the address and birth-date data it has collected over the past few days (over which period of time banknotes are likely not yet to have changed hands). In cases where Bar X and Merchant Y hold common serial numbers, Merchant Y can send mailings directly to customers indeed, even to those customers who merely enter or pass by Merchant Y's shops without buying anything. Merchant Y can even tailor mailings according to the ages of targeted customers. Patrons of Bar X and Merchant Y might be entirely unaware of the information harvesting described in this example."

EXAMPLE 2 *"A private detective wishes to know whether Bob is conducting large-value cash transactions at Carl's store. She surreptitiously intercepts the serial numbers on banknotes withdrawn by Bob and also records the serial numbers of those brought by Carl out of his store. If there is any overlap between sets of numbers, she concludes that Bob has given money to Carl. The private detective might reach the same conclusion if Bob leaves without banknotes that he carried into Carl's store. The private detective might also try to reduce her risk of detection by reading the banknotes of Bob and Carl at separate times, e.g., en route to or from the bank."*

3.3 Description of the method

We explain in this section the operations that should be performed on the banknote. Let $\mathsf{Sign}(k, m)$ be the signature on a message m with a key k and $\mathsf{Enc}(k, m, r)$ the encryption of m under the key k with the random number r. We note $\|$ the concatenation of two bit-strings.

Setup. Central bank $\mathcal{B}$ and law enforcement agency $\mathcal{L}$ respectively own a pair of public/private keys $(PK_{\mathcal{B}}, SK_{\mathcal{B}})$ and $(PK_{\mathcal{L}}, SK_{\mathcal{L}})$. $PK_{\mathcal{B}}$ and $PK_{\mathcal{L}}$ are published as well as a collision-resistant hash function h.

Banknote creation. For every banknote i, $\mathcal{B}$ selects (according to its own rules – which can be assumed to be public) a unique serial number S_i and computes its signature $\Sigma_i = \mathsf{Sign}(SK_{\mathcal{B}}, S_i \| den_i)$ where den_i is the banknote denomination. $\mathcal{B}$ then computes an access-key D_i such that $D_i = h(\Sigma_i)^1$, prints S_i and Σ_i on the banknote, and computes

$$C_i = \mathsf{Enc}(PK_{\mathcal{L}}, \Sigma_i \| S_i, r_i)$$

where r_i is a random number. C_i is written into γ and r_i is written into δ. Note that the access-keys D_i is not stored in the databases of $\mathcal{B}$. In order to keep in mind the values stored on/in the banknote, we give in Tab. 1, established from [8], the content of the optical information as well as those of cells γ and δ.

---

[1] Juels and Pappu point out that it is important that the hash function be applied on Σ_i rather than on S_i because an attacker who knows a serial number would be able to compute the corresponding access-key without any optical contact with the banknote.

RFID	
Cell γ	Cell δ
universally-readable / keyed-writable	*keyed-readable / keyed-writable*
$C = \mathsf{Enc}(PK_{\mathcal{L}}, \Sigma \| S, r)$	r

Optical	
S	$\Sigma = \mathsf{Sign}(SK_{\mathcal{B}}, S \| den)$

Table 1.   Optical and RFID data

Banknote verification and anonymization. When a merchant $\mathcal{M}$ receives a banknote, he verifies it then re-encrypts it according to the following steps:

1 $\mathcal{M}$ reads the optical data S_i and Σ_i and computes $D_i = h(\Sigma_i)$.

2 $\mathcal{M}$ reads C_i, stored in γ, and keyed-reads r_i which is stored in δ.

3 $\mathcal{M}$ checks that $C_i = \mathsf{Enc}(PK_{\mathcal{L}}, \Sigma_i \| S_i, r_i)$.

4 $\mathcal{M}$ chooses randomly r_i' and keyed-writes it into δ.

5 $\mathcal{M}$ computes $C_i' = \mathsf{Enc}(PK_{\mathcal{L}}, \Sigma_i \| S_i, r_i')$ and keyed-writes it into γ.

If one of these steps fails then the merchant should warn the law enforcement agency.

Banknote tracking. Let us consider a target banknote that the law enforcement agency $\mathcal{L}$ wants to check or track. $\mathcal{L}$ is able to easily obtain the cipher C reading the cell γ and then to compute the plaintext $\Sigma \| S = \mathsf{Dec}(SK_{\mathcal{L}}, C)$. $\mathcal{L}$ can then check whether or not Σ is a valid signature. If Σ is valid then $\mathcal{L}$ obtains the banknote serial number S.

3.4   Cryptographic algorithms

Encryption and signature schemes can be chosen among the existing secure schemes. However they should bring security without involving high overhead. Juels and Pappu suggest to use an El Gamal-based encryption scheme [5] and the Boneh–Shacham–Lynn signature scheme [3], both using elliptic curves. Let $\mathcal{G}$ denote an elliptic-curve-based group with prime order q and let P be a generator of $\mathcal{G}$. Let $SK_{\mathcal{L}} = x \in_R Z_q$ be the law enforcement agency private key and $PK_{\mathcal{L}} = Y = xP$ the corresponding public key. A message $m \in \{0,1\}^n$ where n is reasonable sized, is encrypted with the El Gamal scheme under the random number r as follows:

$$\mathsf{Enc}(PK_{\mathcal{L}}, m, r) = (m + rY, rP).$$

Since El Gamal encryption scheme is not secure against adaptive chosen-ciphertext attacks, Juels and Pappu suggest to use the secure integration method due to Fujisaki and Okamoto [6]; the message m is then encrypted as follows:

$$\mathsf{Enc}^*(PK_{\mathcal{L}}, m, r) = (\mathsf{Enc}(PK_{\mathcal{L}}, r, h_1(r||m)), h_2(r) \oplus m)$$

where h_1 and h_2 are two hash functions from $\{0,1\}^*$ to $\{0,1\}^n$. As explained in [8], signature size could be 154 bits. Assuming than a serial number can be encoded over 40 bits, the plaintext $\Sigma||S$ requires 194 bits. Let us consider a 195 bits order elliptic curve group, the size of $\mathsf{Enc}^*(PK_{\mathcal{L}}, \Sigma||S, r)$ will be 585 bits. The total required size will then be 780 bits (585 bits in γ and 195 bits in δ). As pointed out in [8], current RFID tags can provide such resources. For instance the Atmel TK5552 [4] offers a 1056 bits memory (only 992 bits are usable by the user). [11] suggests that only RFID costing about 50 US cents could supply such a capacity but less expensive tags with a few hundreds bits memory could appear in a few years. RFID tags currently costing 5 US cents supply less capacity, usually 64 or 96 bits of user memory [10].

4   Attacks on the banknote protection system

We introduce in this section several attacks that can be performed on the Juels – Pappu banknote protection scheme. While some of these attacks are proper to that scheme (Sections 4.2, 4.3, and 4.7), some other are more general and could be applied to other RFID-based privacy protection schemes (Sections 4.1, 4.4, 4.5, and 4.6).

4.1   Pickpocketing attack

This attack that Juels and Pappu already mentioned is significant enough to be recalled here. It requires an attacker to test a passer-by in order to detect if he carries some banknotes. Even if the attacker is not able to discover neither the serial number nor the denomination, he is able to establish how many banknotes the passer-by is bearing.

The attacker has less information if banknotes of all denominations are tagged than if only the largest ones are tagged. However, tagging banknotes of all denominations may be dangerous with the Juels – Pappu scheme due to the fact that scanning banknotes takes some time. Merchants would not agree to re-encrypt notes of small denominations; privacy could consequently be threatened.

EXAMPLE 3 *Some criminals want to steal some cars in a car park. During daylight hours, they only break into a few cars so as not to attract attention. Their problem is therefore to determine which cars could be the "best" ones, that is the cars that contain currency. Thanks to the banknote protection system, they can radio-scan numerous cars in order to pinpoint their targets.*

These "pickpocketing" attacks show that the attack described in the second example of Juels and Pappu (See page 39) still occurs even using their banknote protection scheme.

4.2   Data recovery attack

The *data recovery* attack consists of two steps: the first one aims at obtaining the access-key D and then the random number r which is stored in the δ-cell; the second step exploits a misuse of the secure integration method of Fujisaki and Okamoto, in order to recover S and Σ. So, the attacker can obtain the serial number of the banknote without optical access to it. Note that even well-behaving merchants are not supposed to obtain this information from the electronic data.

Step 1: One of the goals of the scheme is to avoid γ-write-access without optical reading of the banknote. This implies that an attacker must have physically access to the banknote to modify the γ-cell. However a merchant who is willing to re-encrypt the banknote sends the access-key $D = h(\Sigma)$ (obtained by optical reading) to the tag in order to receive the value stored in the δ-cell, i.e. the random number r. The attacker can just eavesdrop the communication in order to steal D and then he is able to communicate with the tag and finally obtain the δ-cell value r. To buttress our argumentation, remind that it is usually easier to eavesdrop the forward channel, that is from the reader to the tag, than the backward channel. Note also that the communication range should not be too short since one of the goals of the Juels – Pappu scheme is to enforce banknotes tracking be the law enforcement agency, even in areas of dense traffic, such as airports.

Step 2: The attacker first obtains the value stored in the γ-cell (universally readable). γ-cell contains:

$$\begin{aligned} \mathsf{Enc}^*(PK_{\mathcal{L}}, m, r) &= (\mathsf{Enc}(PK_{\mathcal{L}}, r, h_1(r\|m)), h_2(r) \oplus m) \\ &= (r + h_1(r\|m)PK_{\mathcal{L}},\ h_1(r\|m)P,\ h_2(r) \oplus m) \end{aligned}$$

Notation is defined in Section 3.4. Let us consider $(\epsilon_1, \epsilon_2, \epsilon_3)$ such that

$$(\epsilon_1, \epsilon_2, \epsilon_3) = \mathsf{Enc}^*(PK_{\mathcal{L}}, m, r).$$

So:

$$\epsilon_1 = r + h_1(r||m)PK_{\mathcal{L}}, \ \epsilon_2 = h_1(r||m)P, \text{ and } \epsilon_3 = h_2(r) \oplus m.$$

She obtains therefore

$$m = \epsilon_3 \oplus h_2(r) \text{ where } \epsilon_3, r, \text{ and } h_2 \text{ are known.}$$

Since $m := \Sigma||S$, this proves that an attacker can discover the serial number and the signature of a banknote without having optical access to it, contrary to what Juels and Pappu claim.

The problem arises from the fact that the integration method of Fujisaki and Okamoto is not secure anymore when the random value r is revealed. Indeed, the purpose of the asymmetric encryption $\mathsf{Enc}(PK_{\mathcal{L}}, r, h_1(r||m))$ is to "hide" the random value that is used to generate the key of the symmetric encryption. If this random value is public or can be determine easily by an attacker, the integration method becomes null and void.

4.3   Ciphertext tracking

The protection method that we are discussing in this paper uses re-encryptions to prevent tracking attacks. However, re-encryptions can only be performed in practice by merchants or retail banks[2]. Therefore the time period between two re-encryptions could last long enough to track banknotes.

EXAMPLE 4 *Many supermarkets use massive computing power to analyze the buying patterns of their clients. Identifying these patterns enables merchants to reorganize their store layouts to increase their sales. Data mining consists of using computer-based search techniques to sort through the mounds of transaction data captured through goods barcoding. The frequently cited example is the "beer and diapers" example: a large discount chain discovered by data mining its sales data that there was a correlation between beer and diaper purchases during the evening hours. The discount chain therefore moved the beer next to the diapers and increased sales. Let us now consider a merchant who installs RFID readers in his store departments: now he is able to analyze precisely his client's path and thereby to reorganize his store layout. Since some existing payment systems contain names and addresses, a client who stays during a long time in the bicycle department without buying anything will receive directly advertising literature to his door.*

---

[2]We could imagine a scenario where citizens are able to re-encrypt their own banknotes, but it is an unrealistic assumption.

Ciphertext tracking attacks show that the threat described in the first example of Juels and Pappu (See page 38) still occurs within their banknote protection scheme. Let us first consider a milder version of the attack: bar X cannot read the optical data on the banknotes of his customers (We consider that a *customer* is a person who comes in the shop; he does not necessarily need to buy anything). So, he stores in a database all the γ-values that he is able to collect matched with the name and address of their handlers. Merchant Y also reads the γ-values of his clients and stores them. Bar X and merchant Y can combine their databases: if a γ-value appears in both databases, they are almost sure that it is the same client. Let us now consider a stronger attack: when bar X returns change to a client, he re-encrypts banknotes with a fixed number, denoted r_0 also known by merchant Y. When a customer arrives in Merchant Y's store, the merchant reads the γ-value of the customer's banknotes (universally readable) and computes Σ_0 using r_0 (applying the method described in Section 4.2). He then computes $D_0 = h(\Sigma_0)$ and tries to read δ with D_0; if the tag agrees this means that r_0 was the appropriate random number and that merchant Y can be almost sure that this client comes from Bar X. Note that Merchant does not "touch" the banknote here: he has just to scan the people when they pass through the store door for instance.

This issue is inherent in re-encryption-based privacy protection schemes: since re-encryptions cannot be performed very frequently, it is possible to track tags with their universally readable values (even if these values seem to be some garbage for a person who is not authorized to decrypt them). Note that even with a higher re-encryption frequency, the attack still works if the re-encryptions are performed by the merchants, and not by the users themselves.

4.4   Access-key tracking

The goal of the re-encryptions is to prevent banknotes tracking, as we mentionned. If an attacker does not have optical contact with a given banknote, then he should not be able to track it in the long-term. Unfortunately, we demonstrate here that a side channel can be used to track the banknotes. Indeed, if the attacker can see the banknote once (or even, more simply, if he effects the first step of the attack described in Section 4.2) then thanks to the static access-key D, he will be able to track the banknote by just trying to read the δ-cell: the tag responds if and only if the key D is the good one;

This attack is particularly devastating because it dashes the purpose of the scheme. Actually, when a tag owns a unique access-key and responds if and only if the key sent by the reader is the valid one, this key can be used to track the tag. On may think that the tag could thwart such an attack by replying with some garbage when the access-key is wrong, instead of remaining silent. Unfortunately, sending *static* garbage opens a new way to perform tracking attacks, and requiring the tag to be able to generate *random* garbage is not yet realistic due to the low capability of such devices.

EXAMPLE 5 *Mrs Johnson suspects that her husband is having an affair with his secretary. It seems that he has been giving her money. Mrs Johnson decides to read the optical data on her husband's banknotes - in order to generate the access-key - and to surreptitiously follow his secretary after work. She will soon know whether her suspicions are true or not.*

4.5   Cookies threat

According to [8], the sizes of the δ-cell and γ-cell are 195 bits and 585 bits respectively. Since these values can be modified for everyone having access to the banknote (or using the attack described in Section 4.2), the δ-cell and the γ-cell can be used to hide a certain amount of information. This hidden information channel looks like an HTTP cookie. This cookie will however be detected during the next re-encryption of the tag data (since merchants have to check the current value before performing the re-encryption) because δ and γ are not consistent anymore.

A clever way to abuse the tag is to put the cookie only in the δ-cell: since the value r stored in this cell is a random number, it can be used to store some information. Obviously, the γ-cell value will have to be re-encrypted with the new random number. This kind of cookie will be untraceable and will stay available until the next re-encryption.

4.6   Denial of service attack

We saw that when a merchant finds a discrepancy on a banknote, he cannot accept the payment and should warn the law enforcement agency. This could however be used to harm banknote bearers: all that is required is to input incorrect data into either δ or γ. This could be done not only by a merchant who has access to the optical data but also anyone who is able to perform the first step of the attack described in Section 4.2. Due to its simplicity, this malicious attack may bring about many problems as the law enforcement agency as well as the Central Bank – that has to restore the banknotes – would be flooded.

EXAMPLE 6 *In a store, some hackers are waiting in line for the cash register. The customer in front of them pays for his purchases. The hackers eavesdrop the communication between the reader and the tag, thus obtaining the access-key D; they then replace the data stored in the cell γ with a false value, just for fun. The banknote becomes out of service until it is restored by the Central Bank. They can block all cashiers this way and take advantage of panic scenes between cashiers and complaining customers in order to steal goods.*

4.7   Sleeping and dead banknotes

We present here two attacks that are possible using the native commands of the RFIDs. The first one exploits the `sleep` function of the device, and the second uses the `kill` function.

Juels and Pappu point out that banknotes issued from dirty money could pass, for instance, airport checking, by replacing Σ and S by values issued from clean money. Another solution would be to put banknotes into the sleep mode with the `sleep` function. After having passed through bank policy checking, money launderers are able to "wake up" the banknotes. It is therefore important that this function is not universally available. However, if forgers create fake banknotes (by cloning) they will be able to embed a `sleep` function in their tags: law enforcement agents consequently cannot detect the counterfeit banknotes during checking.

We propose now a stronger denial of service attack than those proposed in Section 4.6, involving the `kill` function of the tags. Obviously, this function is a keyed command but the key may be too short to ensure real security. According to the Auto-ID center's standards, the kill-key should be 24 bits [2] or even 8 bits [1]! This means that it would be so simple to perform an exhaustive search to destroy tags. Tag makers should therefore embed longer kill-keys, but the longer the key, the more expensive the tag.

5   Conclusion

We have outlined in this paper the main aspects of banknote protection and described the Juels - Pappu scheme, which is based on both Optical and Radio Frequency Identification systems. We show that two parties can benefit directly from the use of tags in the banknotes: the central bank and the law enforcement agency, both profiting from this system by enforcing banknote tracking and anti-counterfeiting. What about the other interested parties such as merchants and citizens? The role of merchants here is crucial since privacy relies only on their collaboration. Even if most of the merchants are compliant, one can notice that re-encrypting banknotes implies a loss of time for merchants. Another issue is the attitude of a merchant when faced with a problematic banknote; according to [8], he should warn the law enforcement agency. However he is able to repair the problematic data as he has optical access to the banknote. Will he risk losing a customer by warning the law enforcement agency? From the citizens point of view, it would be difficult to tolerate such a system for three reasons. The first one is that citizens have to travel to the central bank (or perhaps to a retail bank) every time they have a faulty banknote. The second reason is that they will lose confidence in their banknotes: they can no longer be sure that they will be able to use their currency at the cash register! Last but not least, they will be suspicious about the fact that their privacy and their anonymity remain intact.

Beyond the sociological issues brought by this scheme, we proved that, despite what the authors claimed, the proposed banknote protection scheme suffers from some privacy issues and thus compromises the privacy of the banknotes' bearers. We have described many attacks that can be performed on the scheme and so, proved that the only RFID banknote protection scheme published until now is null and void and should not be used in practice. Even if this solution could be partially fixed, some attacks are inherent in re-encryption-based privacy protection schemes, as explained in Section 4.3, what strengthens our feeling in the fact that such an approach is not suitable to privacy protection schemes. Furthermore, some described attacks are beyond the scope of the banknote protection, as the access-key tracking attack, and we think that our contribution should be taken into account in future designs of RFID privacy protection schemes.

Acknowledgments

The work presented in this paper was supported (in part) by the National Competence Center in Research on Mobile Information and Communication Systems (NCCR-MICS), a center supported by the Swiss National Science Foundation under grant number 5005-67322. I would like to thank Serge Vaudenay and Lu Yi for their helpful comments on this work.

References

[1] Auto-ID Center. 860MHz-960MHz class I radio frequency identification tag radio frequency & logical communication interface specification: Recommended standard, version 1.0.0. Technical report http://www.autoidcenter.org, Massachusetts Institute of Technology, MA, USA, November 2002.

[2] Auto-ID Center. 13.56MHz ISM band class 1 radio frequency identification tag interface specification: Recommended standard, version 1.0.0. Technical report http://www.autoidcenter.org, Massachusetts Institute of Technology, MA, USA, February 2003.

[3] Dan Boneh, Ben Lynn, and Hovav Shacham. Short signatures from the weil pairing. In Colin Boyd, editor, *Advances in Cryptology – ASIACRYPT'01*, volume 2248 of *Lecture Notes in Computer Science*, pages 514–532, Gold Coast, Australia, December 2001. IACR, Springer-Verlag.

[4] Atmel Corporation. http://www.atmel.com.

[5] Taher El Gamal. A public key cryptosystem and a signature scheme based on discrete logarithms. *IEEE Transactions on Information Theory*, 31(4):469–472, July 1985.

[6] Eiichiro Fujisaki and Tatsuaki Okamoto. Secure integration of asymmetric and symmetric encryption schemes. In Michael Wiener, editor, *Advances in Cryptology – CRYPTO'99*, volume 1666 of *Lecture Notes in Computer Science*, pages 537–554, Santa Barbara, California, USA, August 1999. IACR, Springer-Verlag.

[7] RFID Journal. Michelin embeds RFID tags in tires. http://www.rfidjournal.com/article/view/269, January 2003.

[8] Ari Juels and Ravikanth Pappu. Squealing euros: Privacy protection in RFID-enabled banknotes. In Rebecca N. Wright, editor, *Financial Cryptography – FC'03*, volume 2742 of *Lecture Notes in Computer Science*, pages 103–121, Le Gosier, Guadeloupe, French West Indies, January 2003. IFCA, Springer-Verlag.

[9] Mark Roberti. The money trail – RFID journal. http://www.rfidjournal.com, August 2003.

[10] Sanjay Sarma. Towards the five-cent tag. Technical Report MIT-AUTOID-WD-006, MIT auto ID center, Cambridge, MA, USA, November 2001.

[11] Sanjay Sarma, Stephen Weis, and Daniel Engels. Radio-frequency identification: security risks and challenges. *Cryptobytes, RSA Laboratories*, 6(1):2–9, spring 2003.

[12] Stephen Weis, Sanjay Sarma, Ronald Rivest, and Daniel Engels. Security and privacy aspects of low-cost radio frequency identification systems. In Dieter Hutter, Günter Müller, Werner Stephan, and Markus Ullmann, editors, *First International Conference on Security in Pervasive Computing – SPC 2003*, volume 2802 of *Lecture Notes in Computer Science*, pages 454–469, Boppard, Germany, March 2003. Springer-Verlag.

[13] Junko Yoshida. Euro bank notes to embed RFID chips by 2005. http://www.eetimes.com/story/OEG20011219S0016, December 2001.

SMARTCARD-BASED ANONYMIZATION

Anas Abou El Kalam, Yves Deswarte,
LAAS-CNRS, 7 avenue du colonel Roche, 31077 Toulouse Cedex 4; {Deswarte, anas}@laas.fr

Gilles Trouessin,
ERNST & 3YOUNG, 1 place Alfonse Jourdain, 31000 Toulouse; gilles.trouessin@fr.ey.com

Emmanuel Cordonnier
ETIAM. 20 Rue du Pr Jean Pecker, 35000 Rennes; emmanuel.cordonnier@etiam.com

Abstract: This paper presents a new technique for anonymizing personal data for studies in which the real name of the person has to be hidden. Firstly, the privacy problem is introduced and a set of related terminology is then presented. Then, we suggest a rigorous approach to define anonymization requirements, as well as how to characterize, select and build solutions. This analysis shows that the most important privacy needs can be met by using smartcards to carry out the critical part of the anonymizaton procedure. By supplying his card, the citizen (e.g., the patient in the medical field) gives his consent to exploit his anonymized data; and for each use, a new anonymous identifier is generated within the card. In the same way, reversing the anonymity is possible only if the patient presents his personal smartcard (which implies that he gives his consent). In this way, the use of the smartcard seems be the most suitable means of keeping the secret as well as the anonymization and the disanonymization procedures under the patient control.

Key words: Privacy, anonymization.

1.   INTRODUCTION

Privacy is becoming a critical issue in many emerging applications in networked systems: E-commerce, E-government, E-health, etc. Of course, the networks development facilitates data communication and sharing, but such data are often personal and sensitive. Consequently, the citizen's privacy could be easily endangered, e.g., by inferring personal information.

International [1] and European legislations are more and more worried about protecting personal data and privacy is nowadays considered as a constitutional right in many countries [2, 3, 4]. To comply with these laws, systems dealing with personal data need more and more security and reliability. But unfortunately, the real privacy requirements are sometimes neglected, and solutions intended to solve some privacy problems are often developed empirically. What we first need is to have a systematic methodology to create and manage securely anonymized data.

The next section presents an analytic approach, based on risks analysis, security requirement identification, security objective definition, requirement formalization, and finally, solutions characterization. We have to reply questions such as: which data must be protected and for how long? Do we need unlinkability, unobservability anonymity, or pseudonimity? Where the cryptographic transformations should be carried out? Which transformation should be used, and do we need complementary technical or organizational measures? What is the best trade-off between robustness and flexibility? Etc.

Following our methodology, we suggest in Section 3 a new generic solution to anonymize and link identities. We will show that the use of a smart card is important to meet better privacy needs.

Let us take for example the healthcare systems (the reasoning remains the same for other applications such as E-commerce or E-government). It is clear that if a patient anonymous identifier is used to link the anonymized medical data coming from different sources or at different periods for the same patient, this identifier has to be kept secret to preserve the patient's privacy. In addition, the patient must rationally gives his consent when his data are anonimized or disanonimized. For all these reasons, and for other reasons that we will explain below, we believe that the use of smartcards could be a suitable solution. We suggest that the most critical part of the anonymization procedure should be carried out in a *personal medical data smart card*. The anonymization procedure should be based on a secret, the *patient anonymous identifier*, which is a randomly generated number, stored in the card, and never transmitted out of the card. This anonymous identifier is used to link uniquely the patient's anonymized medical data to the patient, while preserving his privacy.

In this way, the secret is held under the patient control. Except when it is legally mandatory, patient medical data may appear in a certain database only if, by supplying his card, he gives his consent to exploit his medical data as part of a certain project (e.g., epidemiologic research concerning a rare disease). In the same way, the patient consent is necessary to reverse the anonymity, for example, in order to refine epidemiological studies.

In the last section, we show that our solution presents a large flexibility and efficiency in that the data transformation procedures (e.g., anonymization, impoverishment) depend on the purpose of use (i.e., the reason why the data are collected). Moreover, our solution resists dictionary attacks, respects the least privilege principle and fulfills the European legislation requirements. It is also adaptable to non-medical areas, such as demographic studies, and supports (without compromising the security nor the flexibility) some organizational changes like the merging of hospitals.

2. ANALYTIC APPROACH

Based on previous works, we present a set of concepts that are necessary to cover this topic. *Anonymity* can be defined as the state of being not identifiable within a set of subjects. *Pseudonymity* add accountability to anonymity, more precisely, pseudonymity ensures that a user may use a resource or a service without disclosing its identity, but can still be made accountable for that use. *Unlinkability* between two or more items (operations, users initiating operations, terminals where operations were launched from, etc.) means that, within the system, it is not possible to distinguish if these items are related or not. *Unobservability* ensures that a user may use a resource or service without others, especially third parties, being able to observe that the resource or service is being used; therefore, the intent is not only to hide the identity of the person using a resource or a service, but even to hide if the resource or service is used [7; 9].

Besides, the privacy protection necessitates specifying two major categories of concepts:

the *request* (demand) in the form of needs to be satisfied;

the *response* in the form of security functionalities and solutions.

Like most of security functionalities, anonymization analysis can be expressed with regard to three levels of user expectations:

The *anonymization* needs represent the user expectations; generally, their form is neither very explicit nor very simple to formalize.

The *anonymization* objectives specify the information to protect, the threats (against privacy) to avoid, the security level to reach, etc.

The *anonymization* requirements represent how to express the needs (and the threats to counter) with a non-ambiguous semantics and, as far as possible, with a formal system.

The anonymization needs depend on the studied system. For instance in the healthcare systems, some anonymization needs could be:

1. both directly nominative information (identities, sex, addresses) and indirectly nominative information (through a characterization of the corresponding persons) should be anonymized;

2. a patient appears in a certain database (e.g., for a medico-commercial study), only if it is obligatory or if he gives his consent;
3. in the same way, it is necessary to have the patient's consent when reversing this anonymity.
4. the anonymization procedure is based on a secret, the *patient anonymous identifier*, which is used to link uniquely the patient's anonymized medical data to the patient, while preserving his privacy.

Considering the identified needs, the use of a smart card is the most suitable means, and this is the most novel contribution of this paper.

At this step, we can easily suggest how the use of smartcards can help to meet these needs:

1. the smartcards are considered as sufficiently tamper-resistant to prevent, for instance, forging fake patient cards or cloning existing patient cards (in any case, the cost to forge or clone cards would be much higher than the expected profit);
2. with current smartcard technology, it is possible to anonymize nominative data;
3. by providing his smartcard, the patient gives his consent (at least implicitly); in other words, the use of a smartcard is suitable to guarantee the patient consent;
4. to keep secret the patient anonymous identifier, we suggest to generate it randomly, and to store it in the card; if in addition, we carry out the cryptographic transformations (anonymization and disanonymization) in the card, we guarantee that this secret (the identifier) is never transmitted out of the card. Section 4 gives details of our proposition.

In the next subsection we will first present what we mean by anonymization objectives and requirements. We will then refine our requirement analysis by giving details of some typical scenarios. Subsequently, for each scenario, we will identify some anonymization objectives and requirements. As suggested by our approach, our analysis will result in proposing a new anonymization procedure.

2.1   Anonymization objectives

An anonymization objective can be defined according to one of the three following properties, applied to the anonymization function [8]:

Reversibility: this corresponds to hiding data by encryption. In this case, from encrypted data, it is always possible to calculate the corresponding original nominative data (decryption), and conversely (encryption), by using the corresponding keys and algorithms.

Irreversibility: this is the property of anonymization. A typical example is to use a one-way hash function. Once replaced by anonymous codes, the original nominative data is no longer recoverable. However, in some cases, attacks by inference (or by dictionary) are able to re-identify the person if the anonymization is imperfect.

Inversibility: this is the case where it is, in practice, impossible to re-identify the person, except by applying an exceptional procedure. This procedure must be controlled by a highly trustworthy authority such as the medical examiner, the inspector-doctor or a trustworthy advisory committee. This authority can be seen as the privacy guarantor. Actually, it is a matter of a pseudonymization rather than anonymization, according to the common criteria terminology [9].

2.2 Anonymization requirements

The analysis of the requirements must take into account the system environment (users categories, attacks types, etc.). For instance, even if the information is anonymous, a malicious user can infer confidential information by using illegitimate deduction from external data. In this respect, two kinds of requirements must be imposed to any anonymization system: the "*linking*" requirements and the "*robustness*" requirements [8].

Linking allows associating (in time and in space) different anonymous data to the same person, without necessarily disclosing the real identity of that person. As mentioned in Figure 1, linking can be temporal (always, sometimes, never) or geographic (international, national, local, etc.).

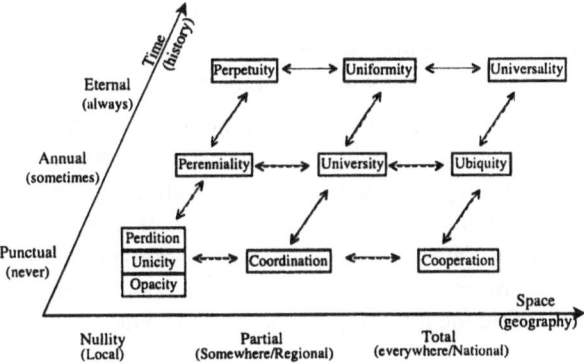

Figure 1. Network of the anonymization cases

The robustness requirements, concerning exclusively illicit disanonymization, can be divided into two distinct cases: robustness to reversion and to inference.

The *reversion robustness* concerns the possibility to inverse the anonymization function, for example if the used cryptographic techniques are not strong enough.

The *inference robustness* concerns data disanonymization by means of unauthorized recovery of nominative data. Generally, an inference can be:

   deductive: it consists in inferring, mainly by first-order logic calculation, unauthorized information on the only basis of publicly available data;

   inductive: if the reasoning that uses data explicitly stored in the information system is not sufficient to infer information, this reasoning can be completed by making some hypothesis on certain information;

   probabilistic: it consists in inferring, by stating a set of various plausible assumptions, a secret information from valid available data.

This list is not exhaustive, and naturally, we can imagine other types of inference channels based on other types of reasoning.

2.3   Solution characterization

For a given scenario, once the privacy needs, objectives and requirement are defined, we have to characterize the most suitable solutions. In particular, we have to identify:

   the *type of the solution* to develop: is it an organizational procedure, a cryptographic algorithm, a one-way function, or a combination of subsets of these solutions?

   the *plurality of the solution*: do we need simple, double or multiple anonymization? Rationally, the choice is related to the type of threats considered against the anonymization function;

   the *interoperability of the* solutions that are to be combined: *transcoding* (manually, for some continuity reasons) or *translating* (mathematically, for some consistency reasons) an anonymization system of anonymous identifiers into another anonymization system of anonymous identifiers; or *transposing* (automatically) several anonymization systems of anonymous identifiers into a unique anonymization system, in order to authorize or forbid the matching of anonymized data.

3. APPLICATION OF OUR METHODOLOGY TO TYPICAL HEALTHCARE EXAMPLES

3.1 Medical data transmission

The sensitivity of the information exchanged between healthcare providers (e.g., between the biology laboratories and the physicians) emphasizes the needs of confidentiality and integrity on transmitted data. Moreover, we need that only the legitimate addressee can receive and read the transmitted data. The use of an asymmetric (or hybrid) cryptographic system seems suitable [17].

The technique used should be reversible when duly authorized (*objective*) and robust to illegitimate reversion (*requirement*).

3.2 Professional unions

In France, for evaluation purpose, the physicians have to send to the professional unions data related to their activity. At first sight, a requirement is to hide patient's and physician's identities. However, when the purpose of use is to evaluate the physician's behavior (to assess care quality), it should be possible to re-identify the concerned physician. Our study of the French law [10] allowed us to identify the following anonymization objectives:

Inversible anonymization (*pseudonymization*) of the physician's identities: only an official body duly authorized to evaluate the physician's behavior can re-establish the real identities.

Inversible anonymization (*pseudonymization*) of the patient's identifiers: only welfare consulting doctors can reverse this anonymity.

In this way, the following risks are avoided:

attempts by a dishonest person to get more details than those necessary to his legitimate task in the system. For example, if the purpose is to study the global functioning of the system, it is not necessary to know the real identities (in accordance with the least privilege principle);

considering that the French law gives to patients the right to forbid the sharing of their information between several clinicians, the identified objectives aim to avoid privacy violation (inasmuch as patients could confide in some clinicians, and only in these clinicians).

3.3   PMSI framework

The *Information System Medicalization Program* (PMSI) aims at evaluating hospital information systems. Actually, it analyses the healthcare establishments activities in order to allocate resources while reducing budgetary inequality [11]. Given that the purpose is purely and simply medico-economic (and not epidemiologic), it is not necessary to know to whom a given medical information belongs (*anonymization*). On the other hand it is important to recognize that different data are related to the same, anonymous person even if they come from different sources at different times (*linkability*). Having said that, every patient must (always) have the same irreversible, anonymous identifier for the PMSI.

3.4   Statutory notification of disease data

Some diseases have to be monitored, through statutory notification, to evaluate the public healthcare policy (e.g., for AIDS) or to trigger an urgent local action (e.g., for cholera, rabies, etc.). Originally, patient records are nominative, but they are irreversibly anonymized before any transmission.

Various needs can be identified: prevention, care providing, epidemiological analysis, etc. The main objective is *anonymization* and *linkability*. Furthermore, universal linking, robustness to inversion, and robustness to inference are the main requirements.

In this respect, the choice in terms of protection must depend on these objectives. In fact, would we like to obtain an exhaustive registry of HIV positive persons? In this case, the purpose would be to know the epidemic evolution, and to globally evaluate the impact of prevention actions. Inversely, would we like to institute a fine epidemiological surveillance of the HIV evolution, from the infection discovery to the possible manifestation of the disease? In this case, the objective is to finely evaluate the impact of therapeutic actions, as well as a follow-up of certain significant cases.

This choice of objectives has important consequences on the nature of data to be collected, on the links with other monitoring systems, and consequently, on the access control policy and mechanisms.

Currently, we identify the following findings related to data impoverishment, to reduce inference risks:

Instead of collecting the zip code, it is more judicious to collect a region code. Obviously, a zip code could allow a precise geographic localization, resulting in identifying a small group of persons.

Instead of collecting the profession, we think that a simple mention of the socio-professional category is sufficient.

Instead of mentioning the country of origin it is sufficient to know if the HIV positive person has originated from a country where the heterosexual transmission is predominant.

3.5   Processing of medical statistical data

Nominative medical data should never be accessible, except when expressly needed for a course of treatment. This applies, in particular, to purely statistical processing and scientific publications. In this respect, not only such data should be anonymized, but also it should be impossible to re-identify the concerned person. Therefore, anonymization inversibility and robustness to inference are essential. Of course, everybody knows that, even after anonymization, identities could be deducted by a malicious statistician if he can combine several well-selected queries, and possibly, by complementing the reasoning by external information.

The problem of statistical database inference has been largely explored in previous works [12, 13]. In the reference [14, ch3] we list some example and solutions, but we believe that it is difficult to decide which solution is the most satisfying. In some cases, the solution could be to exchange the attributes values (in a certain database) so that the global precision of the statistics is preserved, while the precise results are distorted. The inherent difficulty in this solution is the choice of values to be permuted. Another solution could modify the results (of statistical requests) by adding random noise. The aim is to make request cross-checking more difficult.

3.6   Focused epidemiological studies

As mentioned earlier in the introduction, it is sometimes desirable to re-identify patients in order to improve care quality, especially in some focused studies such as cancer research protocols, genetic disease follow-up, etc.

To make this clearer, let us consider a simple example. Suppose that the patients of the category C, having undergone a certain treatment *Tbefore*; and that a certain study concludes that if these patients does not take the treatment *Tafter*, they will have a considerably reduced life expectancy. In such situations, it is necessary to re-identify the patients so that they take advantage of these new results. Having said that, we can conclude that this is a matter of an inversible anonymization. Of course, only authorized persons should be able to reverse the anonymity (e.g., consulting physician, medical examiner), and only when it is necessary.

In the case of cancer research protocols, the process starts by identifying the disease stage, then the protocol corresponding to the patient is identified, and finally, according to this protocol, the patient is registered in a regional, national or international registry. The epidemiological or statistical studies of theses registries could bring out new results (concerning patients following a certain protocol). In order to refine these studies and improve the scientific research, it is sometimes useful to re-identify the patients, to link some data already collected separately, and finally complement the results.

4.   OUR SOLUTION

4.1   General scheme

Previously, we recommended that every anonymization needs a judicious prior analysis. This study must clearly and explicitly identify the security needs, objectives and requirements. After that, we have identified some scenarios and we have applied our approach to these scenarios. Now, we give shape to our analysis by developing a new solution that uses smartcards to better meet privacy needs and to satisfy the identified requirements.

First, in order to decide which view (specific form of data) is accessible by which user, our solution takes into consideration some parameters: the user's role, his establishment, and the purpose of use. Our main aims are to respect the least privilege principle as well as to make use of the legislation related to privacy, especially the European norms [15].

So as to do, before distributing anonymous data to end users, our solution suggests to cascade cryptographic processing (one after the other), carried out in different organizations. Of course, in each step, the transformation to carry out depends on the purpose of the use that follows. The outlines of our solution are first represented in Figure 2, then detailed and discussed in the following sections.

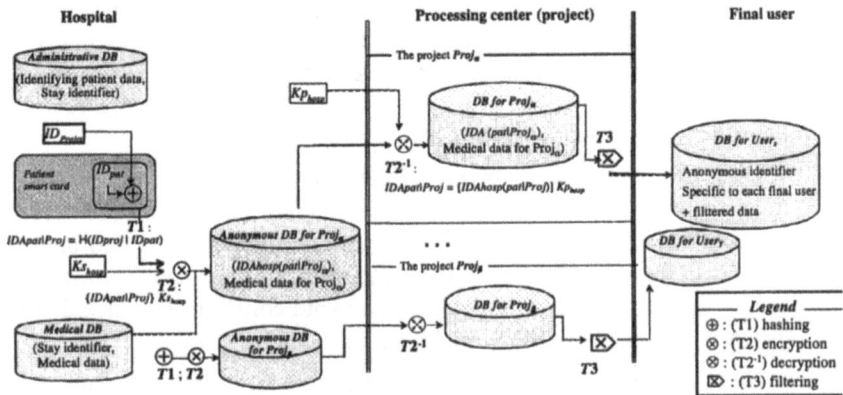

Figure 2. The suggested anonymization procedure.

4.1.1   Transformations processed in healthcare establishments

In healthcare establishments (hospitals, clinics, etc.), three kinds of databases can be distinguished:

an administrative database, accessible to administrative staff (e.g., secretaries, reception staff), according to their role;

a medical database, accessible to clinical staff in charge of the patients;

several anonymous databases, each one containing the necessary and sufficient information for a certain project. A project is an entity that makes statistical or medico-economical data processing such as the PMSI, healthcare insurance companies, associations of diabetic persons, offices for medical statistics, research centers, etc.

The system must possess a well-defined access control policy, and this policy must be implemented by suitable security mechanisms. In the reference [16], we present the *Or-BAC (Organization-Based Access Control)* security model. Or-BAC offers the possibility to handle several security policies associated with different organizations. It is not restricted to permissions, but it also includes the possibility to define prohibitions, obligations and recommendations. In this respect, Or-BAC is able to specify policies developed for a large range of complex and distributed applications.

In our proposal, the transition from a medical database to an anonymized one (dedicated to a certain project) needs the application of two transformations, **T1** and **T2**.

T1: consists in calculating "*IDA$_{pat|Proj}$*", an anonymous identifier per person and per project. *IDA$_{pat|Proj}$* is computed from the two identifiers "*ID$_{proj}$*" and "*ID$_{pat}$*", and characterizes the pair (patient, project):

ID$_{proj}$ is the project identifier; it is known by the healthcare establishments that collaborate with this project;

ID$_{pat}$ is the individual anonymous identifier of the patient; we state that this identifier should be kept under the patient control, on his personal *medical data smart card*; *ID$_{pat}$* is a random number generated uniquely for this patient, and is totally independent from the social security number; a length of 128 bits is sufficient to avoid collisions (the risk that two different persons have the same identifier).

In the healthcare establishment, at the time of supplying data to the anonymous databases (per project), the user (i.e., the hospital employee) transmits *ID$_{proj}$* (the project identifier) to the card. The card already contains *ID$_{pat}$* (the patient anonymous identifier). By supplying his card, the patient gives his consent to transmit his medical data as part of this project. The *T1* procedure, run by the smart card, consists in applying a one-way hash function (i.e., SHA) to the concatenated set (*ID$_{proj}$* | *ID$_{pat}$*):

$$\textbf{(T1)} \qquad IDA_{pat|Proj} = H(ID_{proj} \mid ID_{pat})$$

By generating the fingerprint $H(ID_{proj} \mid ID_{pat})$, *T1* aims at the following objectives:

the patient data appear in a certain database only if it is obligatory or if he has given his consent by producing his patient data card;

IDA$_{pat|Proj}$ does not use any secret whose disclosure would undermine the privacy of other persons (as opposed to the use of a secret key that would be used for all the patients). In addition, since *IDA$_{pat|Proj}$* calculation is run into the card, *ID$_{pat}$* remains into the card; it is never transmitted outside, and it is only used for creating an anonymous database entry (in the hospitals);

since *ID$_{proj}$* is specific to each project, the risks of illicit linkage of data belonging to two different projects are very low; moreover, the anonymous databases (per project) are isolated from external users, and so, can be protected by strict measures of access control;

knowing that *IDA$_{pat|Proj}$* is always the same for the pair (patient, project), every project can *link* data concerning the same patient, even if they are issued by different establishments or at different times, as long as they concern the project.

Nevertheless, the transformation *T1* does not protect against attacks where attackers try to link data held by two different hospitals. To make this clearer, let us take an example where a certain patient Paul has been treated in the hospitals $Hosp_A$ and $Hosp_B$. In each of these two hospitals, Paul has consented to give his data to the project $Proj_\alpha$. Let us assume that Bob, an $Hosp_B$ employee, knows that the fingerprint X (=$IDA_{Paul|Proj\alpha}$) corresponds to Paul, and that Bob obtains (illicitly) access to the anonymous database held by $Hosp_A$ and concerning $Proj_\alpha$. In this case, the malicious user Bob can easily establish the link between Paul and his medical data (concerning $Proj_\alpha$) held by $Hosp_A$ and $Hosp_B$.

In order to face this type of attacks, a cryptographic asymmetric transformation (**T2**) is added. Thus, before setting up the anonymous databases (specific to each project), the hospital encrypts (using an asymmetric cipher) the fingerprint $IDA_{pat|Proj}$ with the key Ks_{hosp} specific to the hospital; (the notation "{M}ᴋ" indicates that M is encrypted with key K):

$$\textbf{(T2)} \qquad IDAhosp_{(pat|Proj)} = \{IDA_{pat|Proj}\}\, Ks_{hosp}$$

If we take again the previous scenario, the malicious user Bob cannot re-identify the patients because he does not know the decryption key Kp_A. In fact, each hospital holds its key Ks_{hosp}, while Kp_{hosp} is held only by the projects.

It is easy to observe that the two transformations ($T1$ and $T2$) allow having a great robustness against attacks attempting to reverse the anonymity (in particular by linking) in an illicit manner.

The procedure, carried out in the hospitals, remains very flexible. Indeed, if two hospitals ($Hosp_a$ and $Hosp_b$) decide to merge someday, it is easy to link data concerning every patient that has been treated in these hospitals. In fact, each hospital decrypts its data with its key Kp_{hosp}, and then encrypts the result by $Ks_{hosp_{ab}}$ the new hospital private key. If $IDAhosp_a{(pat|Proj)}$ (respectively $IDAhosp_b{(pat|Proj)}$) designates an anonymous identifier in $hosp_a$ (respectively $hosp_b$), and "[]ᴋ" designates a decryption with K:

The processing carried out on the former data of $hosp_a$ is:

$$\{\,[IDAhosp_a{(pat|Proj)}]\, Kp_{hosp_a}\,\}\, Ks_{hosp_{ab}}\;;$$

The processing carried out on the former data of $hosp_b$ is:

$$\{\,[IDAhosp_b{(pat|Proj)}]\, Kp_{hosp_b}\,\}\, Ks_{hosp_{ab}};$$

In this way, the resulting fingerprints are the same in the two hospitals (for each anonymous database associated to a certain project).

4.1.2   Transformations carried out when projects receive data

Data contained in the anonymous databases (in the hospitals) undergoes transformations that depend on $IDA_{proj|pat}$ and on Ks_{hosp}. Every processing center (project) decrypts received data by using Kp_{hosp}:

$$[IDAhosp_{(pat|Proj)}]\, Kp_{hosp}$$

according to **(T2)**, $= [\{IDA_{pat|Proj}\} Ks_{hosp}] Kp_{hosp} = IDA_{pat|Proj}$

The processing center thus retrieves, for each patient, the persistent anonymous identifier dedicated to the project, $IDA_{pat|Proj}$, i.e., an information that is sufficient and necessary to link data corresponding to each patient, even if they come from different establishments at different times.

4.1.3   Transformations carried out before the distribution to the end users

Before their distribution to the end users (scientist researchers, web publishing, press, etc.) the information can undergo a targeted filtering. For instance, this can be done by applying data aggregation or data impoverishment.

If an additional security objective is to control if end users can link information, it is advisable to apply another anonymization (e.g., by MD5) with a secret key $Kutil_{|proj}$ generated randomly.

$$IDA_{pat|util} = H(IDA_{pat|Proj} | Kutil_{|proj})$$

In accordance to needs, this transformation corresponds to two different cases:

If the aim is to allow the full time linking (per project for that particular user), the key $Kutil_{|proj}$ has to be stored by the processing center, so that it can reuse it each time it transmits information to this end user.

Inversely, if the center wishes to forbid users to link data transmitted at different times, the key $Kutil_{|proj}$ is randomly generated just before each distribution.

4.2   Discussion

By using the personal smartcard, the solution that we suggest guarantees the following benefits:

The smartcards are sufficiently tamper-resistant.

The secret as well as the anonymization and the disanonymization procedures are held under the patient control. Not only the patient identifier is never transmitted out of the card, but also the generation of project-specific anonymous identifiers is carried out within the card.

The patient's consent must be provided for each non-obligatory, but desirable, utilization of his anonymized data. Indeed, the patient medical data can appear in a certain database only if, by supplying his card, the patient gives his consent to exploit his medical data as a part of a certain project (e.g., epidemiological research concerning a rare disease). The same goes when reversing the anonymity: the patient consent is necessary. Let us take the example where the end user (e.g., researcher in rare diseases) discovers important information that necessitates re-identifying the patients. At first, it sends back results to the hospitals participating to the concerned project (e.g., a given orphan disease study). Only if the patient produces his medical data card (which implies that he gives explicitly his consent), it is possible to calculate $IDApat|Proj = H(IDproj \mid IDpat)$ and $IDAhosp(pat|Proj) = \{IDApat|Proj\}_{Kshosp}$, and so, to establish the link between the patient, his anonymous identifiers, and his medical data. A simple (and automatic) comparison between the anonymous identifier and the inversion list, would allow triggering an alarm. This alarm asks the patient if he wants to consult the results.

The solution resists dictionary attacks that could be led in different organizations: hospitals, processing centers and end users.

The combination of the suggested anonymization procedure with access control mechanisms satisfies the non-inversibility requirement as well as the least privilege principle.

It is possible to merge data belonging to several establishments without compromising security nor flexibility;

In accordance with European legislation, our solution takes the purpose of use into account. Moreover its fine-grain analysis allows to easily adapt it to different sector needs (e.g., social domain, demographic projects, other scientific research areas, etc.).

Currently, French hospitals anonymize patient identities by using a one-way hash coding based on the standard hash algorithm (SHA). The principle is to ensure an irreversible transformation of a set of variables that identify an individual (last name, first name, date of birth, sex). In order to link all the information concerning the same patient, the anonymous code obtained is always the same for the given individual (and for all uses). Two keys have been added before applying SHA. The first pad, k1, is used by all senders of information as follow "$Code_1 = H(k_1 \mid Identity)$"; and the second, k2, is applied by the recipient "$Code_2 = H(k_2 + Code_1)$". Nominal information is therefore hashed twice, consecutively with these two keys. The aim of pad k1 (resp. k2) is to prevent attacks by a recipient (resp. a sender) (Figure 3).

However, this protocol is both complex and risky: the secret key should be the same for all information issuers (clinicians, hospitals) and stay the same over time. Moreover, this key must always remain secret: if this key is corrupted, the security level is considerably reduced. It is very difficult to keep secret during a long time a key that is largely distributed. This means that new keys have to be distributed periodically. The same applies when the hash algorithm (or the key length) is proven not sufficiently robust any more. But, how can we link all the information concerning the same patient before and after changing the algorithm or the key? If this problem occurs, the only possible solution consists in applying another cryptographic transformation to the entire database, which may be very costly.

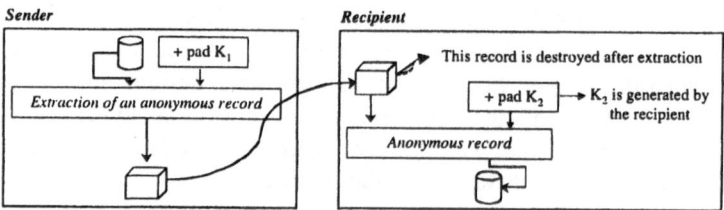

Figure 3. Outline of the current French hashing procedure

Inversely, in our solution, the identifiers (*IDproj*, *IDpat*, *IDApat|Proj* and *IDApat|util*) used in the various transformations are located in different places. Similarly, the keys (*Kshosp*, *Kphosp*) are held by different persons. Indeed, *IDproj* concerns a unique project; the pair (*Kshosp*, *Kphosp*) is specific to one hospital; *IDApat|util* is dedicated to a single end user. It is therefore practically impossible to make illicit disanonymization. Moreover, *IDpat* is specific to one patient, and only held on his card. So even if a certain *IDpat* (corresponding to Paul, for example) is disclosed (which is not easy!), only Paul's privacy could be endangered (but not all the patients' privacy, as it is the case in the current French procedure) and only for certain projects.

Furthermore, according to the security needs of the studied cases, we suggest to complement our solution by other technical and organizational security mechanisms:

the access to data has to be strictly controlled; a well-defined security policy must be implemented by appropriate security mechanisms (hardware and/or software);

sometimes, it is advisable to use thematic anonymizations, so that even if a certain user succeeds in breaking the anonymity, risks (particularly inference risks) are limited;

in some particular contexts, it is more efficient to completely remove identifying data from medical data.

as complementary deterrent or repressive measures, it is recommended to control the purpose of use by implementing intrusion detection mechanisms; in particular, these mechanisms should easily detect sequences of malicious requests (illicit inferences, abuse of power, etc.).

CONCLUSION

In an electronic dimension that becomes henceforth omnipresent, this paper responds to one of the major recent concerns, fathered by the new information and communication technologies: the respect of privacy.

In this framework, we firstly analyzed the anonymization in the medical area, by identifying and studying some representative scenarios. Secondly, we have presented an analytic approach putting in correspondence anonymization functionalities and adequate solutions. Finally, we suggested a new procedure adapted to privacy needs, objectives and requirements of healthcare information and communication systems. This fine-grain procedure is generic, flexible and could be adapted to different sectors. The use of the smartcards in this procedure responds to some security needs.

Although this solution suggests successive anonymizations, the cryptographic mechanisms that it uses are not expensive in terms of time and computation resources, and are compatible with current smartcard technology. We are currently implementing a prototype of this solution, using Java Cards, and we will soon be able to measure the performance and complexity of a real application.

REFERENCES

[1] The resolution A/RES/45/95 of the General assembly of United Nations: *"Guidelines for the Regulation of Computerized Data Files"*; 14 December 1990.

[2] Directive 2002/58/EC of the European Parliament on: *"the processing of personal data and the protection of privacy in the electronic communications sector"*; July 12, 2002; Official Journal L 201, 31-7-2002, p. 37-47.

[3] Directive 95/46/CE of the European Parliament and the Council of the European union: *"On the protection of individuals"*; October 24, 1995.

[4] Recommendations R(97)5 of the Council of Europe, *On The Protection of Medical Data Banks*, Council of Europe, Strasbourg, 13 February 1997.

[5] Loi 78-17 du 6 janvier 1978 relative à l'Informatique, aux fichiers et aux libertés, Journal officiel de la République française, pp. 227-231, décret d'application 78-774 du 17 juillet 1978, pp. 2906-2907.

[6] Loi 94-43 du 18 janvier 1994 relative à la santé publique et à la protection sociale, art. 8.

[7] A. Pfitzmann, M. Köhntopp, "Anonimity, Unobservability, and Pseudonymity – A Proposal for Terminology", *International Workshop on Design Issues in Anonymity and Unobservability*, Berkley, CA, USA, July 25-26, 2000, Springer.

[8] Trouessin, G (1999). "Dependanility Requirements and Security Architectures for Healthcare/Medical Sector", 18th *International Conference SAFECOMP'99*, Toulouse, France, September 1999, Springer, pp. 445-458.

[9] *Common Criteria for Information Technology Security Evaluation, Part 1: Introduction and general model*, 60 p., ISO/IEC 15408-1 (1999).

[10] Loi 94-43 du 18 janvier 1994 relative à la santé publique et à la protection sociale, art. 8.

[11] Circulaire n° 153 du 9 mars 1998 relative à la généralisation dans les établissements de santé sous dotation globale et ayant une activité de soins de suite ou de réadaptation d'un recueil de RHS, ministère de l'emploi et de la solidarité, France.

[12] D. Denning et P. Denning, "Data Security". *ACM Computer Survey*, vol. 11, n° 3, September 1979, ACM Press, ISBN : 0360-0300, pp. 227-249.

[13] S. Castano, M. G. Fugini, G. Martella, P. Samarati, "*Database Security*", 1995, ACM press, ISBN: 0201593750, 456 pp.

[14] A. Abou El Kalam, "*Modèles et politiques de sécurité pour les domaines de la santé et des affaires sociales*", Thèse de doctorat, Institut National Polytechnique de Toulouse, 190 pp., 4 December 2004.

[15] CEN/TC 251/WG I, *Norme prENV 13606-3: Health Informatics - Electronic Healthcare Record Communication*, n° 99-046, Comité Européen de Normalisation, 27 May 1999.

[16] A. Abou El Kalam, P. Balbiani, S. Benferhat, F. Cuppens, Y. Deswarte, R. El-Baida, A. Miège, C. Saurel, G. Trouessin "Organization-Based Access Control", *4th International Workshop on Policies for Distributed Systems and Networks (Policy'03)*, Como, Italy, 4-6 June 2003, IEEE Computer Society Press, pp. 120-131.

[17] A. *Menezes*, P. C. Van Oorshot, S. A. Vanstone, "*Handbook of Applied Cryptography*", 1997, CRC press, ISBN : 0849385237, pp. 780.

PRIVACY PROTECTING PROTOCOLS
FOR REVOKABLE DIGITAL SIGNATURES

István Zsolt BERTA
Laboratory of Cryptography and Systems Security,
Department of Telecommunications,
Budapest University of Technology and Economics
istvan.berta@crysys.hit.bme.hu

Levente BUTTYÁN
Laboratory of Cryptography and Systems Security,
Department of Telecommunications,
Budapest University of Technology and Economics
buttyan@hit.bme.hu

István VAJDA
Laboratory of Cryptography and Systems Security,
Department of Telecommunications,
Budapest University of Technology and Economics
vajda@hit.bme.hu

Abstract   Consider an application where a human user has to digitally sign a message. It is usually assumed that she has a trusted computer at her disposal, however, this assumption does not hold in several practical cases, especially if the user is mobile. Smart cards have been proposed to solve this problem, but they do not have a user interface, therefore the user still needs a (potentially untrusted) terminal to authorize the card to produce digital signatures. In order to mitigate this problem, we proposed a solution based on conditional signatures to provide a framework for the repudiation of unintended signatures. Our previous solution relies on a trusted third party who is able to link the issuer of the signature with the intended recipient, which may lead to severe privacy problems. In this paper we extend our framework and propose protocols that allow the user to retain her privacy with respect to this trusted third party.

1.   Introduction

Cryptographic protocols are often described informally in terms of message passing between hypothetic principals usually called *Alice* and *Bob*. These principals perform cryptographic computations and communicate through a network. When a protocol participant is supposed to be a human, it is implicitly assumed that she uses a *terminal* (e.g., a PC), which stores cryptographic keys, performs cryptographic computations, and handles network connections on behalf of her. It is also implicitly assumed that the terminal is trusted by the user for behaving as expected, and in particular for not compromising the security of the user (e.g., by leaking her keys).

Unfortunately, many terminals (especially public ones) should not be trusted by the user. For instance, a terminal that is installed at a public place (e.g., a PC in a hotel or in an airport lounge), or a terminal operated by an untrusted principal (e.g., a POS terminal of an unknown merchant in a foreign country) fall into this category. In the former case, anyone can access the terminal and install malicious software on it that alters its behavior; in the latter case, the operator of the terminal can easily do the same.

In order to defend against untrusted terminals, the user may carry a trusted computer with her everywhere. A convenient solution is a smart card, because it is small and users are already used to carry them in their pockets. However, smart cards lack user interface, which enables various new attacks, that are not possible in case of traditional computers. This is why Schneier and Schostack [Schneier and Shostack, 1999] called smart cards 'handicapped computers'. A typical example for such an attack is the following man-in-the middle attack: If the user wishes to compute a digital signature using her smart card, the terminal can send a different message to the card, a message the user would never want to sign. Unfortunately, the user at an untrusted terminal cannot prevent this attack, so the malicious terminal may obtain the user's digital signature for an arbitrary message.

In [Berta et al., 2004], we proposed a solution that could be used to solve or at least alleviate the above problem. Our solution relies on a concept called *conditional signature* [Lee and Kim, 2002]. In our protocol, whenever the user wants to sign a message at an untrusted terminal, the card adds a condition to it before the message is signed. Thus, the terminal does not obtain the user's regular signature, but it obtains her conditional signature (a signature on the message together with the condition). A conditional signature is equivalent with a regular signature if and only if the condition is fulfilled and the signature is valid. In our model, we assume that from time to time the user has access to trusted terminals too. Whenever she approaches a trusted terminal, she may review messages she signed at untrusted ones, and – depending on the exact protocol – she may fulfill or prevent their appropriate conditions, and thus confirm or revoke the signatures. If a user does not confirm or revoke a conditional signature before a certain deadline, the condition obtains a default truth value. In a nutshell, our solution is a framework for the controlled revocation of digital signatures.

From a practical point of view, those conditional signature protocols seem to be most usable, that make signatures automatically confirmed, unless they are explicitly revoked by the user. We assume, that most terminals – though untrusted – do not actually mount attacks on the user. Similarly, most users are not criminals and do not repudiate messages they did intend to sign. Protocols that make signatures automatically confirmed allow well-behaving merchants (terminals) to do business successfully even with those users who tend to forget to confirm their signatures.

Unfortunately, all of these protocols that automatically confirm signatures unless revoked require a trusted third party to operate. (see Section 3.4) In these protocols, the trusted third party is able to link the user with the recipient of the digital signature, which may lead to severe privacy problems in some applications. The user may trust the TTP for assisting her in revoking unin-tended signatures, but she may not want the TTP to know when, where and on what messages she generated signatures (especially intended ones). Requiring the TTP to be trusted only for the revocation of unintended signatures makes users less reluctant to use its services, and thus allow more service providers to get into the signature revocation service provider business. If the TTP would need to know less about the user, perhaps more organizations could fulfill this task, and conditional signature schemes protecting the user from the attacks of malicious terminals could become more widespread.

In this paper, we propose protocols based on smart cards and conditional signatures, that allow the user to retain her privacy (and often provide even unconditional privacy) with respect to the trusted third party.

The organization of the paper is the following: In Section 2, we report on related work. In Section 3, we define our system model and present our solution based on conditional signatures. In Section 4 we show three protocols that extend our previous solution by allowing the user to retain her privacy with respect to the trusted third party. Finally, in Section 5, we conclude the paper.

2. Related work

2.1 Untrusted terminals

Terminal identification is perhaps the most basic problem addressed in the literature. In this most simple model, terminals are categorized into two main groups: terminals trusted by the user and untrusted terminals. Naturally, trusted terminals have to be tamper resistant, otherwise tampered terminals could serve an attacker while still being able to authenticate themselves. Asokan et al.

[Asokan et al., 1999] and Rank and Effing [Rankl and Effing, 1997] show a simple protocol, that – using smart cards and one-time passwords – enables the authentication of terminals.

Other papers consider the terminal untrusted, and propose solutions for its secure usage. The problem of man-in-the-middle attacks of untrusted terminals was addressed by Abadi et al. [Abadi et al., 1992] first, who analyzed the dangers of delegation of rights to a terminal. They examined, what additional peripherals a card needs to have to solve the above problem. Clarke et al. [Clarke et al., 2002] also propose a solution based on a super smart card. Other authors tried to come up with solutions that are workable with cards that exist today. [Stabell-Kulo et al., 1999], [Berta and Vajda, 2003]

According to Rivest [Rivest, 2001] there is a fundamental conflict between having a secure device and having a 'reasonable customizable user interface' that supports downloading of applications. He suggests, that digital signatures should not be considered non-repudiable proofs, but simply plausible evidence. Thus, users should be given well-defined possibilities for repudiating such signatures.

Since smart cards did not solve the problem of untrusted terminals, other ideas emerged. Pencil-and-paper cryptography (or human-computer cryptography) tries to give the user methods to protect the secrecy or authenticity of the message without the help of a smart card. The most notable solutions based on human-computer cryptography are [Schneier, 1999] and [Matsumoto, 1996], and those, that rely on visual cryptography ([Naor and Shamir, 1995], [Naor and Pinkas, 1997]). Unfortunately, most of these solutions are rather awkward, and require the user to perform complex or unusual operations.

None of the above solutions try to protect the privacy of the user.

2.2   Anonymous payment systems

In case of any protocol that protects privacy, anonymous communication is a key element. Mixes yield a typical solution for anonymous communication, they were introduced in [Chaum, 1981]. A Syverson et al. [Syverson et al., 1997] describe a famous solution that uses a network of mixes for communication on the WWW. A simpler solution is [Anonymizer Inc., 1999], that provides a proxy for its users.

Electronic payment systems require a balance between anonymity and traceability. Chaum introduced an anonymous payment system (the late DigiCash), it is based on blind signatures. [Chaum, 1982] The foundations of some other famous anonymous payment systems are introduced in [Brands, 1994] and [Franklin and Yung, 1992]. Jakobsson et al. [Jakobsson and Raïhi, 1998] also propose an electronic payment system in which anonymity is based on a mix-network.

The trusted third party in this paper is in a position very similar to that of a bank or trustee in the above papers. However, in contrast to the above papers, we discuss the privacy of a user who is only able to perform cryptographic operations using her smart card beyond the untrusted terminal. We also rely on the existence of methods for anonymous communication.

3. A protocol without privacy

3.1 Model and assumptions

We consider a system with human users who want to generate digital signatures at untrusted terminals. *User U is a human* and has limited memory and computational power. She is able to memorize some passwords or PIN codes, but cannot memorize cryptographic keys, neither can she perform cryptographic computations. For this reason, the private key of U is stored and the signatures are generated by a smart card C in possession of user U.

Essentially, *smart card C* is a trusted personal microcomputer without direct interfaces towards U. C is connected to the terminal in front of U, and all messages between C and U, must pass through the terminal. Smart card C is assumed to be trustworthy and tamper-resistant. The card is *tamper-resistant*, so it is impossible to alter its behavior, reverse engineer it or extract information from it. Card C is *trustworthy*, because it is manufactured by a trusted manufacturer. Since smart cards undergo extremely rigorous evaluation and certification, and not even the manufacturer can alter their behavior after issuance, we consider this assumption to be justified. Thus, if properly authenticated (e.g. by a challenge and response method), smart card C is assumed to be a trusted party.

We assume that the *untrusted terminal T* in front of U is fully under the control of an attacker, who may have installed all kinds of malicious software on the terminal before U started to use it. Since T is able to modify messages that U sends to C for signing, the attacker can obtain a signature from the smart card for an arbitrary message.

However, we assume, that from time to time, U has access to C from trusted terminals too. Such a trusted terminal could be the home PC of U, but it can also be a terminal operated by a trusted organization and believed to be tamper resistant (e.g., an ATM machine). Of course, in order to use a terminal for this purpose, it must be properly authenticated first.

We denote by M *the intended recipient of the digital signature* generated by C. M could be a service provider, a merchant, another user, etc. It is a well-spread scenario that terminal T is operated by service provider M. So, we assume, that M and T may cooperate against the user.

We are going to assume that there is *trusted third party TTP* in the system that both U and M trust. However, user U would still like to retain her privacy with

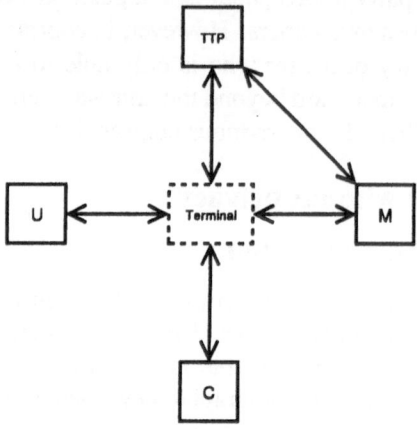

Figure 1.   Physical connections

respect to TTP. This means, that while U trust TTP only for the revocation of unintended signatures, but U would like to prevent TTP from knowing where, when and what messages she signed. Thus, she would like to prevent TTP from knowing, which merchants or service providers she does business with. TTP follows the given protocols, and does not try to cheat by breaking into the terminal or intercepting messages for other parties. Neither does TTP collaborate with T or M to reveal the identity of the user.

A note on the background of this latter assumption: The identity of the user can neither be hidden from M (otherwise M would not be able to verify the signature of U), nor from T (because U is physically present at T who can even take a photograph of her). This means, in case TTP collaborates with T or M, the user is unable to retain her privacy with respect to TTP.

The physical connections defined so far are illustrated in Figure 1.

3.2   Conditional signatures

Conditional signatures were introduced by Lee and Kim [Lee and Kim, 2002], who used this concept for solving fair exchange problems without expensive cryptographic primitives like verifiable escrow. A conditional signature of U on a message m is U's ordinary signature $sig_U(m, c)$ on m and a description of a condition c. If $sig_U(m, c)$ is correct and condition c is true, then $sig_U(m, c)$ is considered to be equivalent with $sig_U(m)$, U's ordinary digital signature on m. However, if c is false, then U is not responsible for m. Intuitively, U's conditional signature is U's commitment: *"I signed m, but if c is not true, then my signature on m is not valid."*

We proposed the use of conditional signatures against malicious terminals in [Berta et al., 2004]. Since it seems to be impossible to prevent the terminal from obtaining signature from the card on an arbitrarily chosen document, instead of generating an ordinary signature, we proposed that C generates a conditional signature.

In our proposed schemes, it is guaranteed that the condition cannot become true before a certain amount of time has passed. This leaves time for the user to access a trusted terminal for checking the signatures generated by the card, and to ensure that the conditions of the fake signatures can never become true.

3.3 The protocol

In this subsection we present a protocol we proposed in [Berta et al., 2004], that allows the user to repudiate signatures computed at untrusted terminals, but does not allow her to retain her privacy with respect to TTP.

User U approaches an untrusted terminal:

Step 1: $U \rightarrow T$: m
U types message m using the keyboard of the terminal.

Step 2: $T \rightarrow C$: m

Step 3: $C \rightarrow T$: $c, sig_U(m, c)$
The card logs m and, computes the conditional signature $sig_U(m, c)$ of U on m. In the most simple case, condition c is the following string:
"My signature on m is valid if and only if deadline t has passed and TTP countersigned this conditional signature."

Step 4: $T \rightarrow M$: $(m, c, sig_U(m, c))$

Later, at a trusted terminal:

Step 5: $C \rightarrow U$: M, m, c
Before deadline t, U reviews the list of messages logged by C at a trusted terminal. This can be done, for instance, by U returning to her home and inserting C into the smart card reader of her home PC. Before outputting its log, C authenticates the terminal to be sure that it is a trusted one.

If user U did not intend to sign m, then:

Step 6: $U \rightarrow TTP$: *I revoke my signature $sig_U(m, c)$*.

Otherwise, the signature becomes confirmed after deadline t:

Step 7: $M \rightarrow TTP$: $c, sig_U(m, c)$

If deadline t has passed and user U did not revoke her signature in Step 6:

Step 8: $TTP \rightarrow M$: $sig_{TTP}(sig_U(m, c))$
TTP countersigns the conditional signature.

A third party needs to check if the digital signature $sig_U(m, c)$ of the card is correct and condition c is true in order to verify a conditional signature.

We assume, that the card is able to log messages. While smart cards have severe memory limitations, it is possible for them to outsource logging to an external log server. In [Berta et al., 2004] we show a protocol that solves this problem.

However, the above protocol does require the user to reveal her identity to TTP. Naturally, the easiest way to protect the privacy of the user from TTP would be to eliminate TTP from the protocol. In the next subsection, we examine the possibility of this option.

3.4   Need for a trusted third party

In most of the applications, it is desirable that the status of a digital signature does not vary in time. In our scheme, this is not fully supported, since every signature is invalid until t, and then it may become valid. There is a good reason to allow this, namely to mitigate the untrusted terminal problem. We note, however, that any conditional signature scheme must guarantee that after t, the status of the signature becomes stable. In particular, once the user reviewed and accepted a signature, it cannot be revoked.

It seems to be a good idea to define a default truth value for c after t that cannot be changed later, because this ensures that the status of each signature will indeed become stable after t independently of the negligence of the involved parties. In other words, if user U does not confirm or revoke the signature before deadline t, then its status will take a default value and U can no longer do anything about it.

Depending on the default truth value, we can distinguish between two classes of protocols. Protocols in the first class support the *default deny* approach, where a signature remains invalid after t (and forever), unless it is explicitly confirmed by the user before t. Such protocols may operate without TTP, we demonstrate an example for this in [Berta et al., 2004]. However, protocols of this class might not be suitable in certain applications, because users may tend to forget to confirm conditional signatures, which means that the signatures are automatically revoked. Protocols in the second class support the *default accept* approach, where a signature automatically becomes valid after t (and remains so forever) unless it is explicitly revoked by the user before t. The majority of terminals do not mount attacks on the user, so forgetting to review a signature rarely has severe consequences. We consider it a critical issue, not to prevent terminals from doing business with those users who tend to forget to confirm signatures. In these protocols it is the interest of the user to review signatures made on untrusted terminals, so she is less likely to forget this important protocol step.

While protocols supporting the *default accept* approach seem more practical, all of them require a TTP. In these protocols, after Step 3, the conditional signature is in the hands of the untrusted terminal (or the untrusted merchant). Condition c (and the signature of the user) will become valid after a certain deadline automatically, unless U revokes it. If U has to revoke it at T or M, either of T or M could simply repudiate the receipt of such a revocation. While smart card C is also a trusted party, C is not online continuously. Thus, it is impossible for C to maintain a signature revocation list that is trusted and can be checked by all other parties. Thus, TTP is needed to enforce a default value for the signature and to handle revocation.

It seems that all of the practical protocols require the help of a TTP. However, if the TTP is able to log all the contracts a user signs, it is in a very critical position. Few organizations would be trusted enough to be a TTP in such protocols. We reckon, that if the protocol prevented the TTP from linking the user with the merchant or service provider, more organizations would qualify to be a TTP. In the next section we propose three protocols that would allow user U to retain her privacy with respect to TTP.

4. Protocols to protect the user's privacy

Our goal is to develop a protocol for signature revocation, that allows U *to retain her privacy with respect to TTP*. While U may trust TTP for signature revocation, U does not want TTP to know, where, when and what messages she wanted to sign. Thus, apart from hiding the message from the eyes of TTP, U should hide her identity too, and the fact that she sent a message to M. Thus, she needs to protect the *confidentiality of message m and digital signature* $sig_U(x)$ she computes on any x.

It is clear, that she does not want to protect them against M, because she intends to send message m to service provider M. Moreover, she cannot protect m against T, because she types the message using the keyboard of the terminal in Step 1.

All three protocols we propose follow the concept of the protocol in Section 3.3. In Step 3, the smart card sends a cryptogram encrypted by the public key of TTP that contains condition c along with revocation token r. This cryptogram is forwarded to the merchant and later to TTP in Step 7. The user receives revocation token r from the card via a trusted terminal, and may repudiate her signature by submitting r to TTP in Step 6 via an *anonymous channel*. We assume, that such an anonymous channel exists. (Naturally, anybody may repudiate the signature using r, so it is advisable not to let the untrusted terminal compute it. Otherwise, T would be able to repudiate messages in the name of U, and could spoil the reputation of the user.)

TTP decrypts the cryptogram that was sent by the merchant in Step 7, and enforces condition c to become true (in Step 8) if the revocation token r inside the cryptogram was not submitted before. Revocation token r is a random number, statistically independent from the identity of U, the contents of message m or the value of conditional signature $sig_U(m, c)$. Based on r, TTP is unable to link U with M. (Note, that the identity of M is not hidden from TTP.)

While TTP needs to store revocation token r, it may not be necessary to store it forever. This problem could be solved e.g. by introducing a lapse time, so TTP could refuse to validate very ancient conditional signatures.

4.1   Bit commitment

Our first protocol follows the spirit of bit commitment protocols. [Schneier, 1996] User U commits herself to her signature to M. However, U does not reveal her signature to M immediately, only after deadline t contained in c. In this case, condition c is the following string: *"My signature on the above message is not valid before deadline t."*

In contrast to the classical bit commitment, the "reveal" phase is not performed by user U (because of reasons described in 3.4), but by trusted third party TTP in Step 8. Moreover, in this case not even TTP is allowed to read the bits U committed herself to. Thus, in this case not all known bit commitment methods can be used (e.g., the solution proposed in [Naor, 1991] cannot be used in this case), only those that do not require sending bits (the user committed herself to) in cleartext when they are revealed.

We propose the following protocol to protect the privacy of U with respect to TTP:

Step 1: $U \to T$: m

Step 2: $T \to C$: m

C generates random symmetric key k and revocation token r.

Step 3: $C \to T$: c, $E_k[sig_U(m)]$, $E_{TTP}(r, k, c)$

Step 4: $T \to M$: m, $c, E_k[sig_U(m)]$, $E_{TTP}(r, k, c)$
   Unlike in the protocol described in Section 3.3, terminal T is unable to verify the signature in this step. However, C is a device trusted by T, so T may assume, that C follows the protocol, and is not sending garbage.

Later, at a trusted terminal:

Step 5: $C \to U$: M, m, c, r

If user U would like to repudiate the signature on message m then

   Step 6: $U \to TTP$: r (via an anonymous channel)

After deadline t:

Step 7: $M \rightarrow TTP$: $E_{TTP}(r, k, c)$

If deadline t has passed, and r was not submitted to TTP, then:

> **Step 8:** $TTP \rightarrow M$: k
>
> **Step 9:** M decrypts $E_k[sig_U(m)]$ using k and obtains $sig_U(m)$.

It is an important merit of this protocol, that a third party needs to have m and $sig_U(m)$ only in order to verify the conditional signature of U. Since, this conditional signature is not different from a regular one, its verification requires the same procedure too. Note, that in this protocol TTP does not have to perform a digital signature operation.

In this protocol, U is able to retain a *provable degree of privacy* with respect to TTP. In case everyone behaves honestly, r is not sent in Step 6. In this case TTP only receives cryptogram $E_{TTP}(r, k, c)$. Since r and k are random numbers, and c is the same string in case of all users in the system, each of them is independent from U, m and $sig_U(m)$. Thus, if everyone behaves honestly, U has unconditional privacy.

On the other hand, if U chooses to revoke the signature, TTP receives random number r too. Naturally, r is independent from U, m and $sig_U(m)$, so in this case, the privacy of user U is equal to the one provided by the channel that is used in Step 6 for signature revocation.

4.2   Blind signatures

Our next protocol relies on the concept of blind signatures introduced in [Chaum, 1982]. In this scheme, the cryptogram the card outputs in Step 3 contains the conditional signature of the user. TTP has to countersign the conditional signature in order to validate it, without being able to read it. To prevent TTP from reading the conditional signature, the signing process of TTP is blinded by the card. Smart card C also releases a token that M will be able to use to unblind the signature.

In this case, condition c if the following: *"My signature on the above message is valid if and only if deadline t has passed and TTP countersigned it."*

We are going to use the following notation:

- The signature of the user on message x is denoted as $sig_U(x)$.
- TTP has two key pairs. One of them is an RSA key pair for digital signatures, the other keypair is for an arbitrary algorithm for encryption/decryption.

  - The RSA keypair of TTP for digital signature is the following: the public key is e, the public modulus is m, and the private exponent d. According to the RSA algorithm, the signature of TTP on message x is $x^d \bmod m$.

Note, that an attacker may obtain signature (or decryption) from this keypair on an arbitrary message, so TTP should not use this keypair for any other purpose.

– Encryption of message x with the public encryption key of TTP is denoted as $E_{TTP}(x)$. Decryption of message y with the private decryption key of TTP is denoted as $D_{TTP}(y)$.

■ Symbol "$*$" stands for multiplication modulo m. Modulus m should be larger than the largest possible value of $sig_U(x)$.

The proposed protocol is as follows:

Step 1: $U \rightarrow T$: m

Step 2: $T \rightarrow C$: m

Step 3: $C \rightarrow T$: $c, b, E_{TTP}(c, r, sig_U(m, c) * b^e)$
C generates random numbers r and b, where r is a repudiation token and b is going be used to blind the signature of TTP.

Step 4: $T \rightarrow M$: $m, c, b, E_{TTP}(c, r, sig_U(m, c) * b^e)$

Later, at a trusted terminal:

Step 5: $C \rightarrow U$: M, m, c, r

If user U would like to repudiate the signature, then:

Step 6: $U \rightarrow TTP$: r (via an anonymous channel)

After the t deadline:

Step 7: $M \rightarrow TTP$: $E_{TTP}(c, r, sig_U(m, c) * b^e)$

If deadline t has passed and r was not submitted to TTP, then:

Step 8 : $TTP \rightarrow M$: $[sig_U(m, c)) * b^e]^d = [sig_U(m, c)]^d * b$
Step 9: M acquires $[sig_U(m, c)]^d$ using b.

A third party needs to have m, c and $(sig_U(m, c))^d$ in order to verify the conditional signature of U.

Again, this protocol provides a *provable degree of privacy for U with respect to TTP*, since TTP receives only r (which is just a random number) and $E_{TTP}(c, r, sig_U(m, c) * b^e)$. In this latter cryptogram only $sig_U(m, c) * b^e$ carries information that could be connected to user U. However, according to [Chaum, 1982], the blind signature is unconditionally secure, so no algorithm exists that can compute $sig_U(m, c)$ based on $sig_U(m, c) * b^e$ (with a probability better than $1/2^{|sig_U(m,c)|}$) without knowing parameter b. Thus, the degree of privacy user U has is the same as the one provided by the channel that is used in Step 6 for signature revocation. Moreover, if everyone behaves honestly, U obtains unconditional privacy.

Note, that in the above protocol TTP signs an incoming message without being able to see what it is. Although TTP may authenticate M to prevent denial of service attacks, M may still obtain the signature of TTP on an arbitrary message. Since TTP does not use this private key for anything else, such a signature is useful only if a conditional signature of the user was signed. Although M can compute the cryptogram and repeat Step 7 at will, M can only gain countersignatures on signatures U did not revoke.

4.3 Halving the digital signature

In our final protocol we make use of the fact that even if TTP obtains one half of the bits of the conditional signature of U, TTP is still unable to compute the other half. Meanwhile, if TTP countersigns the half of the bits of the conditional signature, it is authenticated not much less securely than if TTP countersigned the whole signature.

In this case, condition c looks as follows: *"My signature on the above message is valid if and only if TTP countersigned its right half and deadline t has passed."*

According to our notation, $left(x)$ means the left half of the bits of bitstring x, and $right(x)$ means its right half. Naturally, $left(x)\|right(x) = x$ where operation $\|$ is concatenation. The proposed protocol is as follows:

Step 1: $U \to T$: m

Step 2: $T \to C$: m

Step 3: $C \to T$: c, $left[sig_U(c, m)]$, $E_{TTP}(r, c, right[sig_U(c, m)])$
  C generates random number r. Again, the terminal cannot verify the conditional signature, but since C is a trusted device, T may believe that the conditional signature on m was encrypted by the public key of TTP.

Step 4: $T \to M$: m, c, $left[sig_U(c, m)]$, $E_{TTP}(r, c, right[sig_U(c, m)])$

Later, at a trusted terminal:

Step 5: $C \to U$: M, m, c, r

If user U would like to repudiate the signature, then

  Step 6: $U \to TTP$: r (via an anonymous channel)

After the deadline t:

Step 7: $M \to TTP$: $E_{TTP}(r, c, right[sig_U(c, m)])$

If deadline t has passed and r was not submitted to TTP, then:

  Step 8: $TTP \to M$: $right[sig_U(c, m)]$, $sig_{TTP}(right[sig_U(c, m)])$

Step 9: M computes: $left[sig_U(c, m)] \parallel right[sig_U(c, m)] = sig_U(c, m)$

A third party needs to have m, c, $sig_U(c, m)$, $sig_{TTP}(right[sig_U(c, m)])$ in order to verify the conditional signature of U.

Again, M may obtain multiple countersignatures on signatures not revoked by U. However, this is not a problem. See the note at the end of Section 4.2. In contrast to the other solutions we proposed, the smart card does not need to perform any complex computation (like blinding or encryption) in this one, apart from generating the digital signature.

5.   Conclusion and future work

If a user at a malicious terminal would like to perform sensitive operations like digital signature, she is subject to man-in-the attacks from the terminal. The user may rely on a conditional signature scheme to gain time to review signatures from a trusted terminal and revoke signatures on messages she did not intend to sign. Unfortunately, those conditional signature schemes that allow this and are most suitable for commercial use, have to rely on the help from a trusted third party.

In this paper we have shown three protocols that allow the user to retain her privacy with respect to this third party when signing messages with conditions on untrusted terminals.

In our future work we intend to examine if a user is able to identify the untrusted terminals that mount attacks against her, and in what extent she is able to prove the attack to third party.

References

Abadi, M., Burrows, M., Kaufman, C., and Lampson, B. (1992). Authentication and Delegation with Smart-cards. Theoretical Aspects of Computer Software: Proc. of the International Conference TACS'91, Springer, Berlin, Heidelberg.

Anonymizer Inc. (1999). . http://www.anonymizer.com.

Asokan, N., Debar, Hervé, Steiner, Michael, and Waidner, Michael (1999). Authenticating Public Terminals. Computer Networks, 1999.

Berta, I. Zs. and Vajda, I. (2003). Documents from Malicious Terminals. SPIE Microtechnologies for the New Millenium 2003, Bioengineered and Bioinspired Systems, Maspalomas, Spain.

Berta, István Zsolt, Buttyán, Levente, and Vajda, István (2004). Mitigating the Untrusted Terminal Problem Using Conditional Signatures. Proceedings of International Conference on Information Technology ITCC 2004, IEEE, 2004, IEEE, Las Vegas, NV, USA, April.

Brands, S. A. (1994). Untraceable off-line cash in wallets with observers. In Crypto'93 Springer-Verlag, LNCS 773 pp. 302-318.

Chaum, David (1981). Untraceable electronic mail, return addresses and digital pseudonyms. Communications of the ACM, v24, n.2 pp.84-88.

Chaum, David (1982). Blind signatures for untraceable payments. Advances in Proceedings of Crypto 82, D. Chaum, R.L. Rivest, & A.T. Sherman (Eds.), Plenum, pp. 199-203.

Clarke, Dwaine, Gassend, Blaise, Kotwal, Thomas, Burnside, Matt, Dijk, Marten van, Devadas, Srinivas, and Rivest, Ronald (2002). The Untrusted Computer Problem and Camera-Based Authentication.

Franklin, M. and Yung, M. (1992). Towards provably secure efficient electronic cash. Columbia Univ. Dept. of CS TR CSUCS-018-92.

Jakobsson, M. and Raïhi, D. (1998). Mix-based electronic payments. Fifth Annual Workshop on Selected Areas in Cryptography (SAC'98), Queen's University, Kingston, Ontario, Canada.

Lee, B and Kim, K (2002). Fair Exchange of Digital Signatures using Conditional Signature. SCIS 2002, Symposium on Cryptography and Information Security.

Matsumoto, T (1996). Human-Computer cryptography: An attempt. In ACM Conference on Computer and Communications Security, pp 68-75.

Naor, Moni (1991). Bit Commitment Using Pseudo-Randomness. Journal of Cryptology: the journal of the International Association for Cryptologic Research, volume 2, pp 151-158.

Naor, Moni and Pinkas, Benny (1997). Visual Authentication and Identification. Lecture Notes in Computer Science, volume 1294.

Naor, Moni and Shamir, Adi (1995). Visual Cryptography. Lecture Notes in Computer Science, vol 950, pp 1–12, 1995, http://citeseer.nj.nec.com/naor95visual.html.

Rankl, W. and Effing, W. (1997). Smart Card Handbook. John Wiley & Sons, 2nd edition, ISBN: 0471988758.

Rivest, R (2001). Issues in Cryptography. Computers, Freedom, Privacy 2001 Conference http://theory.lcs.mit.edu/~rivest/Rivest-IssuesInCryptography.pdf.

Schneier, B. and Shostack, A. (1999). Breaking up is Hard to do: Modelling security threats for smart cards. USENIX Workshop on Smart Card Technology, Chicago, Illinois, USA, http://www.counterpane.com/smart-card-threats.html.

Schneier, Bruce (1996). Applied Cryptography. John Wiley & Sons, ISBN: 0471117099.

Schneier, Bruce (1999). The Solitaire Encryption Algorithm. http://www.counterpane.com/solitaire.htm.

Stabell-Kulo, Tage, Arild, Ronny, and Myrvang, Per Harald (1999). Providing Authentication to Messages Signed with a Smart Card in Hostile Environments. Usenix Workshop on Smart Card Technology, Chicago, Illinois, USA, May 10-11, 1999.

Syverson, Paul, F., Goldschlag, David M., and Reed, Michael G. (1997). Anonymous Connections and Onion Routing. IEEE Symposium on Security and Privacy, Oakland, California.

ANONYMOUS SERVICES USING SMART CARDS AND CRYPTOGRAPHY

Sébastien Canard
France Telecom R&D
42 rue des Coutures - BP6243
14066 CAEN Cedex
France
sebastien.canard@francetelecom.com

Jacques Traoré
France Telecom R&D
42 rue des Coutures - BP6243
14066 CAEN Cedex
France
jacques.traore@francetelecom.com

Abstract   More and more services provided by Internet pose a problem of privacy
and anonymity. One cryptographic tool that could be used for solving
this problem is the group signature [1, 5, 8]. Each member of the group
is able to anonymously produce a signature on behalf of the group and
a designated authority can, in some cases, revoke this anonymity.

During the last decade, many anonymous services using this concept
have been proposed: electronic auctions [11], electronic cash systems
[12, 10, 7], anonymous credentials [2]. But for some other services where
the anonymity is essential (such as electronic voting or call for tenders),
group signature schemes cannot be applied as they are. For this reason,
the authors of [6] proposed a variant that is partially linkable and not
openable, called list signature scheme.

In this paper, we first improve the cryptographic tool of [6] by proposing
some optional modifications of list signature schemes such as anonymity
revocation. We then propose more efficient list signature schemes, by
using a smart card to produce the signature. We finally propose some
concrete implementations of our proposals. As a result, we obtain more
efficient solutions that are useful in many more services.

Keywords: Cryptography, group signature schemes, voting, call for tenders, subscription tickets.

Introduction

Group signature schemes have been introduced in 1991 by Chaum and van Heyst [8]. They allow members to sign a document on behalf of the group in such a way that the signatures remain anonymous and untraceable for everyone but a designated authority, who can recover the identity of the signer whenever needed (this procedure is called "signature opening"). Moreover, this type of signature must be unlinkable for everybody but the authority (no one else can decide whether two different valid signatures were computed by the same group member or not). Currently, the most secure group signature scheme is the one of Ateniese et al. [1].

In most cases, it is desirable to protect group member's signing keys in a secure device, such as a smart card. However, these devices are restricted in terms of processing power and memory. To solve this problem, Canard and Girault [5] proposed a smart card based solution where all computations can be efficiently done inside the (tamper-resistant) smart card.

In some applications, group signature schemes can be used as they are. For example, an electronic auction system [11] can be based on any group signature scheme without any modification (other than completing it with some extra cryptographic mechanisms for auction purposes). However, in some other services, it is not possible to use group signatures as they are. It is often desirable to make possible for everybody (and without opening every signature) to link signatures produced by group members. For example, in [6], the authors, in order to develop an electronic voting system, need to introduce a variant of group signature schemes called list signature scheme. These signatures are similar to group signatures except that they are partially linkable and not openable. It permits them to introduce a new electronic voting system that directly uses list signature schemes and that satisfies the fundamental needs of security in electronic voting. In some other cases, we need that the number of anonymous signatures produced by the same member be limited, even in a multi-verifier setting and, again, without opening every signature.

In this paper, we improve the work of [6] in two ways. First, we generalise these list signature schemes by making them optionally openable. Second we propose various implementations of these list signatures (with optionally anonymity revocation) by assuming that a signature is produced by a tamper-resistant device (typically a smart card). Our concrete solutions permit us to use list signatures with optionally anonymity revocation in various applications: electronic voting system or opinion poll, call for tenders and anonymous subscription tickets. Thus, we present some services that can be developed in a mobility context, using a device that is restricted in terms of processing power and memory.

The paper is organised as follows. In the next section, we recall existing solutions and we explain why they are not applicable in some context. Section 2 presents our technical implementations of list signatures with optionally anonymity revocation. Finally, Section 3 shows some examples of services that can be implemented in a smart card using our solutions.

1. Existing Solutions

This section presents some cryptographic tools for anonymous services in Internet. This paper does not consider blind signature schemes or mixnets, which are other tools that provide anonymity. Here, we restrict ourselves to group signature schemes and their variants.

1.1 Group Signature Schemes

There are many entities that are involved in a group signature scheme. A member of the group is denoted by $\mathcal{M}$. The Group Manager $\mathcal{GM}$ is the authority in charge of generating the keys of new members. The Opening Manager $\mathcal{OM}$ is the authority that revokes the anonymity of a group signature, whereas the Revocation Manager $\mathcal{RM}$ is the authority that has the capability of revoking the right of signing of a group member (also called member deletion). We do not consider this kind of revocation in this paper since generic solutions using smart cards (such as [5]) already exist for this problem. Consequently, we will not refer to $\mathcal{RM}$ anymore.

Currently, the most secure group signature scheme is the one of Ateniese, Camenisch, Joye and Tsudik [1] (ACJT for short). In this proposal, if someone wants to become a group member, he computes a secret key x and interacts with $\mathcal{GM}$ in order to obtain a certificate (A, e) (such that $A^e = a_0 a^x \pmod{n}$ where n, a and a_0 are public and where the factorization of n is only known by $\mathcal{GM}$). An ACJT group signature consists in performing

1 an El Gamal encryption of A, that is computing $T_1 = Az^w$ and $T_2 = g^w$ where z is the encryption public key and w is a random number. $\mathcal{OM}$ is able to decrypt it in case of anonymity revocation by using the discrete logarithm of z in base g as the decryption key.

2 a zero-knowledge proof of knowledge of (A, e, x) (a zero-knowledge proof of knowledge on a message M of a value α that verifies the predicate f is denoted by $PK(\alpha : f(\alpha))(M)$).

The group signature scheme of Canard and Girault [5] is based on the use of a tamper-resistant device (such as a smart card) to produce a group signature. Consequently, each member of a group owns a smart card. This proposal makes use of a signature scheme (such as RSA, DSA, Schnorr or GPS) and an encryption scheme that can be symmetric (such as AES) or asymmetric (such as RSA or El Gamal). It is however necessary that this encryption scheme be probabilistic.

During a setup phase, the group manager $\mathcal{GM}$ computes a signature private key sk_G such that he can distribute it without knowing it. This can be done, for example, by using a secret sharing protocol: sk_G is shared by various entities and the associated public key, denoted by pk_G, is computed by the cooperation of all these entities. The signature of the message M with sk_G is denoted by $\textbf{Sign}_G(M)$.

$\mathcal{OM}$ generates a decryption private key $sk_{\mathcal{OM}}$ and the associated encryption public key $pk_{\mathcal{OM}}$. He keeps the first one secret. If the chosen encryption algorithm is symmetric, then $pk_{\mathcal{OM}} = sk_{\mathcal{OM}}$ and the encryption key must also be kept secret. The encryption of a message M is denoted by $\textbf{Encrypt}_{\mathcal{OM}}(M)$.

When someone wants to become a new group member, he interacts with the group manager $\mathcal{GM}$ that sends him an identifier $Id_{\mathcal{M}}$ and the signature private key sk_G (consequently, this key is shared by all group members). It is important that $\mathcal{GM}$ knows the link between the identifier $Id_{\mathcal{M}}$ and the identity of the group member $\mathcal{M}$. $\mathcal{M}$ also obtains the encryption key $pk_{\mathcal{OM}}$ from the opening manager $\mathcal{OM}$.

After that, a group member can sign on behalf of the group by using his smart card. Then the algorithm in figure 1 is executed inside the smart card. The verification of a group signature only consists in verifying the

```
GSign  (M ,Id_M ,pk_OM ,sk_G):
  C  :=  Encrypt_OM (Id_M);
  M̃  :=  Concatenate (M,C);
  S  :=  Sign_G (M̃);
  return (C,S).
```

Figure 1.   Group Signature of [6].

signature S by using the public key pk_G. If this signature is correct, then the group signature is correct. Finally, if the anonymity needs to be revoked, $\mathcal{OM}$ uses his private key $sk_{\mathcal{M}}$ to decrypt C to obtain $Id_{\mathcal{M}}$. The Group Manager knows the link between this value and the identity of the group member.

1.2 Openable versus Non-openable Signatures

The previous section has shown that anonymity revocation is made possible by the encryption of an identifier. An ACJT group signature includes an El Gamal encryption of a value A that is known by $\mathcal{GM}$. In the group signature scheme of [5], a signature is made of an encryption of the identifier $Id_{\mathcal{M}}$ that is shared by the group member and $\mathcal{GM}$. Consequently, if one wants to make a group signature scheme non openable, it is (almost) sufficient to remove this encryption.

In ACJT, it is then unnecessary to compute $T_1 = Az^w$ and $T_2 = g^w$ anymore. But, to prove his membership, the group member has to make a commitment on A. This can be done by computing $T_1 = Az^w$ and $T_2 = g^w h^{w_1}$ where w and w_1 are random and where the discrete logarithm of z in base g is unknown (which is different from the ACJT group signature scheme where the discrete logarithm of z in base g is known by $\mathcal{OM}$ to open signatures). This is the approach chosen in [6] (see above for more details).

In the group signature of [5], it is sufficient to remove the value $C := \text{Encrypt}_{\mathcal{OM}}(Id_{\mathcal{M}})$ during the group signature process.

1.3 List Signature Schemes

List signatures have been introduced by Canard and Traoré [6]: they correspond to partially linkable and non openable group signatures.

In a list signature scheme, the time is divided into distinct sequences, each of them being a fixed time period (one hour, one day, one month, the time of an election day, ...). A list signature scheme involves the same protagonists as a group signature scheme, except the Opening Manager. Moreover, in some cases, $\mathcal{GM}$ needs to create a public value that is representative of a sequence: this is executed at the begin of each sequence. Furthermore, everybody is able, taking as input two valid signatures produced during a particular sequence, to test whether or not they have been produced by the same list member or not. However, two signatures produced during two different sequences are unlinkable.

In [6], $\mathcal{GM}$ firstly randomly computes a $2l_p$ bits safe RSA modulus $n = pq = (2p'+1)(2q'+1)$ and chooses random elements $a, a_0, g, h \in_R QR(n)$. When someone becomes a new list member, he obtains a secret key

$sk_{\mathcal{M}} = x$ and a list certificate (known by $\mathcal{GM}$) $c_{\mathcal{M}} = (A, e)$ such that $A^e = a_0 a^x \pmod{n}$.

Before the beginning of a new sequence, $\mathcal{GM}$ generates a representative of the sequence, that is a random integer $f \in QR(n)$ (where $QR(n)$ denotes the group of quadratic residues modulo n). A simple solution is to compute $f = (\mathcal{H}(date))^2 \pmod{n}$ where $date$ is the date of the beginning of the sequence and $\mathcal{H}$ is a collision-resistant hash function. Each list member can then sign on behalf of the group by doing the following:

- choosing random values $w, w_1 \in \{0,1\}^{2l_p}$.

- computing $T_1 = Az^w \pmod{n}$, $T_2 = g^w h^{w_1} \pmod{n}$, $T_3 = g^e h^w$ $\pmod{n}$ and $T_4 = f^x \pmod{n}$.

- making a proof of knowledge $U = PK(\alpha, \beta, \gamma, \delta, \zeta, \eta : a_0 = T_1^\alpha/(a^\beta z^\gamma) \wedge$ $1 = T_2^\alpha/(g^\gamma h^\eta) \wedge T_2 = g^\delta h^\zeta \wedge T_3 = g^\alpha h^\delta \wedge T_4 = f^\beta)(M)$ where M is the message to be signed.

The couple (T_1, T_2) is a commitment of the value A (as explained in Section 1.2) and the signature on M is finally (T_1, T_2, T_3, T_4, U).

The verifier only has to check the validity of the proof of knowledge U to be sure that the signer is actually a member of the list. Finally, since a signer is constrained to compute T_4 by using the representative f of the sequence and his secret key x, he will always compute the same T_4 for a particular sequence. It is consequently possible to know if two or more signatures produced during a particular sequence come from the same member or not.

1.4   Limitations of these Proposals

Group signature is a convenient tool for anonymous auctions, as explained in [11]. List signature schemes can be used to enhance the security of electronic voting system since they have been designed for this purpose in [6]. In some other interesting services (such as call for tenders for example), it can be desirable to have a cryptographic scheme in which:

- each group member can anonymously sign a message on behalf of the group,

- a designated authority can revoke this anonymity and

- it is important to know when two (or more) signatures have been produced by the same user.

In this case, existing solutions (group and list signatures) are not suitable and we have to design list signatures with revocable anonymity.

In some other cases, the number of anonymous signatures produced by the same member needs to be limited. We must be sure that each member is unable to produce one more signature per sequence. An anonymous subscription ticket requires this property since the user of the ticket cannot use it more than the number of uses he has subscribed for. Again, group signatures and list signatures are not suitable for this kind of service.

Moreover, the existing list signature schemes are too expensive in terms of processing power and memory since they involve a lot of modular exponentiations. Consequently, it is not possible to use them in a mobility context and a smart card (possibly embedded in a mobile phone or a PDA) cannot be used.

2. List Signature Schemes and Smart Cards

In this section, our aim is to solve the problems identified in the previous one. We consequently design some list signatures with optionally anonymity revocation that can be produced by a tamper-resistant device which is restricted in terms of processing power and memory. In the following, we first propose two solutions, called "Testing and Deciding" and "Semi-Probabilistic Encryption Scheme". We then improve the second one to obtain the "Pseudo Randomizing" method.

2.1 Testing and Deciding

Our first solution implies that we let decide the smart card about its capacity to sign on behalf of the group. In this solution, it is not possible to link two signatures made by the same user during the same sequence. In fact, each member has the possibility to anonymously sign a fixed number of times during a particular sequence.

Our list signature scheme relies on a group signature scheme (that can be the ACJT one or the one described in [5]). Each smart card can sign on behalf of the group by using this group signature scheme. The group signature of a message M is denoted by **GSign**(M). Furthermore, we assume that each smart card has a tamper-resistant memory space denoted by *Mem* and manages a counter *cpt*.

At the beginning of a new sequence, the group manager $\mathcal{GM}$ randomly creates a representative of this sequence, denoted by *RepSeq*. This element is not simply public but also authenticated and timestamped by $\mathcal{GM}$.

The group manager also manages an integer that represents the number of signatures that a list member's smart card can produce during one sequence. This integer, denoted by $NbSig$, can be the same for every list member but can also be different for each smart card: this depends on the service. It must also be authenticated by $\mathcal{GM}$. We denote by $Sig_{\mathcal{GM}}$ the signature on $RepSeq$ and $NbSig$[1].

When a list member wants to sign a message M, he uses his smart card that receives the message M, the representative of the sequence $RepSeq$, the signature $Sig_{\mathcal{GM}}$ and the number $NbSig$ of signatures that it is allowed to produce during this sequence. Then, the smart card executes the algorithm presented in figure 2. For each new signature during the

```
LSign_{T&D} (M, RepSeq, Sig_{GM}, NbSig):
  if ( Verify(Sig_{GM}) == OK)
  {
    if (RepSeq is not recorded in Mem);
    {
      Record RepSeq in Mem;
      cpt := 0;
    }
    cpt := cpt + 1;
    if (cpt > NbSig)
    {
      return Error;
    }
    return GSign(M);
  }
  else return Error;
```

Figure 2.   "Testing and Deciding" List Signature: **LSign**$_{T\&D}$.

sequence $RepSeq$, the smart card increments its proper counter cpt and, consequently, can test whether or not its owner has exceeded the number of authorized signatures for this particular sequence.

A verifier of a correct list signature only knows that the card that produced this signature belongs to a valid list member without knowing which one. He knows that if this signature was produced it means that the list member has the right to do it and that he has not exceeded the number of authorized signatures.

Our list signatures are openable, due to the property of the underlying group signature scheme. To create a non openable list signature, it is sufficient to transform the group signature into a non openable one, as explained in Section 1.2. It is also possible to revoke the anonymity of a signature when the user tries to produce more list signatures than he is entitled to do. Instead of the signature, the smart card can then send an identifier (possibly encrypted with $\mathcal{OM}$'s public key) as $Id_\mathcal{M}$ that can be linked to the identity of the cheat.

This list signature scheme can be used in call for tenders, in e-voting or opinion poll and in anonymous subscription tickets.

2.2 Semi-Probabilistic Encryption Scheme

In our second solution, we propose a list signature scheme that necessary supports revocable anonymity. As for [6], it is also possible to link two signatures produced during a particular sequence. Moreover, the list signature will be produced by a smart card since we do not trust the list member.

As for the previous list signature scheme, the following list signature scheme relies on a group signature scheme (that can be the ACJT one or the proposal in [5]). Each smart card can sign on behalf of the list by using this group signature scheme. The latter one must use a probabilistic encryption scheme for anonymity revocation (this is the case for both group signature schemes presented in Section 1.1). The group signature of the message M is denoted by $\mathbf{GSign}(M, C)$ where C denotes the ciphertext derived from the group signature scheme (in ACJT, $C = (T_1, T_2)$ and in [5], $C = \mathbf{Encrypt}_{\mathcal{OM}}(Id_\mathcal{M})$).

Roughly, a probabilistic encryption scheme takes as input the message to be encrypted and a freshly generated random number. This random number is not used for the decrypting phase. The core of our second solution is that, for a particular sequence, this random number will be fixed. Consequently, the encryption scheme is deterministic during a sequence and probabilistic for two different sequences: this is what we call a pseudo probabilistic encryption scheme.

At the beginning of a new sequence, the group manager $\mathcal{GM}$ randomly creates a representative of this sequence. This representative is denoted by *RepSeq*. This element is public, authenticated and timestamped by $\mathcal{GM}$ but it may also be required that this number be publicly verifiable: everybody must be convinced that $\mathcal{GM}$ generated it in a truly random manner. For this purpose, we suggest to use the proposal of [6] and to define *RepSeq* as follows: $RepSeq = \mathcal{H}(date)$ where *date* is, for example, the date of beginning of the sequence and where $\mathcal{H}$ is a hash function.

Each list member's smart card holds, from the used group signature scheme, an identifier denoted by $Id_\mathcal{M}$. In ACJT, this identifier corresponds to the certificate (A, e) whereas in [5], it corresponds to the value also denoted $Id_\mathcal{M}$.

We assume that the smart card can produce a group signature and that the group signature scheme relies on a probabilistic encryption scheme **Encrypt$_G$** (that takes as input a random number and a message). It also has access to a pseudo-random number generator **PRNG** that depends on two variables. Each smart card also holds a secret key K that is only known by it. The pseudo random number generator can be, for example, the ANSI X9.17 pseudo random bit generator (it is also possible to take the FIPS 186 generator family, that is SHA-1 or DES, or any other suitable pseudo random number generator). It takes as input the number of output bits, a seed and a 3DES key. In our scheme, when a list member wants to sign a message M, the seed is the representative of the sequence and the 3DES key is the secret key of the smart card. The size of the output depends on the used probabilistic encryption scheme and is not specified in this paper.

Using this pseudo random number generator, the smart card can then use the output as the random input of the encryption scheme (for example, in the ACJT group signature scheme the input of **Encrypt$_G$** corresponds to w). It finally computes a group signature of the message M by using the output of the encryption scheme. The production of a list signature can be summarized as explained in figure 3. The verifi-

```
LSign_SPES (M, RepSeq, Sig_gM, K):
  if ( Verify(Sig_gM) == OK)
  {
     r := PRNG(K, RepSeq);
     C := Encrypt_G(r, Id_M);
     return GSign(M, C);
  }
  else return Error;
```

Figure 3.   "Semi-Probabilistic Encryption Scheme" List Signature: **LSign$_{SPES}$**.

cation of such a list signature only consists in verifying the validity of the group signature. The opening of a valid list signature corresponds to the opening procedure of the used group signature scheme.

For a given sequence (and consequently a given representative $RepSeq$) and a given smart card (and consequently a given secret key K), the output r of the pseudo random generator will always be the same, as well as the ciphertext C. Using C, it will be possible to link various signatures made by the same list member during a given sequence. But, for two distinct sequences, the two corresponding representatives $RepSeq1$ and $RepSeq2$, and consequently the two corresponding outputs r_1 and r_2 of the pseudo random generator, will be different. Since r_1 and r_2 will be different, the two corresponding ciphertexts C_1 and C_2 will be different and unlinkable.

This second list signature scheme is suitable for call for tenders and opinion polls applications and supports anonymity revocation. We can modify this proposal to make it optionally with anonymity revocation and this is the purpose of the next section, that develops a solution also based on the use of a pseudo random number generator.

2.3   Pseudo Randomizing

Our third solution is close to the previous one. In this list signature scheme, it is also possible to link two signatures produced during a particular sequence. Again, we do not trust the list member but we trust his smart card.

Our solution is related to the one of [5]. Consequently, it needs an ordinary public-key signature scheme (such as RSA, DSA, GPS, GQ, Schnorr, etc.) with a public key sk_G which is shared by all list member's smart cards while the associated public key pk_G is public. The signature of a message M is denoted by $\mathbf{Sign}_G(M)$. If we require for the list signature scheme to be openable, then our solution requires a (symmetric or asymmetric) encryption scheme. The corresponding encryption key $pk_{\mathcal{OM}}$ is sent to all smart cards (secretly if the scheme is symmetric, and publicly if it is asymmetric) whereas the decryption key $sk_{\mathcal{OM}}$ is kept secret by $\mathcal{OM}$. The encryption of the message M is denoted by $\mathbf{Encrypt}_{\mathcal{OM}}(M)$. In this case, each list member is known by the group manager by an identifier denoted by $Id_{\mathcal{M}}$. This data is stored in the smart card.

The core of our solution is, as for the second proposal, the use of a pseudo random number generator **PRNG** that takes as input a seed denoted by s and a key K that is secret and different for all smart cards.

At the beginning of a new sequence, the group manager randomly creates a representative of this sequence denoted by $RepSeq$ and which is used as the seed of **PRNG**. This can be done as for the second proposal (using the date and a hash function).This element is public, authenticated and dated by $\mathcal{GM}$.

When a list member wants to sign a message on behalf of the list, he uses his smart card. The algorithm executed by the smart card depends on the "openability" of the scheme. If the list signature is not openable, the smart card performs what is described in figure 4. In case the list

```
LSign_NOPR (M, RepSeq, Sig_GM, K, pk_OM, sk_G):
   if ( Verify(Sig_GM) == OK)
   {
      R := PRNG(RepSeq, K);
      M̃ := Concatenate(M, R);
      S := Sign_G(M̃);
      return (S, R);
   }
   else return Error;
```

Figure 4.   Non-Openable "Pseudo Randomizing" List Signature: **LSign**$_{NOPR}$.

signature scheme supports anonymity revocation[2], the smart card needs to execute the algorithm presented in figure 5. The verification of a list

```
LSign_OPR (M, RepSeq, Sig_GM, K, pk_OM, sk_G):
   if ( Verify(Sig_GM) == OK)
   {
      R := PRNG(RepSeq, K);
      C := Encrypt_OM(Id_M);
      M̃ := Concatenate(M, R, C);
      S := Sign_G(M̃);
      return (S, R, C);
   }
   else return Error;
```

Figure 5.   Openable "Pseudo Randomizing" List Signature: **LSign**$_{OPR}$.

signature produced by **LSign**$_{OPR}$ or **LSign**$_{NOPR}$ can simply be done by verifying the signature S, using the public key pk_G. If this signature is correct, then the list signature is considered as valid.

If the signature is openable, $\mathcal{OM}$ can revoke the anonymity by decrypting C to obtain the identifier Id_M. The group manager, who knows the link between this data and the actual identity of the list member, can then identify the member who produces this signature.

Finally, everybody is able to link list signatures produced by the same list member during a specified sequence, using the value R. Indeed, for a given sequence (and consequently a given representative $RepSeq$) and a given smart card (and consequently a given secret key K), the output R of the pseudo random generator will be always the same. However, for two distincts sequences, the two corresponding representative $RepSeq1$ and $RepSeq2$ will be different and, consequently, the corresponding outputs of the pseudo random generator will be different and unlinkable. Our third solution is suitable for call for tenders and opinion polls.

3. Examples of Applications

We have seen in Section 1.4 that the state of the art is not suitable for some services such as opinion polls, call for tenders and anonymous subscription tickets. We explain in the sequel why the tools introduced in the previous sections are more convenient for the above services.

3.1 Opinion Polls

In [6], the authors propose an off-line electronic voting system based on list signature schemes. To vote, each voter produces a list signature of his vote and sends an encryption of the vote and the signature to a ballot box. This system is also directly applicable for an opinion poll service. But, in a mobility context, such as an opinion poll by the means of a mobile phone, the list signature of [6] in not useful since it is too expensive in terms of processing power and memory.

In our proposal, each user can have the possibility to produce a list signature with the help of his mobile phone. For each opinion poll, as for the voting system of [6], the embedded smart card produces a list signature of the voting option of his owner, then encrypts the choice and the signature, and finally sends the encrypted value to the ballot box. It is important to prevent someone from voting many times for a given election and our list signature schemes suit very well.

In this context, it is possible to use $\textbf{LSign}_{SPES}$, $\textbf{LSign}_{NOPR}$ or $\textbf{LSign}_{OPR}$. The list signature can be either openable or not, depending on the organizer of the opinion poll (it can be useful to revoke the anonymity in case of fraud). It can also be possible to use $\textbf{LSign}_{T\&D}$ with $NbSig = 1$.

For a choice c from the user, the embedded smart card executes the algorithm described in figure 6, where $\textbf{Encrypt}_{OP}$ is the encryption algorithm related to the opinion poll. The organizer of the opinion poll creates a new sequence and a corresponding representative $RepSeq$ of it. He also signs it. This algorithm, and more particularly the list signature, also needs some more data (such as K, $NbSig$, etc.), denoted by $data$, that are not described in this section. During the counting phase, it is

OpinionPoll (c) :
$\quad S := \mathbf{LSign}(c, RepSeq, Sig_{\mathcal{GM}}, data)$;
$\quad M := \mathbf{Concatenate}(S, c)$;
$\quad C := \mathbf{Encrypt}_{OP}(M)$;
$\quad$ return (C) ;

Figure 6.   Opinion Poll Service.

sufficient to decrypt each ballot and then to verify the validity of each list signature (to verify that the ballot is valid). Finally, one can publish the result.

3.2   Call for Tenders

In an anonymous call for tenders service, each tenderer can propose an anonymous offer. Moreover, the winner must be identified at the end of the call. In some cases, it can be desirable to prevent a participant from proposing several tenders[3]. Sometimes, we can let the tenderer propose several offers and take the better one, or the last one. In these two cases, it is important to be able to link the offers which come from the same tenderer. For these reasons, our list signature schemes are useful for this purpose.

Our proposal is close to the opinion poll system described above. It first consists in supplying each participant with a smart card. When someone wants to make an offer for a call for tenders, he produces a list signature of his proposal and then encrypts the signature along with his proposal. The call for tenders authority can decrypt each proposal and then test its validity by verifying the list signature scheme. The jury can identify the winner by opening the list signature.

The signing algorithm can be **LSign**$_{SPES}$ or **LSign**$_{OPR}$. It is also possible to use **LSign**$_{T\&D}$ with $NbSig = 1$ if the authority does not want a participant to propose several offers to a given call for tenders.

For a proposal p from a participant, the embedded smart card executes the algorithm described in figure 7, where **Encrypt**$_{CT}$ is an encryption algorithm related to the call for tenders (necessary to prevent a fraudulent user from learning something about the proposal of a rival). The authority creates a new sequence and a corresponding representative *RepSeq*. This algorithm, and more particularly the list signature, needs some *data* that depend on the used scheme (see previous section).

```
CallTenders (p):
  S := LSign(p, RepSeq, Sig_{GM}, data);
  M := Concatenate(S, p);
  C := Encrypt_{CT}(M);
  return (C);
```

Figure 7. Call for Tenders Service.

3.3 Anonymous Subscription Tickets

An anonymous subscription ticket permits someone to use a service he had paid for without being traced by the supplier. This is useful for example for cinema tickets, subway or bus tickets, etc.

In our solution, everybody can (non anonymously) obtain a subscription ticket by simply buying it. This one is represented by a numeric data that can be stored in a mobile phone, a PDA, a dongle or a specific smart card. This numeric data corresponds to the number $NbSig$ of "entries" that the customer has bought and is signed, with the identity of the customer, by the issuer. The embedded smart card is able to produce a "Testing and Deciding" list signature with a number of signatures $NbSig$.

When the user wants to go to cinema or to take the bus, he only has to exhibit his mobile phone (for example), which produces a list signature. The embedded smart card knows when all tickets are used and, in this case, refuses to sign.

In this context, the sequence can either correspond to the end of validity of the ticket or the lifetime of the system. In the first case, the representative of the sequence $RepSeq$ corresponds to an identifier of the end date. In the latter case, we can imagine that an embedded smart card can be used, for example, in various cinemas. Thus, $RepSeq$ represents one cinema.

When a smart card is presented to a verification machine, this last one sends it a commitment c and the representative $RepSeq$ (in the first case, this representative is sent after the reception of the date of validity from the smart card). The smart card then produces a list signature $S := \text{LSign}_{T\&D}(c, RepSeq, Sig_{GM}, NbSig)$ that can easily be verified by the verification machine.

In some cases, in can be useful to revoke the anonymity of a signature (for example, if it is necessary to know who was on the cinema at a particular date).

Notes

1. For sake of simplicity, we assume that there is only one signature for both values *RepSeq* and *NbSig*, even if, in practice, it can have two differents signatures, since these two values can be obtained at different times.

2. In terms of efficiency, this solution is not very interesting in comparison with the "Semi-Probabilistic Encryption Scheme" solution since there are two computations instead of one. We nevertheless present both solutions to be as general as possible.

3. A tenderer can try to propose various times the same offer to leave nothing to chance.

References

[1] G. Ateniese, J. Camenisch, M. Joye, G. Tsudik. A Practical and Provably Secure Coalition-Resistant Group Signature Scheme. In L. Bellare, editor, Advances in Cryptology-Crypto'2000, volume 1880 of LNCS, pages 255-270. Springer-Verlag, 2000.

[2] J. Camenisch, A. Lysyanskaya. An Efficient System for Non-transferable Anonymous Credentials with Optional Anonymity Revocation. In B. Pfitzmann, editor, Advances in Cryptology-Eurocrypt'2001, volume 2045 of LNCS, pages 93-118. Springer-Verlag, 2001.

[3] J. Camenisch, M. Michels. A Group Signature Scheme based on an RSA-variant. Technical Report RS-98-27, BRICS, Dept. of Comp. Sci., University of Arhus, preliminary version in Advances in Cryptology-EUROCRYPT'98, volume 1514 of LNCS.

[4] J. Camenisch, M. Stadler. Efficient Group Signature Schemes for Large Groups. In B. Kaliski, editor, Advances in Cryptology-CRYPTO'97, volume 1296 of LNCS, pages 410-424. Springer-Verlag, 1997.

[5] S. Canard, M. Girault. Implementing Group Signature Schemes with Smart Cards. Proc. of CARDIS 2002.

[6] S. Canard, J. Traoré. List Signature Schemes and Application to Electronic Voting. Proc. of WCC'03.

[7] S. Canard, J. Traoré. On Fair E-cash Systems based on Group Signature Schemes. Proc. of ACISP 2003, volume 2727 of LNCS, pages 237-248. Springer-Verlag, 2003.

[8] D. Chaum, E. van Heyst. Group Signatures. In D. W. Davies, editor, Advances in Cryptology-Eurocrypt'91, volume 547 of LNCS, pages 257-265. Springer-Verlag, 1991.

[9] G. Maitland, C. Boyd. Co-operatively Formed Group Signatures. In B. Preneel, editor, CT-RSA 2002, volume 2271 of LNCS, pages 218-235. Springer-Verlag, 2002.

[10] G. Maitland, C. Boyd. Fair Electronic Cash Based on a Group Signature Scheme. ICICS 2001, volume 2229 of LNCS, pages 461-465. Springer-Verlag, 2001.

[11] K.Q. Nguyen, J. Traoré. An Online Public Auction Protocol Protecting Bidder Privacy. Information Security and Privacy, 5th Australasian Conference-ACISP 2000, pages 427-442. Springer-Verlag, 2000.

[12] J. Traoré. Group Signatures and Their Relevance to Privacy-Protecting Off-Line Electronic Cash Systems. ACISP'99, volume 1587 of LNCS, pages 228-243. Springer-Verlag, 1999.

EFFICIENT COUNTERMEASURES AGAINST POWER ANALYSIS FOR ELLIPTIC CURVE CRYPTOSYSTEMS

Kouichi Itoh, Tetsuya Izu, and Masahiko Takenaka
FUJITSU LABORATORIES Ltd.
{kito,izu,takenaka}@labs.fujitsu.com

Abstract   The power analysis on smart cards is a real threat for cryptographic applications. In spite of continuous efforts of previous countermeasures, recent improved and sophisticated attacks against Elliptic Curve Cryptosystems are not protected. This paper proposes two new countermeasures, the Randomized Linearly-transformed Coordinates (RLC) and the Randomized Initial Point (RIP) against the attacks including the Refined Power Analysis (RPC) by Goubin and the Zero-value Point Analysis (ZPA) by Akishita-Takagi. Proposed countermeasures achieve notable speed-up without reducing the security level.

Keywords: Smart cards, power analysis, Elliptic Curve Cryptosystems, countermeasure

1. Introduction

Smart cards are becoming a new infrastructure in the coming IT society for their applications such as the SIM cards for mobile phones, identification cards for entrance systems and electronic tickets for movies. However, the power analysis attacks against these devices are real threats for these applications. In these attacks, an adversary observes traces of the power consumption of the device, and then, he detects a correlation between this information and some secret information hidden in the device. The simple power analysis (SPA) and the differential power analysis (DPA) are classical but typical examples [17, 18, 22]. Fortunately, various countermeasures which is not only secure but also efficient, have been proposed before 2002 [5, 15].

Recently, improved and sophisticated power analysis on Elliptic Curve Cryptosystems (ECC) have been proposed. In 2003, Goubin presented a new analysis, the Refined Power Analysis (RPA), which detects special points with 0-coordinate on the curve by chosen messages [9]. Then, Akishita-Takagi extended RPA to the Zero-value Point Analysis (ZPA), which detects 0 value in additions and doublings [2]. Some previous countermeasures resist RPA and ZPA [1, 5, 6], however, they require larger amount of processing time. A practical countermeasure was proposed by Ciet-Joye [7]. Note that Smart's countermeasure resists only RPA [32]; it does not always resist ZPA [3].

In this paper, we propose two practical countermeasures for ECC, the Randomized Linearly-transformed Coordinates countermeasure (RLC) and the Randomized Initial Point countermeasure (RIP), which resist power analysis including above newer attacks, and provide efficient processing speed for scalar multiplications. Proposed countermeasures achieve notable speed-up without reducing the security level.

The rest of this paper is organized as follows: we briefly review Elliptic Curve Cryptosystems (ECC) in section 2. Side channel attacks and countermeasures are in section 3. Then section 4 describes our proposed countermeasures. A comparison of countermeasures are described in section 5. Concrete algorithms of proposed countermeasures are in the appendix.

2. Elliptic Curve

This section describes a brief review of elliptic curves defined over finite fields with prime elements. This is just for a simlicity and basic ideas of most attacks and countermeasures can be applied to other finite fields.

Elliptic Curve Cryptosytems (ECC)

Let K be a finite field with prime elements $p > 3$. An elliptic curve over K is represented by an equation $E(K) := \{(x, y) \in K \times K \mid y^2 = x^3 + ax + b,\ a, b \in K,\ 4a^3 + 27b^2 \neq 0\} \cup \{\mathcal{O}\}$. The special point $\mathcal{O}$ is called the point at infinity. In this representation, a point on the curve (different from $\mathcal{O}$) is represented as a pair of K-elements (affine coordinates). An elliptic curve $E(K)$ forms an additive group by the following addition law: the point $\mathcal{O}$ is a neutral element, and an inverse point of $P = (x, y)$ is given by $-P = (x, -y)$. We call $P_1 + P_2$ $(P_1 \neq P_2)$ the elliptic curve addition (ECADD) and $P_1 + P_2$ $(P_1 = P_2)$, that is $2P_1$, the elliptic curve doubling (ECDBL). Let $P_1 = (x_1, y_1)$, $P_2 = (x_2, y_2)$ be two elements of $E(K)$ different from $\mathcal{O}$ and satisfying $P_1 \neq -P_2$. Then the addition $P_3 = P_1 + P_2 = (x_3, y_3)$ is defined by $x_3 = \lambda^2 - x_1 - x_2$, $y_3 = \lambda(x_1 - x_3) - y_1$, where $\lambda = (y_2 - y_1)/(x_2 - x_1)$ for $P_1 \neq P_2$, and $\lambda = (3x_1^2 + a)/(2y_1)$ for $P_1 = P_2$.

Elliptic Curve Cryptosystems (ECC) are one of the standard technologies in the area of cryptography [24]. The biggest advantage of ECC is the key length; it is currently chosen much shorter than that of existing other cryptosystems such as RSA and ElGamal. This feature is quite suitable for smart card implementations. Let d be an integer (scalar) and P be a point on the elliptic curve $E(K)$. The point P is called the base point. A *scalar multiplication* is an operation to compute a point $dP = P + \cdots + P$ $(d - 1$ additions). Computing d

Table 1. Add-and-double method

```
INPUT: d, P,    OUTPUT: dP
1:   T[0] = 0, T[1] = P
2:   for i=0 upto n-1
3:     if(d[i]==1){
4:        T[0] = ECADD(T[0],T[1])
5:     }
6:     T[1] = ECDBL(T[1])
7:   }
8:   return T[0]
```

Table 2. Add-and-double-always method

```
INPUT: d, P,    OUTPUT: dP
1:   T[0] = 0, T[2] = P
2:   for i=0 upto n-1 {
3:      T[1] = ECADD(T[0],T[2])
4:      T[2] = ECDBL(T[2])
5:      T[0] = T[d[i]]
6:   }
7:   return T[0]
```

from dP and P is called the elliptic curve discrete logarithm problem, and the hardness of this problem assures the security of ECC. In most ECC schemes, d is used as a secret key. A dominant computation of all encryption/decryption and signature generation/verification algorithms of ECC is the scalar multiplication. Efficiency of a scalar multiplication strongly depends on a choice of addition chains and coordinate systems for the elliptic curve representation.

Addition Chain and Coordinate System

Let $d = d_{n-1}2^{n-1} + \cdots + d_1 2^1 + d_0$ be a binary expression of a scalar d with $d_{n-1} = 1$. Then the add-and-double (AD) method from the least significant bit (LSB) for a scalar multiplication is shown in Table 1, which requires $n - 1$ ECDBLs and $n/2$ ECADDs in average. The number of ECADDs can be reduced by using the signed-binary expression of d, however, the technique is vulnerable to side channel attacks and we are not interested in it in this paper. A similar algorithm can be constructed from the most significant bit (MSB), however, we omit it because of the space limitation.

Coordinate systems for the elliptic curve representation effect on the processing speed of ECADD/ECDBL and hence a scalar multiplication [8]. In the affine coordinates $\mathcal{A}$, the addition formula always requires an inverse in the definition field K. In some environments, especially in prime fields (our situation), computing inverses are expensive. Since the projective coordinates and the Jacobian coordinates provide division-free formulas, these coordinates are widely used in practice. In the *projective coordinates* $\mathcal{P}$, a point is represented as a tuple (X, Y, Z) where two points (X, Y, Z) and $(\lambda X, \lambda Y, \lambda Z)$ $(\lambda \in K^*)$ are identical. A curve equation is obtained by setting $x = X/Z$, $y = Y/Z$ in the curve equation. In the *Jacobian coordinates* $\mathcal{J}$, a point is also represented as a tuple (X, Y, Z) where two points (X, Y, Z) and $(\lambda^2 X, \lambda^3 Y, \lambda Z)$ $(\lambda \in K^*)$ are identical. A curve equation is obtained by setting $x = X/Z^2$, $y = Y/Z^3$ in the curve equation. The addition formulas for general cases (namely, $P_1, P_2 \neq \mathcal{O}$ and $P_1 \neq -P_2$ for ECADD) in the Jacobian coordinates are given in Table 3. Note that in these coordinates, the point at infinity is represented as a tuple, which is the only point whose Z-coordinate value equals 0.

Table 3.   Addition formulas in the Jacobian coordinates (general cases)

ECADD (in the Jacobian coordinates)	ECDBL (in the Jacobian coordinates)
Input: $P_1 = (X_1, Y_1, Z_1), P_2 = (X_2, Y_2, Z_2)$	Input: $P_1 = (X_1, Y_1, Z_1)$, a
Output: $P_3 = P_1 + P_2 = (X_3, Y_3, Z_3)$	Output: $P_4 = 2P_1 = (X_4, Y_4, Z_4)$
$U_1 \leftarrow X_1 Z_2^2, \ U_2 \leftarrow X_2 Z_1^2, \ S_1 \leftarrow Y_1 Z_2^3$	$W \leftarrow a Z_1^4, \ M \leftarrow 3X_1^2 + W$
$S_2 \leftarrow Y_2 Z_1^3, \ W \leftarrow U_2 - U_1, \ R \leftarrow S_2 - S_1$	$S \leftarrow 4 X_1 Y_1^2, \ T \leftarrow 8 Y_1^4$
$T \leftarrow U_2 + U_1, \ M \leftarrow S_2 + S_1$	$X_4 \leftarrow M^2 - 2S$
$X_3 \leftarrow R^2 - TW^2, \ V \leftarrow 3TW^2 - 2R^2$	$Y_4 \leftarrow M(S - X_4) - T$
$Y_3 \leftarrow (VR - MW^3)/2, \ Z_3 \leftarrow Z_1 Z_2$	$Z_4 \leftarrow 2 Y_1 Z_1$

Table 4.   Processing times of ECADD and ECDBL

Coordinates	ECADD		ECDBL	
	$Z \neq 1$	$Z = 1$	$a \neq -3$	$a = -3$
$\mathcal{A}$	$2M + 1S + 1I$	—	$2M + 2S + 1I$	
$\mathcal{P}$	$12M + 2S$	$9M + 2S$	$7M + 5S$	$7M + 3S$
$\mathcal{J}$	$12M + 4S$	$8M + 3S$	$4M + 6S$	$4M + 4S$

Required processing time of ECADD/ECDBL is summarized in Table 4, where M, S, I denotes the processing time of a multiplication, a squaring, and an inverse in the definition field K, respectively. These processing time can be improved when one of Z-coordinate values in $\mathcal{P}, \mathcal{J}$ equals to 1 or the coefficient of the definition equation satisfies $a = -3$.

3.   Power Analysis

The power analysis is a powerful attack against cryptographic schemes on some devices such as smart cards. An adversary observes leaked power consumption, and detects a correlation between this information and some secret information hidden in the device. The simple power analysis (SPA) [17] and the differential power analysis (DPA) [18, 22] are typical examples.

Simple Power Analysis

The binary methods compute ECADD only when $d_i = 1$. Hence an adversary easily detects this irregular procedure by observing power consumption and obtains the value of d_i. The attack is called the *simple power analysis* (SPA) [17]. A standard approach to resist SPA is to use so called the *add-and-double-always* method (Table 2) [5], in which both ECADD and ECDBL are computed repeatedly (in step 3 and step 4). Since a pattern of the power trace becomes fixed, the adversary cannot detect the bit information d_i by SPA. As a variant of this countermeasure, the Montgomery ladder (Table 5) [21] is sometimes used [4, 11, 12, 16, 26, 27], because it provides good processing speed by using the specialized addition formula, the x-coordinate-only addition formula [4, 12, 11] which will be discussed later.

Table 5. Montgomery ladder

INPUT: d, P,   OUTPUT: dP
1:   T[0] = P
2:   T[1] = ECDBL(T[0])
3:   for i=n-2 downto 0 {
4:      T[2] = ECDBL(T[d[i]])
5:      T[1] = ECADD(T[0],T[1])
6:      T[0] = T[2-d[i]]
7:      T[1] = T[1+d[i]]
8:   }
9:   return T[0]

Table 6. ADA and RPC

INPUT: d, P,   OUTPUT: dP
1:   T[0] = 0, T[2] = RPC(P)
2:   for i=0 upto n-1 {
3:      T[1] = ECADD(T[0],T[2])
4:      T[2] = ECDBL(T[2])
5:      T[0] = T[d[i]]
6:   }
7:   return invRPC(T[0])

Note that some SPA-countermeasures use window-based methods for efficiency [14, 19, 20, 30, 31]. However, the memory requirement is severe in most cases on smart cards and we do not follow these approaches in this paper.

Differential Power Analysis

In the *differential power analysis* (DPA) [18, 22], an adversary guesses $d_i = 0$, for example, and simulates the computation repeatedly. Then he/she divides the power consumption into two sets depending on the assumption, in order to make a bias of hamming weights of the internal information between these sets. If the assumption is correct, a difference of the power consumption of two sets can be observed (as a spike) in the trace. Note that even if a scheme is SPA-resistant, it is not always DPA-resistant.

In order to resist DPA, a randomization of parameters is a well-known technique to make the simulation impossible [5]. One approach is to randomize a scalar d or an addition chain for d. Coron proposed the scalar blinding countermeasure, in which d is replaced by $d + r\phi$ for a random integer r and the order of the point ϕ [5]. Original r (20-bit) does not provide sufficient security [27]. Larger r will provide higher security and less speed. Coron also proposed the base point blinding countermeasure, in which dP is computed by $dP = d(P + R) - dR$ for a random point R. However biases are pointed out [27]. Oswald-Aigner used the randomized addition-subtraction chain [25], but effect of the randomization is not enough [28, 29, 33].

Clavier-Joye splits the scalar randomly, in which dP is computed by $dP = rP + (d - r)P$ for a random integer r (the exponent splitting, ES) [6]. As no security problems have been pointed out, ES requires at least twice of the processing time than without it. Ciet-Joye proposed another splitting method, in which dP is computed by $\lfloor d/r \rfloor (rP) + (d \bmod r)P$ for a random integer r [7]. Here r is as long as half of d and, with Shamir's trick, it provides practical processing speed without reducing the security.

Another approach for resisting DPA is to randomize a point representation. Coron proposed the randomized projective coordinates (RPC) countermeasure

[5]. Let $P = (X, Y, Z)$ be a base point in the projective coordinates. Then (X, Y, Z) equals to (rX, rY, rZ) for all $r \in K^*$ mathematically; but they all are different data as bit sequences. The power consumption of a scalar multiplication is randomized if (X, Y, Z) is randomized to (rX, rY, rZ) for a randomly chosen integer $r \in K^*$. An example of RPC for the add-and-double-always method is shown in Table 6, where a function RPC outputs a randomized point by RPC and a function invRPC denotes its inverse function. Joye-Tymen proposed another DPA-countermeasure based on the randomization of point representation, the randomized curve (RC) countermeasure [15]. RC uses an isomorphism of elliptic curves with which a curve equation and a base point are transformed with holding the group structure. Again, let $P = (X, Y, Z)$ be a base point in the projective coordinates. Then (X, Y, Z) and (r^2X, r^3Y, Z) correspond to each other under the isomorphism. Two points are the same (under the isomorphism) mathematically; but different as bit sequences. Thus the power consumption will be randomized if the curve and the base point is randomized by a randomly chosen integer $r \in K^*$. A sample algorithm is easily obtained similarly to Table 6 by changing functions RPC, invRPC to RC, invRC. RPC and RC provide efficient scalar multiplications because the conversions are quite cheap compared to the main multiplications.

Note that all countermeasures of ES, RPC, RC does not resist SPA. These countermeasures should be used combined with SPA-countermeasures.

Refined Power Analysis and Zero-value Point Attack

In 2003, Goubin presented a new power analysis against ECC, the *refined power analysis* (RPA), which reveals the secret scalar by detecting special points with 0-coordinates on the curve by chosen messages [9]. Let $E(K)$ be an elliptic curve which has the special point $Q_y = (0, y)$, for example. Suppose scalar multiplications are computed by add-and-double-always method from LSB (Table 2). When an adversary knows the bit information $d_0, \dots, d_{j-1}$ of the secret key d, he/she tries to reveal the next bit d_j. At the end of the $(j-1)$-th loop in Table 2, we have $T[2] = 2^j P$, $T[0] = \sum_{i=0}^{j-1} d_i 2^i P$. Then, in the j-th loop, we will compute

$$T[1] \leftarrow \left(\sum_{i=0}^{j-1} d_i 2^i + 2^j \right) P, \quad T[0] \leftarrow \begin{cases} T[0] = \sum_{i=0}^{j-1} d_i 2^i P & \text{if } d_j = 0, \\ T[1] = (\sum_{i=0}^{j-1} d_i 2^i + 2^j) P & \text{if } d_j = 1. \end{cases}$$

Then the adversary pre-computes $\overline{P} = \left(\sum_{i=0}^{j-1} d_i 2^i + 2^j \right)^{-1} Q_y$, where the inverse is computed modulo the order of the curve, and computes a scalar multiplication on input d and $\overline{P}$. If $d_i = 1$, at the end of the j-th loop, we have $T[0] = \left(\sum_{i=0}^{j-1} d_i 2^i + 2^j \right) \overline{P} = Q_y = (0, y)$ and the adversary easily detects d_i, because the power consumption of "0" is quite distinctive. This is a basic idea of RPA. The attack can be applied to the add-and-double-always method from MSB and the Montgomery ladder [9]. Note that the pre-computation is very easy for the adversary because elliptic curve parameters are public.

In the same year, Akishita-Takagi extended Goubin's RPA to the *zero-value point attack* (ZPA) [2]. A basic idea of ZPA is to observe 0 in the functions ECADD and ECDBL rather than in the addition chain where RPA paid attention. Akishita-Takagi discussed conditions when such 0 can be observed. In order to resist ZPA, implementers should avoid operations which possibly output 0 in ECADD and ECDBL.

Countermeasures against RPA/ZPA

A key of RPA is to observe 0 in the process. The most noteworthy property of RPA is that RPC and RC cannot resist RPA, because in RPC and RC, the special point $Q_x = (x, 0)$, $Q_y = (0, y)$ are converted to point $(X, 0, Z)$, $(0, Y, Z)$ by the conversion with keeping the specialty "0". Because of the same reasons as RPA, RPC and RC cannot resist ZPA.

Smart proposed an RPA-countermeasures using isogeny of the curve [32]. However, the countermeasure does not resist ZPA [3]. Avanzi analyzed the resistance of Coron's scalar blinding and base point blinding countermeasure against RPA, and concluded that they are resistant to RPA [1]. Similarly, they also resist ZPA. However, these countermeasures require larger amount of computing time than without them. The exponent splitting countermeasure [6] and its improved version [7] also resist both RPA and ZPA. Especially, Ciet-Joye's countermeasure has almost no penalty from a viewpoint of efficiency.

4. Proposed Countermeasures

This section proposes two practical countermeasures for ECC against power analysis including DPA, RPA and ZPA: the *randomized linearly-transformed coordinates* (RLC) and the *randomized initial point* (RIP) countermeasures.

Randomized Linearly-transformed Coordinates (RLC)

Before proposing the countermeasure, we introduce the linearly-transformed coordinates (LC) in the following:

DEFINITION 1 *In the linearly-transformed coordinates (LC) with a parameter set* $(\lambda_X(\lambda), \lambda_Y(\lambda), \lambda_Z(\lambda), \mu_X, \mu_Y, \mu_Z)$, *a point* (X, Y, Z) *and* (X', Y', Z') *are identified if there exists* $\lambda \in K^*$ *such that*

$$X - \mu_X = \lambda_X(\lambda)(X' - \mu_X), \quad Y - \mu_Y = \lambda_Y(\lambda)(Y' - \mu_Y),$$
$$Z - \mu_Z = \lambda_Z(\lambda)(Z' - \mu_Z),$$

where $\lambda_X(\lambda)$, $\lambda_Y(\lambda)$, $\lambda_Z(\lambda)$ *are functions of* λ *and* $\mu_X, \mu_Y, \mu_Z \in K$.

Table 7.　Addition formulas in LC for a parameter set $S_1 = (\lambda^2, \lambda^3, \lambda, \mu, 0, 0)$ (general case)

LCECADD for $S_1 = (\lambda^2, \lambda^3, \lambda, \mu, 0, 0)$	LCECDBL for $S_1 = (\lambda^2, \lambda^3, \lambda, \mu, 0, 0)$
Input: $P_1 = (X_1, Y_1, Z_1)$,	Input: $P_1 = (X_1, Y_1, Z_1)$, a, μ
$\quad P_2 = (X_2, Y_2, Z_2)$, μ	
Output: $P_3 = P_1 + P_2 = (X_3, Y_3, Z_3)$	Output: $P_4 = 2P_1 = (X_4, Y_4, Z_4)$
$U_1 \leftarrow (X_1 - \mu)Z_2^2,\ U_2 \leftarrow (X_2 - \mu)Z_1^2$	$W \leftarrow aZ_1^4$
$S_1 \leftarrow Y_1 Z_2^3,\ S_2 \leftarrow Y_2 Z_1^3$	$M \leftarrow 3(X_1 - \mu)^2 + W$
$W \leftarrow U_2 - U_1,\ R \leftarrow S_2 - S_1$	$S \leftarrow 4(X_1 - \mu)Y_1^2$
$T \leftarrow U_2 + U_1,\ M \leftarrow S_2 + S_1$	$T \leftarrow 8Y_1^4$
$X_3 \leftarrow R^2 - TW^2 + \mu$	$X_4 \leftarrow M^2 - 2S + \mu$
$V \leftarrow 3TW^2 - 2R^2$	$Y_4 \leftarrow M(S - X_4 + \mu) - T$
$Y_3 \leftarrow (VR - MW^3)/2,\ Z_3 \leftarrow Z_1 Z_2 W$	$Z_4 \leftarrow 2Y_1 Z_1$

The projective coordinates $\mathcal{P}$ and the Jacobian coordinates $\mathcal{J}$ are examples of the linearly-transformed coordinates; the parameter set for $\mathcal{P}$ is $(\lambda, \lambda, \lambda, 0, 0, 0)$ and for $\mathcal{J}$ is $(\lambda^2, \lambda^3, \lambda, 0, 0, 0)$. One of the advantage of LC is a variation of representations. Especially, if we set $\mu_X, \mu_Y \neq 0$, the special point $Q_y = (0, y)$ and $Q_x = (x, 0)$ are converted to some points whose coordinate values are different from 0. However, the processing speed of ECADD/ECDBL become slower in general than previous coordinates.

As an example, we give explicit addition formulas LCECADD, LCECDBL which compute ECADD and ECDBL in the linearly-transformed coordinates for a parameter set $S_1 = (\lambda^2, \lambda^3, \lambda, \mu, 0, 0)$ in Table 7. Here the coordinates are very similar to the formulas for the Jacobian case (Table 3).

Proposed Countermeasure.　An idea of our countermeasure is quite simple. For given functions $\lambda_X(\lambda)$, $\lambda_Y(\lambda)$, $\lambda_Z(\lambda)$, we randomly choose μ_X, μ_Y, μ_Z and convert the base point into the transformed coordinates with the parameter set $(\lambda_X(\lambda), \lambda_Y(\lambda), \lambda_Z(\lambda), \mu_X, \mu_Y, \mu_Z)$ in the beginning of a scalar multiplication. After processing the scalar multiplication in LC, we re-convert the result to the original coordinates (*randomized linearly-transformed coordinates, RLC*). A generic algorithm of RLC is obtained by changing RPC, invRPC to RLC, invRLC, where a function RLC converts a point P into a point with a chosen parameter set, and invRLC re-converts to the affine coordinates. Note that RLC can be applied to all addition chains mentioned in section 2.

Let us consider the security of RLC. Basically, we have to use the add-and-double-always methods or the Montgomery ladder for the addition chain, for RLC does not resist SPA itself. On the other hand, RLC resists DPA if a parameter set is properly chosen, because RLC is a natural extension of RPC. But RPA will be successful when we choose $\mu_X = 0$ or $\mu_Y = 0$ (for speeding up), because points with 0 coordinate values can appear. In RLC, by choosing $\mu_X, \mu_Y \neq 0$, such special points will not appear in the process of scalar multiplications. This is the same against ZPA. Consequently, RLC resists SPA, DPA, RPA, and ZPA if the parameter set is chosen suitably (namely $\mu_X, \mu_Y \neq 0$).

Table 8. ADA and RLC

```
INPUT: d, P.    OUTPUT: dP
1:  T[0] = 0, T[2] = RLC(P)
2:  for i=0 upto n-1 {
3:      T[1] = LCECADD(T[0],T[2])
4:      T[2] = LCECDBL(T[2])
5:      T[0] = T[d[i]]
6:  }
7:  return invRLC(T[0])
```

Table 9. ADA and RIP

```
INPUT: d, P.    OUTPUT: dP
1:  R = randompoint()
2:  T[0] = R, T[2] = P
3:  for i=0 upto n-1 {
4:      T[1] = ECADD(T[0],T[2])
5:      T[2] = ECDBL(T[2])
6:      T[0] = T[d[i]]
7:  }
8:  return T[0]-R
```

However, a straight-forward implementation of RLC may be vulnerable to RPA. For example, when we implement LCECADD in Table 7, direct computations of $X_1 - \mu$ and $X_2 - \mu$ should be avoided, because these values equal to 0 for a chosen base point $\overline{P}$ by RPA. Thus we have to compute W, T without computing $X_i - \mu$ and U_i. Realizations of concrete secure algorithms are shown in the following sections.

In the followings, we combine RLC and three addition chains concretely, and then discuss the security against possible attacks. As we will consider over a finite field with prime elements, the condition $\mu_X \neq 0$ should be satisfied.

Combination with Add-and-double-always Method from MSB. Applying RLC to the add-and-double-always method from MSB is easy (so we omit a concrete algorithm because of the space limitation). Here both LCECADD and LCECDBL should be implemented not to compute $X_i - \mu_X$ as mentioned before. Remember that if μ_X is chosen randomly (different from 0), the probability that $X_i - \mu_X = 0$ is negligible. Concrete secure algorithms LCECADD1 and LCECDBL1 for a parameter set $S_1 = (\lambda^2, \lambda^3, \lambda, \mu, 0, 0)$, $\mu \in K^*$ are shown in Table A1 and Table A2 in the appendix. Note that these algorithms are constructed not to output 0 values after each operation in order to resist ZPA. The processing speed of LCECADD1 is $14M + 4S$ (Table A1) and LCECDBL1 is $8M + 7S$ (Table A2). An average processing speed for each bit is $T^{\mathrm{MSB}} = (14M + 4S) + (8M + 7S) = 22M + 11S$.

Combination with Add-and-double-always Method from LSB. Applying RLC to the add-and-double-always method from LSB is also easy. A concrete algorithm is shown in Table 8. In this case, a variable T[2] has no relation to the secret key d. Hence there is no need for T[2] to be protected. In other words, we are allowed to compute $X_2 - \mu_X$ in LCECADD, and allowed not to use secure algorithm for LCECDBL. Thus we only take care not to compute $X_1 - \mu_X$ in LCECADD. A concrete algorithm LCECADD1' for a parameter set $S_1 = (\lambda^2, \lambda^3, \lambda, \mu, 0, 0)$, $\mu \in K^*$ is shown in Table A3 in the appendix. Note that the algorithm is constructed not to output 0 values after each operation in order to resist ZPA. The processing speed of LCECADD1' is $13M + 4S$ (Table

A1) and LCECDBL1 is $4M + 6S$ (Table 3). An average processing speed for each bit is $T^{LSB} = (13M + 4S) + (4M + 6S) = 17M + 10S$.

When the secret key d has successive 0 in lowest bits, the number of 0 can be detected, because variables T[0],T[2] are not randomized by insecure LCECDBL. Once 1 appears in d, T[0] is randomized in LCECADD and no information will be leaked moreover. One solution for the leakage is to use secure LCECDBL and compute $X - \mu_X$ directly. Since the effect of λ's are remained, the security is improved without reducing the processing speed.

In the Jacobian coordinates, Itoh et al. proposed an efficient algorithm called the *iterated ECDBL* [13], which computes $2^k P$ for a given integer $k \geq 1$ more efficiently than computing ECDBL k times by recycling intermediate values; the iterated ECDBL requires $4kM + (4k + 2)S$, while each ECDBL requires $4M + 6S$ (Table 4). In our countermeasure for the parameter set S_1, a variable T[2] only concerns a sequence $P, 2P, 2^2 P, \ldots$ and the iterated ECDBL can be applied efficiently. In this case, an averaged processing speed for each bit is optimized to $T_o^{LSB} = (13M + 4S) + (4M + 4S) = 17M + 8S$.

Combination with Montgomery Ladder.   Applying RLC to the Montgomery ladder is also easy. In this case, all variables T[0], T[1], T[2] depend on the secret key and both LCECADD and LCECDBL should be implemented not to compute $X_i - \mu_X$.

When we use secure algorithms LCECADD1, LCECDBL1 for the parameter set S_1 (Table A1 and Table A2 in the appendix), an average processing speed is $T^{Mon} = (14M + 4S) + (6M + 6S) = 20M + 10S$.

The Montgomery ladder provides good processing speed by using the specialized addition formula, the x-coordinate-only addition formula [4, 12, 11], which computes ECADD and ECDBL with only x-coordinate value in the affine coordinates, or X- and Z-coordinate values in the projective or the Jacobian coordinates. We can easily establish similar formulas xLCECADD, xLCECDBL, which compute ECADD and ECDBL in our linearly-transformed coordinates by the x-coordinate-only formula.

Moreover, Izu-Takagi proposed an encapsuled formula ECADDDBL, which takes P_1, P_2 on input and outputs $P_1 + P_2$ and $2P_1$, and xECADDDBL computed by the x-coordinate-only formula. In this case, an improved version of the Montgomery ladder is required [11]. A concrete algorithm xLCECADDDBL2, which computes xECADDDBL in LC by the x-coordinate-only method for a parameter set $S_2 = (\lambda, \lambda, \lambda, \mu, 0, 0)$, is shown in Table A4 (where the additive ECADD is used [12]). Here the algorithm is constructed not to output 0 values after each operation in order to resist ZPA, and the processing speed for each bit is $T_o^{Mon} = 20M + 5S$.

Randomized Initial Point Countermeasure (RIP)

Next, we propose another countermeasure against power analysis.

Proposed Countermeasure. In the add-and-double always method from LSB (Table 2), an initial value of T[0] is set to $\mathcal{O}$. If this initial value is changed to a randomized point R, intermediate information will be randomized and thus it resists power analysis. (Of course, R should be subtracted from a result of a scalar multiplication.) This countermeasure is called the *randomized initial point* countermeasure (RIP). A generic algorithm of RIP combined with Table 2 is shown in Table 9. Here a function randompoint() generates a random point on the curve. Note that RIP can be applied only to the add-and-double always method from LSB.

Selection of Initial Point. The security of RIP depends on how to select an initial point R. The simplest way is to generate the x-coordinate value of R and compute corresponding y-coordinate value. One can obtain such y with probability $1/2$, namely it might require large amount of time to obtain R. However, R can be obtained in short time practically.

The other way is to convert a fixed and stored point Q into a randomized point R by using RPC. As the conversion is very cheap, it provides an efficient generation of R.

Security Consideration. Firstly, as RIP does not resist SPA (similarly to RLC), we have to use the add-and-double always method for the addition chain (Table 9). When an initial point R is chosen properly, as a representation of R would be randomized, RIP resists DPA (contrary, RIP is vulnerable to DPA if R is fixed). Here there is no need to apply RPC or RC. On the other hand, RIP also resists RPA and ZPA, because the all intermediate values are randomized by the initial point R.

Processing Speed. When R is set from Q by RPC, the processing time of generating R is negligible. Thus, the processing time of RIP per bit equals to a sum of that of ECADD and ECDBL. It is $(12M+2S)+(7M+5S) = 19M+7S$ for the projective coordinates, and $(12M+4S)+(4M+6S) = 16M+10S$ for the Jacobian coordinates.

When the Jacobian coordinates are used, the iterated ECDBL can be also applied similarly to RLC. In this case, the processing time of RIP per a bit equals $(12M+4S)+(4M+6S)-2S = 16M+8S$.

Table 10. Comparison of secure countermeasures

Method	Coordinates	Speed in each bit	
iES [7]	$\mathcal{J}$	$16M + 10S$	$(24.0M)$
iES with Montgomery's trick	$\mathcal{J}$	$12M + 9S$	$(19.2M)$
RLC with MSB	$\mathcal{J}$	$22M + 11S$	$(30.2M)$
RLC with LSB	$\mathcal{J}$	$17M + 10S$	$(25.0M)$
RLC with LSB and iECDBL	$\mathcal{J}$	$17M + 8S$	$(23.4M)$
RLC with Montgomery ladder	$\mathcal{J}$	$20M + 5S$	$(24.0M)$
RIP with LSB	$\mathcal{P}$	$19M + 7S$	$(24.6M)$
RIP with LSB	$\mathcal{J}$	$16M + 10S$	$(24.0M)$
RIP with LSB and iECDBL	$\mathcal{J}$	$16M + 8S$	$(22.4M)$

5.   Comparison

Let us compare countermeasures against power analysis that resist all of SPA, DPA, RPA and ZPA. We provide practical processing time for the improved exponent splitting (iES) [7] and our proposed two countermeasures (RLC and RIP). Processing speed are given averaged for each bit of the scalar d and is estimated with the assumption that $1S = 0.8M$ [27, 12].

As in the table, iES with Shamir's trick and Montgomery's trick provides the most efficient processing speed. However, our proposed countermeasures provide comparable speed. A difference between iES and RIP is the number of registers for points on the curve; iES requires 5 registers while our RIP requires only 4 registers for a scalar multiplication. Note that, in RLC, the seed of an initial point R is required. That is, RLC requires extra memory compared to other methods. On the other hand, RLC requires no extra memory.

6.   Concluding Remarks

In this paper, two new practical countermeasures against power analysis are proposed, the randomized linearly-transformed coordinates countermeasure and the randomized initial point countermeasure. These methods improves the processing speed without reducing the security.

Since RPA and ZPA are newer attacks, the property of the attacks and of the countermeasures are not well studied yet. Theoretically, iES provides a better processing speed, however, because it uses a small table of pre-computed points, the resistance against the address-bit DPA [10] and the 2nd order DPA [28] should be analyzed. On the other hand, applying RIP to other addition chains will be for future work. Very recently, Morimoto-Mamiya-Miyaji attempted to apply RIP to the binary method from MSB [23], for which further analysis will be required.

Acknowledgments

The authors would like to thank Toru Akishita, JJean-Bernard Fischer and Naoya Torii for their valuable and helpful comments on the preliminary version of this paper. We also thank anonymous reviewers of the conference for introducing required references.

References

[1] R. Avanzi, "Countermeasures against Differential Power Analysis for Hyperelliptic Curve Cryptosystems", *CHES 2003*, LNCS 2779, pp.366-381, Springer-Verlag, 2003.

[2] T. Akishita, and T. Takagi, "Zero-value Point Attacks on Elliptic Curve Cryptosystem", *ISC 2003*, LNCS 2851, pp.218-233, Springer-Verlag, 2003.

[3] T. Akishita, and T. Takagi, "On the Optimal Parameter Choice for Elliptic Curve Cryptosystems Using Isogeny", *PKC 2004*, LNCS 2947, pp.346-359, Springer-Verlag, 2004.

[4] E. Brier, and M. Joye, "Weierstraß Elliptic Curves and Side-Channel Attacks", *PKC 2002*, LNCS 2274, pp.335-345, Springer-Verlag, 2002.

[5] J. Coron, "Resistance against Differential Power Analysis for Elliptic Curve Cryptosystem", *CHES'99*, LNCS 1717, pp.292-302, Springer-Verlag, 1999.

[6] C. Clavier, and M. Joye, "Universal exponentiation algorithm – A first step towards provable SPA-resistance –", *CHES 2001*, LNCS 2162, pp. 300-308, Springer-Verlag, 2001.

[7] M. Ciet, and M. Joye, "(Virtually) Free Randomization Technique for Elliptic Curve Cryptography", *ICICS 2003*, LNCS 2836, pp. 348-359, Springer-Verlag, 2003.

[8] H. Cohen, A. Miyaji, and T. Ono, "Efficient Elliptic Curve Exponentiation Using Mixed Coordinates", *Asiacrypt'98*, LNCS 1514, pp.51-65, Springer-Verlag, 1998.

[9] L. Goubin, "A Refined Power-Analysis Attack on Elliptic Curve Cryptosystems", *PKC 2003*, LNCS 2567, pp.199-210, Springer-Verlag, 2003.

[10] K. Itoh, T. Izu, M. Takenaka, "Address-bit Differential Power Analysis of Cryptographic Schemes OK-ECDH and OK-ECDSA", *CHES 2002*, LNCS 2523, pp.129-143, 2003.

[11] T. Izu, B. Möller, and T. Takagi, "Improved Elliptic Curve Multiplication Methods Resistant against Side Channel Attacks", *Indocrypt 2002*, LNCS 2551, pp.296-313, Springer-Verlag, 2002.

[12] T. Izu, and T. Takagi, "A Fast Parallel Elliptic Curve Multiplication Resistant against Side Channel Attacks", *PKC 2002*, LNCS 2274, pp.280-296, Springer-Verlag, 2002.

[13] K.Itoh, M.Takenaka, N.Torii, S.Temma, and Y.Kurihara, "Fast Implementation of Public-Key Cryptography on DSP TMS320C6201", *CHES'99*, LNCS 1717, pp.61-72, 1999.

[14] K. Itoh, J. Yajima, M. Takenaka, and N. Torii, "DPA Countermeasures by Improving the Window Method", *CHES 2002*, LNCS 2523, pp.303-317, Springer-Verlag, 2003.

[15] M. Joye, C. Tymen, "Protections against Differential Analysis for Elliptic Curve Cryptography", *CHES 2001*, LNCS 2162, pp.377-390, Springer-Verlag, 2001.

[16] M. Joye, and S-M. Yen, "The Montgomery Powering Ladder", *CHES 2002*, LNCS 2523, pp.291-302, Springer-Verlag, 2003.

[17] C. Kocher, "Timing Attacks on Implementations of Diffie-Hellman, RSA, DSS, and Other Systems", *Crypto'96*, LNCS 1109, pp.104-113, Springer-Verlag, 1996.

[18] C. Kocher, J. Jaffe, and B. Jun, "Differential Power Analysis", *Crypto '99*, LNCS 1666, pp.388-397, Springer-Verlag, 1999.

[19] B. Möller, "Securing Elliptic Curve Point Multiplication against Side-Channel Attacks", *ISC 2001*, LNCS 2200, pp.324-334, Springer-Verlag, 2001.

[20] B. Möller, "Parallelizable Elliptic Curve Point Multiplication Method with Resistance against Side-Channel Attacks", *ISC 2002*, LNCS 2433, pp.402-413, Springer-Verlag, 2002.

[21] P. Montgomery, "Speeding the Pollard and Elliptic Curve Methods for Factorizations", *Math. of Comp*, vol.48, pp.243-264, 1987.

[22] T. Messerges, E. Dabbish, and R. Sloan, "Power Analysis Attacks of Modular Exponentiation in Smartcards", *CHES'99*, LNCS 1717, pp. 144-157, Springer-Verlag, 1999.

[23] H. Morimoto, H. Mamiya, and A. Miyaji, "Elliptic Curve Cryptosystems Secure against ZPA"(in Japanese), Technical Report of the Institute of Electronicas, Information and Communication Engineers (IEICE), ISEC 2003-103, March, 2004. English version is to appear in the proceedings of *CHES 2004*.

[24] Recommended Elliptic Curves for Federal Government Use, in the appendix of FIPS 186-2, National Institute of Standards and Technology (NIST).

[25] E. Oswald, and M. Aigner, "Randomized Addition-Subtraction Chains as a Countermeasure against Power Attacks", *CHES 2001*, LNCS 2162, pp.39-50, Springer-Verlag, 2001.

[26] K. Okeya, H. Kurumatani, and K. Sakurai, "Elliptic curves with the Montgomery form and their cryptographic applications", *PKC 2000*, LNCS 1751, pp.446-465, Springer-Verlag, 2000.

[27] K. Okeya, and K. Sakurai, "Power analysis breaks elliptic curve cryptosystem even secure against the timing attack", *Indocrypt 2000*, LNCS 1977, pp.178-190, Springer-Verlag, 2000.

[28] K. Okeya, and K. Sakurai, "On Insecurity of the Side Channel Attack Countermeasure Using Addition-Subtraction Chains under Distinguishability between Addition and Doubling", *ACISP 2002*, LNCS 2384, pp.420-435, Springer-Verlag, 2002.

[29] K. Okeya, and K. Sakurai, "A Multiple Power Analysis Breaks the Advanced Version of the Randomized Addition-Subtraction Chains Countermeasure against Side Channel Attacks", to appear in the proceedings of *2003 IEEE Information Theory Workshop*.

[30] K. Okeya, and T. Takagi, "The Width-w NAF Method Provides Small Memory", *CT-RSA 2003*, LNCS 2612, pp.328-342, Springer-Verlag, 2003.

[31] K. Okeya, and T. Takagi, "A More Flexible Countermeasure against Side Channel Attacks using Window Method", *CHES 2003*, LNCS 2779, pp. 397-410 Springer-Verlag, 2003.

[32] N. Smart, "An Analysis of Goubin's Refined Power Analysis Attack", *CHES 2003*, LNCS 2779, pp.281-290, Springer-Verlag, 2003.

[33] C. Walter, "Security Constraints on the Oswald-Aigner Exponentiation Algorithm", Cryptology ePrint Archive, Report 2003/013, 2003.

Appendix: Algorithms for Proposed Countermeasures

The appendix shows detailed concrete algorithms for LCECADD1, LCECDBL1, LCECADD1', and xLCECADDDBL2 for RLC introduced in section 4.1. The parameter set for LCECADD1,

Table A1. ECADD in LC with a parameter set $S_1 = (\lambda^2, \lambda^3, \lambda, \mu, 0, 0)$

LCECADD1 (in 14M+4S)

Input: $P_1 = (X_1, Y_1, Z_1)$,
$\quad\quad\; P_2 = (X_2, Y_2, Z_2)$, μ
Output: $P_3 = P_1 + P_2 = (X_3, Y_3, Z_3)$

$T_1 \leftarrow Z_1^2 \quad (= Z_1^2)$
$T_2 \leftarrow Z_1 \times T_1 \quad (= Z_1^3)$
$T_3 \leftarrow Z_2^2 \quad (= Z_2^2)$
$T_4 \leftarrow Z_2 \times T_3 \quad (= Z_2^3)$
$T_5 \leftarrow X_1 \times T_3 \quad (= X_1 Z_2^2)$
$T_6 \leftarrow \mu \times T_3 \quad (= \mu Z_2^2)$
$T_7 \leftarrow X_2 \times T_1 \quad (= X_2 Z_1^2)$
$T_8 \leftarrow \mu \times T_1 \quad (= \mu Z_1^2)$
$T_9 \leftarrow Y_1 \times T_4 \quad (= Y_1 Z_2^3 = S_1)$
$T_{10} \leftarrow Y_2 \times T_2 \quad (= Y_2 Z_1^3 = S_2)$
$T_{11} \leftarrow T_7 - T_5 \quad (= X_2 Z_1^2 - X_1 Z_2^2)$
$T_{12} \leftarrow T_{11} - T_8 \quad (= U_2 - X_1 Z_2^2)$
$T_{13} \leftarrow T_{12} + T_6 \quad (= U_2 - U_1 = W)$
$T_{14} \leftarrow T_7 + T_5 \quad (= X_2 Z_1^2 + X_1 Z_2^2)$
$T_{15} \leftarrow T_{14} - T_8 \quad (= U_2 + X_1 Z_2^2)$
$T_{16} \leftarrow T_{15} - T_6 \quad (= U_2 + U_1 = T)$
$T_{17} \leftarrow T_{10} - T_9 \quad (= S_2 - S_1 = R)$
$T_{18} \leftarrow T_{10} + T_9 \quad (= S_2 + S_1 = M)$
$T_{19} \leftarrow T_{17}^2 \quad (= R^2)$
$T_{20} \leftarrow T_{13}^2 \quad (= W^2)$
$T_{21} \leftarrow T_{16} \times T_{20} \, (= TW^2)$
$T_{22} \leftarrow T_{19} + \mu \quad (= R^2 + \mu)$
$X_3 \leftarrow T_{22} - T_{21} \, (= R^2 - TW^2 + \mu)$
$T_{23} \leftarrow 2 * X_3$
$T_{24} \leftarrow T_{23} - T_{21}$
$T_{25} \leftarrow 2 * \mu \quad (= 2\mu)$
$T_{26} \leftarrow T_{25} - T_{24} \, (= V)$
$T_{27} \leftarrow T_{25} \times T_{17} \, (= VR)$
$T_{28} \leftarrow T_{18} \times T_{13} \, (= MW)$
$T_{29} \leftarrow T_{28} \times T_{20} \, (= MW^3)$
$2Y_3 \leftarrow T_{27} - T_{29}$
$T_{30} \leftarrow Z_1 \times Z_2 \quad (= Z_1 Z_2)$
$Z_3 \leftarrow T_{30} \times T_{13} \quad (= Z_1 Z_2 W)$

Table A2. ECDBL in LC with a parameter set $S_1 = (\lambda^2, \lambda^3, \lambda, \mu, 0, 0)$

LCECDBL1 (in 8M+7S)

Input: $P_1 = (X_1, Y_1, Z_1)$, a, μ
Output: $P_4 = 2P_1 = (X_4, Y_4, Z_4)$

$T_1 \leftarrow Z_1^2 \quad\quad (= Z_1^2)$
$T_2 \leftarrow T_1^2 \quad\quad (= Z_1^4)$
$T_3 \leftarrow a \times T_2 \quad (= aZ_1^4 = W)$
$T_4 \leftarrow X_1^2 \quad\quad (= X_1^2)$
$T_5 \leftarrow 2 * X_1 \quad\quad (= 2X_1)$
$T_6 \leftarrow \mu - T_5 \quad\quad (= \mu - 2X_1)$
$T_7 \leftarrow \mu \times T_6 \quad (= \mu(\mu - 2X_1))$
$T_8 \leftarrow 3 * T_4 \quad\quad (= 3X_1^2)$
$T_9 \leftarrow 3 * T_7 \quad\quad (= 3\mu(\mu - 2X_1))$
$T_{10} \leftarrow T_3 + T_8 \quad (= W + 3X_1^2)$
$T_{11} \leftarrow Y_1^2 \quad\quad (= Y_1^2)$
$T_{12} \leftarrow X_1 \times T_{11} \, (= X_1 Y_1^2)$
$T_{13} \leftarrow \mu \times T_{11} \quad (= \mu Y_1^2)$
$T_{14} \leftarrow T_{11}^2 \quad\quad (= Y_1^4)$
$T_{15} \leftarrow 8 * T_{14} \quad\quad (= 8Y_1^4 = T)$
$T_{16} \leftarrow T_9^2 \quad\quad (= (3\mu(\mu - 2X_1))^2)$
$T_{17} \leftarrow T_9 \times T_{10}$
$T_{18} \leftarrow 2 * T_{17}$
$T_{19} \leftarrow T_{10}^2 \quad\quad (= (W + 3X_1^2)^2)$
$T_{20} \leftarrow \mu + T_{16}$
$T_{21} \leftarrow T_{20} + T_{18}$
$T_{22} \leftarrow T_{21} + T_{19}(= \mu + M^2)$
$T_{23} \leftarrow 8 * T_{12} \quad (= 8X_1 Y_1^2)$
$T_{24} \leftarrow 8 * T_{13} \quad (= 8\mu Y_1^2)$
$T_{25} \leftarrow T_{22} - T_{23}(= M^2 - 8X_1 Y_1^2 + \mu)$
$X_4 \leftarrow T_{22} + T_{24}(= M^2 - 2S + \mu)$
$T_{26} \leftarrow 4 * T_{12} \quad (= 4X_1 Y_1^2)$
$T_{27} \leftarrow 4 * T_{13} \quad (= 4\mu Y_1^2)$
$T_{28} \leftarrow T_{26} + \mu \quad (= 4X_1 Y_1^2 + \mu)$
$T_{29} \leftarrow X_4 - T_{28}(= 4X_1 Y_1^2 - X_4 + \mu)$
$T_{30} \leftarrow T_{29} - T_{27}(= S - X_4 + \mu)$
$T_{31} \leftarrow T_9 \times T_{30}$
$T_{32} \leftarrow T_{10} \times T_{30}$
$T_{33} \leftarrow T_{31} - T_{15}$
$Y_4 \leftarrow T_{33} + T_{32} \quad (= M(S - X_4 + \mu) - T)$
$T_{34} \leftarrow Y_1 \times Z_1 \quad (= Y_1 Z_1)$
$Z_4 \leftarrow 2 * T_{34} \quad\quad (= 2Y_1 Z_1)$

LCECDBL1, LCECADD1' is $S_1 = (\lambda^2, \lambda^3, \lambda, \mu, 0, 0)$ and that for xLCECADDDBL2 is $S_2 = (\lambda, \lambda, \lambda, \mu, 0, 0)$. In the algorithms, $\times$ denotes a general multiplication, while $*$ denotes a multiplication by a constant. We ignore the processing time for $*$ in the evaluations.

SMART-CARD IMPLEMENTATION OF ELLIPTIC CURVE CRYPTOGRAPHY AND DPA-TYPE ATTACKS

Marc Joye

Gemplus, Card Security Group
La Vigie, Avenue des Jujubiers, ZI Athélia IV, 13705 La Ciotat Cedex, France

marc.joye@gemplus.com

Abstract   This paper analyzes the resistance of smart-card implementations of elliptic curve cryptography against side-channel attacks, and more specifically against attacks using differential power analysis (DPA) and variants thereof. The use of random curve isomorphisms is a promising way (in terms of efficiency) for thwarting DPA-type for elliptic curve cryptosystems but its implementation needs care.

Various generalized DPA-type attacks are presented against *improper* implementations. Namely, a second-order DPA-type attack is mounted against an additive variant of randomized curve isomorphisms and a "refined" DPA-type attack against a more general variant. Of independent interest, this paper also provides an exact analysis of second-order DPA-type attacks.

Keywords: Smart-card implementations, elliptic curve cryptography, side-channel analysis, DPA-type attacks.

1.   Introduction

With shorter key lengths, elliptic curve cryptography has received increased commercial acceptance and is already available in several smart-card products. It is supported by several standardization bodies, including ANSI, IEEE, ISO and NIST.

Because they better fit the constrained environment of smart cards, elliptic curve cryptosystems are particularly relevant to smart-card implementations. Until recently, efficient implementation meant fast running time and small memory requirements. Nowadays, an efficient implementation must also be protected against attacks and more particularly against side-channel attacks (e.g., based on timing analysis (TA) [11] or on simple/differential power analysis (SPA/DPA) [12]).

Two classes of elliptic curves are mainly used in cryptography: (non-supersingular) elliptic curves over binary fields (a.k.a. binary elliptic curves) and elliptic curves over large prime fields. The former class may be preferred for smart card implementations as arithmetic in characteristic two can be made very efficient [6, 14] (especially in hardware). See also [8].

In this paper, we show how to implement in a *proper* yet efficient way countermeasures against DPA-type attacks for binary elliptic curve cryptosystems. Of independent interest, we also provide an exact analysis of second-order DPA [12, 13].

The rest of this paper is organized as follows. In the next section, we briefly review known countermeasures meant to prevent DPA-type attacks. In Section 3, we detail the additive and multiplicative variants of point randomization using curve isomorphisms. We point out that the additive variant may succumb to a DPA-type attack if the slope, resulting in the addition of two elliptic curve points, is implemented in a straightforward way. Next, we mount a second-order DPA-type attack against another additive variant using a randomized slope in Section 4. In Section 5, we describe an attack against the multiplicative variant. This attack also applies to the more general randomized curve isomorphisms, combining both the additive and the multiplicative variants. Finally, we conclude in Section 6.

2.   DPA-type Countermeasures

The basic operation in elliptic curve cryptography is the point multiplication: on input point P and scalar k, point $Q = [k]P$ is returned. DPA-type countermeasures include the randomization of k and/or P.

Let E be a nonsupersingular elliptic curve over $GF(2^n)$ given by the (short) Weierstrass equation

$$E : y^2 + xy = x^3 + a_2 x^2 + a_6 \qquad (1)$$

and let P and $Q = [k]P$ be points on E.

The usual way to randomize k in the computation of $Q = [k]P$ consists in adding a random multiple of the order of E (or of $\mathrm{ord}_E(P)$) [5]:

$$k^* := k + r\#E$$

for a random r and then Q is evaluated as $Q = [k^*]P$. Another option is to split k in two (or several) shares [2, 4] (see also [17]): $k = k_1^* + k_2^*$ with $k_1^* = k - r$ and $k_2^* = r$ for a random r and then $Q = [k_1^*]P + [k_2^*]P$. Further countermeasures dedicated to Koblitz curves (i.e., curves given by Eq. (1) with $a_2 \in GF(2)$ and $a_6 = 1$) are presented in [9, 10].

Input:	k and $P \in E$
Output:	$Q = [k]P$

1. Choose a random curve isomorphism φ;
2. Compute $P^* = \varphi(P)$ on $E^* = \varphi(E)$;
3. Compute $Q^* = [k]P^* \in E^*$;
4. Return $Q = \varphi^{-1}(Q^*) \in E$.

Figure 1. Randomized evaluation of $Q = [k]P$.

Point P can be randomized using a randomized projective representation [5] or a randomized field or curve isomorphism [10]. The use of randomized curve isomorphisms leads to better performances (and is easier to implement) than the use of randomized field isomorphisms. Furthermore, it better suits binary curves as it allows to represent points with affine coordinates (and so runs faster [6, 14]). However, as we will demonstrate in the next section, its implementation needs care.

3. Randomized Curve Isomorphisms

Using randomized curve isomorphisms, a point P on an elliptic curve E is randomized as $P^* = \varphi(P)$ on $E^* = \varphi(E)$, for a random curve isomorphism φ and then $Q = [k]P$ is evaluated as $\varphi^{-1}([k]P^*)$.

More specifically, over $GF(2^n)$, a point $P = (x_P, y_P)$ on the elliptic curve

$$E : y^2 + xy = x^3 + a_2 x^2 + a_6$$

is mapped to point $P^* = (u^2 x_P + r, u^3 y_P + u^2 s\, x_P + t)$ on the isomorphic curve

$$E^* : y^2 + a_1^* xy + a_3^* y = x^3 + a_2^* x^2 + a_4^* x + a_6^* \qquad (2)$$

where

$$\begin{cases} a_1^* &= u \\ a_2^* &= u^2 a_2 + us + r + s^2 \\ a_3^* &= ur \\ a_4^* &= ut + r^2 \\ a_6^* &= u^6 a_6 + u^2 r^2 a_2 + u(r^2 s + rt) + r^3 + t^2 + r^2 s^2 \end{cases} \qquad (3)$$

and $r, s, t, u \in GF(2^n)$ with $u \neq 0$ (see [15, Table III.1.2]).

As already noted in [10], the short Weierstrass equation (i.e., Eq. (2) where $a_1^* = 1$ and $a_3^* = a_4^* = 0$) cannot be used for E^* as this implies $u = 1$ and $r = t = 0$, and hence let unchanged the x-coordinate of point P^*: $x(P^*) = x_P$.

3.1 Additive randomization of P

In [3], Ciet and Joye overcome the above limitation by working on the extended Weierstrass equation

$$y^2 + xy + a_3^* y = x^3 + a_2^* x^2 + a_4^* x + a_6^*,\qquad (4)$$

which, from Eq. (3), corresponds to $u = 1$ (and thus $a_3^* = r$). For simplicity, they also set the value of s to 0. As a result, point $P = (x_P, y_P)$ is randomized into

$$P^* = (x_P + r, y_P + t) .\qquad (5)$$

Although both the x- and y-coordinates of P are now randomized, this technique cannot be used naively in a point multiplication algorithm. Indeed, a closer look at the addition formulas shows that the slope given by the chord-and-tangent law, remains invariant.

Let $P_1^* = (x_1^*, y_1^*)$ and $P_2^* = (x_2^*, y_2^*)$ (with $P_1^* \neq -P_2^*$) denote points on the randomized elliptic curve given by Eq. (4). Then the sum $P_3^* = P_1^* + P_2^*$ is defined as (x_3^*, y_3^*) with

$$x_3^* = \lambda^{*2} + \lambda^* + a_2^* + x_1^* + x_2^* \quad \text{and} \quad y_3^* = \lambda^*(x_1^* + x_3^*) + y_1^* + x_3^* + a_3^*$$

where $\lambda^* = \dfrac{y_1^* + y_2^*}{x_1^* + x_2^*}$ when $x_1^* \neq x_2^*$ and $\lambda^* = \dfrac{x_1^{*2} + a_4^* + y_1^*}{x_1^* + a_3^*}$ otherwise (see [15, III.2.3c]). Recall that (i) we are working in characteristic two, (ii) $x_i^* = x_i + r$ and $y_i^* = y_i + t$ for $i \in \{1, 2\}$, and (iii) $u = 1$, $r = a_3^*$ and $s = 0$. Hence, we see that the slope

$$\lambda^* = \begin{cases} \dfrac{y_1^* + y_2^*}{x_1^* + x_2^*} = \dfrac{y_1 + t + y_2 + t}{x_1 + r + x_2 + r} = \dfrac{y_1 + y_2}{x_1 + x_2} \quad \text{when } x_1^* \neq x_2^* \\[2mm] \dfrac{x_1^{*2} + a_4^* + y_1^*}{x_1^* + a_3^*} = \dfrac{x_1^2 + r^2 + t + r^2 + y_1 + t}{x_1 + r + r} = x_1 + \dfrac{y_1}{x_1} \quad \text{otherwise} \end{cases}$$

does not depend on randoms r and t (we write λ the corresponding value), and consequently may be subject to a DPA-type attack (see e.g. [5]).

Randomly choosing s. The first idea that comes to mind is to randomly choose $s \in GF(2^n)$, leading to the more general randomization,

$$P^* = (x_P + r, y_P + s x_P + t) .\qquad (6)$$

In this case, the slope λ^* involved in the addition of $P_1^* = (x_1^*, y_1^*)$ and $P_2^* = (x_2^*, y_2^*)$ becomes

$$
\lambda^* = \begin{cases}
\dfrac{y_1^* + y_2^*}{x_1^* + x_2^*} = \dfrac{y_1 + sx_1 + t + y_2 + sx_2 t}{x_1 + r + x_2 + r} & \\
\qquad = \dfrac{y_1 + y_2}{x_1 + x_2} + s & \text{when } x_1^* \neq x_2^* \\
\dfrac{x_1^{*2} + a_4^* + y_1^*}{x_1^* + a_3^*} = \dfrac{x_1^2 + r^2 + t + r^2 + y_1 + sx_1 + t}{x_1 + r + r} & \\
\qquad = x_1 + \dfrac{y_1}{x_1} + s & \text{otherwise}
\end{cases}
$$

$$= \lambda + s$$

and hence is masked with the value of s. It should be noted that the evaluation of λ^* needs to be carefully implemented. In particular, field registers cannot contain the values of non-randomized values (e.g., $x_1 + x_2$) as otherwise a DPA-type attack could still be mounted.

Unfortunately, as we will see in Section 4, such an implementation may succumb to a second-order DPA-type attack. It should however be noted that second-order DPA-type attacks are much harder to mount since the attacker needs to know where/when certain operations are done.

3.2   Multiplicative randomization of P

The coordinates of point $P = (x_P, y_P)$ can also be randomized in a multiplicative way. Randomly choosing $u \neq 0$ and $r = s = t = 0$, point P becomes

$$P^* = (u^2 x_P, u^3 y_P) \tag{7}$$

on the isomorphic curve

$$y^2 + u\,xy = x^3 + (a_2 u^2)x^2 + a_6 u^6 . \tag{8}$$

Let $P_1^* = (x_1^*, y_1^*)$ and $P_2^* = (x_2^*, y_2^*)$ (with $P_1^* \neq -P_2^*$) denote points on the randomized elliptic curve given by Eq. (8). Then the sum $P_3^* = P_1^* + P_2^*$ is defined as (x_3^*, y_3^*) with

$$x_3^* = \lambda^{*2} + \lambda^* u + a_2 u^2 + x_1^* + x_2^* \quad \text{and} \quad y_3^* = \lambda^*(x_1^* + x_3^*) + y_1^* + u x_3^*$$

where $\lambda^* = \dfrac{y_1^* + y_2^*}{x_1^* + x_2^*}$ when $x_1^* \neq x_2^*$ and $\lambda^* = \dfrac{x_1^{*2} + u y_1^*}{u x_1^*}$ otherwise. Again, denoting by λ the slope corresponding to the addition of P_1 and P_2 on the initial curve, we see that λ^* is multiplicatively blinded: $\lambda^* = u\lambda$.

4.   A Second Order DPA-type Attack

Higher-order side-channel attacks [12] combine several samples within a single side-channel trace.

To ease the presentation, we consider the simplified Hamming-weight model for power leakage [13]. This model assumes that the instantaneous power consumption (C) is linearly related to the Hamming weight (H):

$$C = \varepsilon H + \ell \qquad\qquad (9)$$

for some constants ε and ℓ.

Using the additive randomization (see Eq. (5) or (6)), point $P = (x_P, y_P)$ on the elliptic curve E is randomized as $P^* = (x_{P^*}, y_{P^*}) = (x_P + r, y_P + sx_P + t)$ on the isomorphic elliptic curve $E^* = \varphi(E)$ given by Eq. (4). Point $Q = [k]P$ is then evaluated as $Q = \varphi^{-1}([k]P^*)$. Recall that the attacker's goal is to recover the value of $k = \sum_{i=0}^{m} k_i\, 2^i$ (or a part thereof) during the evaluation of Q. W.l.o.g., we assume that the point multiplication is carried out with the left-to-right binary algorithm and that the attacker already knows the leading bits of k: $k_m, k_{m-1}, \ldots, k_{t+1}$ with $t < m$. He now wants to know the value of the next bit of k, namely k_t.

Let $C^{(r)}$ represent the instantaneous power consumption when random r is drawn in $GF(2^n)$. Let also $C^{(x_{P^*})}$ represent the instantaneous power consumption when the x-coordinate of point $P^* \in E^*$ is handled in a register. For any point $P^* \in E^*$, we can write $x_{P^*} = x_P \oplus r$ since addition in $GF(2^n)$ is equivalent to a bit-wise XOR operation. From this observation, the attacker guesses that $k_t = 1$ and produces two equal-size sets, S_0 and S_1, of random points, defined as

$$S_b = \{ P \in E \mid g\big(x([2K_{t+1}+1]P)\big) = b \} \quad \text{with } b \in \{0,1\}$$

where $K_{t+1} = \sum_{i=t+1}^{m} k_i\, 2^{i-(t+1)}$ and g is a Boolean selection function returning for $g\big(x([2K_{t+1}+1]P)\big)$ the value of a given bit (in the representation) of the x-coordinate of point $[2K_{t+1}+1]P$. Let $R := [K_t]P = (x_R, y_R)$ and $R^* = \varphi(R) = (x_R + r, y_R + sx_R + t)$. The next step consists in computing the two average differential power consumptions (in absolute value),

$$\Delta_0 := \big\langle |C^{(x_{R^*})} - C^{(r)}| \big\rangle_{S_0} \quad \text{and} \quad \Delta_1 := \big\langle |C^{(x_{R^*})} - C^{(r)}| \big\rangle_{S_1}$$

and the second-order DPA operator

$$\Delta^{(2)} := \Delta_1 - \Delta_0 \ .$$

If $\Delta^{(2)} \not\approx 0$ (i.e., if there are DPA peaks) then the guess of the attacker was right, i.e., $k_t = 1$; otherwise the attacker deduces that $k_t = 0$. See Appendix A for a detailed proof. The attack proceeds iteratively in the same way until the value of k_0 is recovered.

5. A Refinement of Goubin's Attack

In [7], Goubin observes that the x-coordinate of a point $P = (x_P, y_P)$ with $x_P = 0$ is not randomized using the multiplicative randomization (see Eq. (7)). The same holds when considering the y-coordinate of point $P = (x_P, y_P)$ with $y_P = 0$. Over binary fields, elliptic curve points on Weierstrass curve (1) with their x-coordinate equal to 0 are points of order two. They are easily avoided with the cofactor variant in cryptographic protocols [16, Section 3]. Points of large order with their y-coordinate equal to 0 can also be defended against by using the Montgomery ladder as the y-coordinate is not used in this point multiplication algorithm [16, Section 5].

Another way for thwarting Goubin's attack is to use the more general randomization $P^* = (u^2 x_P + r, u^3 y_P + u^2 s\, x_P + t)$ (see e.g. [1, § 2.3]). We will show that this method succumbs to a "refined" power analysis.

We assume that the cofactor variant is not applied and so points with their x-coordinate equal to 0 are valid inputs to the point multiplication algorithm. An elliptic curve over $GF(2^n)$ always possesses a point of order two. Namely, the point $P_2 = (0, a_6^{2^{n-1}})$ satisfies Weierstrass equation

$$y^2 + xy = x^3 + a_2 x^2 + a_6 .$$

As in the attack of the previous section, we assume w.l.o.g. that the point multiplication, $Q = [k]P$, is evaluated with a left-to-right binary algorithm. The attacker's goal is to recover the value of k_t in the binary representation of scalar $k = \sum_{i=0}^{m} k_i 2^i$.

Define $K_t = \sum_{i=t}^{m} k_i 2^{i-t}$. The attacker guesses that $k_t = 1$ and repeatedly feeds the point multiplication algorithm with point $P_2 = (0, a_6^{2^{n-1}})$. As P_2 is of order two, it is worth noting that $R := [K_t]P_2 = [K_t \bmod 2]P_2 = [k_t]P_2$. Hence, when $k_t = 1$, it follows that

$$R^* := \varphi(R) = \varphi(P_2) = (r, u^3 a_6^{2^{n-1}} + t)$$

for some randoms r, t and u.

Next, the attacker computes the average differential power consumption (remember that the computation of $Q = [k]P$ is randomized),

$$\Delta^{(1)} := \left\langle |C^{(x_{R^*})} - C^{(r)}| \right\rangle$$

where $C^{(x_{R^*})}$ and $C^{(r)}$ denote the instantaneous power consumptions when the x-coordinate of point $R^* = \varphi(R)$ is handled and when r is randomly drawn from $GF(2^n)$, respectively. With the (idealized) model given by Eq. (9), this yields $\Delta^{(1)} \approx 0$ when $k_t = 1$. If $\Delta^{(1)} \not\approx 0$ then the attacker can deduce that $k_t = 0$.

6. Conclusion

This paper analyzed the resistance of elliptic curve cryptosystems against DPA-type attacks. Several new attacks were mounted against implementations improperly using randomized curve isomorphisms as a means for thwarting DPA-type attacks.

Since all the attacks presented in this paper require averaging several side-channel traces with the *same* input multiplier, the lesson is that point randomization techniques should be always used in conjunction with multiplier randomization techniques.

Acknowledgments

The author would like to thank to Francis Olivier, Pascal Paillier and Berry Schoenmakers for useful comments.

Appendix: Second-Order DPA Peaks

This appendix explains in further details why second-order DPA peaks reveal the value of the multiplier-bit k_t in a second-order DPA. We use the notations of Section 4. We have to show that $\Delta^{(2)} \not\approx 0$ means that $k_t = 1$ and 0 otherwise.

Proof. Assume that $k_t = 1$. Then we have $K_t = 2K_{t+1} + 1$ and $R = [2K_{t+1} + 1]P$. Hence, we get

$$\langle |C^{(x_{R^*})} - C^{(r)}| \rangle_{S_0} = |\epsilon| \cdot \langle |H^{(x_{R^*})} - H^{(r)}| \rangle_{S_0}$$

$$\approx |\epsilon| \underset{\substack{x_R, r \in \{0,1\}^n \\ g(x_R)=0}}{E} \left[|H^{(x_R \oplus r)} - H^{(r)}| \right]$$

and

$$\langle |C^{(x_{R^*})} - C^{(r)}| \rangle_{S_1} \approx |\epsilon| \underset{\substack{x_R, r \in \{0,1\}^n \\ g(x_R)=1}}{E} \left[|H^{(x_R \oplus r)} - H^{(r)}| \right]$$

provided that the number of power consumption traces is sufficiently large. Moreover, we have

$$\underset{\substack{x_R, r \in \{0,1\}^n \\ g(x_R)=0}}{E} \left[|H^{(x_R \oplus r)} - H^{(r)}| \right]$$

$$= \Pr[g(r) = 0] \underset{\substack{x_R, r \in \{0,1\}^n \\ g(x_R)=0, g(r)=0}}{E} \left[|H^{(x_R \oplus r)} - H^{(r)}| \right] +$$

$$\Pr[g(r) = 1] \underset{\substack{x_R, r \in \{0,1\}^n \\ g(x_R)=0, g(r)=1}}{E} \left[|H^{(x_R \oplus r)} - H^{(r)}| \right]$$

$$= \frac{1}{2} \underset{x_R, r \in \{0,1\}^{n-1}}{E} \left[|H^{(x_R \oplus r)} - H^{(r)}| \right] + \frac{1}{2} \underset{x_R, r \in \{0,1\}^{n-1}}{E} \left[|H^{(x_R \oplus r)} - H^{(r)}| \right]$$

$$= \underset{x_R, r \in \{0,1\}^{n-1}}{E} \left[|H^{(x_R \oplus r)} - H^{(r)}| \right] .$$

We also have

$$\underset{x_R, r \in \{0,1\}^n}{E} \left[|H^{(x_R \oplus r)} - H^{(r)}| \right]$$

$$= \frac{1}{2} \underset{\substack{x_R, r \in \{0,1\}^n \\ g(x_R)=0}}{E} \left[|H^{(x_R \oplus r)} - H^{(r)}| \right] + \frac{1}{2} \underset{\substack{x_R, r \in \{0,1\}^n \\ g(x_R)=1}}{E} \left[|H^{(x_R \oplus r)} - H^{(r)}| \right]$$

$$= \frac{1}{2} \underset{x_R, r \in \{0,1\}^{n-1}}{E} \left[|H^{(x_R \oplus r)} - H^{(r)}| \right] + \frac{1}{2} \underset{\substack{x_R, r \in \{0,1\}^n \\ g(x_R)=1}}{E} \left[|H^{(x_R \oplus r)} - H^{(r)}| \right] .$$

As $E_{x_R, r \in \{0,1\}^n} \left[|H^{(x_R \oplus r)} - H^{(r)}| \right] > E_{x_R, r \in \{0,1\}^{n-1}} \left[|H^{(x_R \oplus r)} - H^{(r)}| \right]$, this implies

$$\underset{\substack{x_R, r \in \{0,1\}^n \\ g(x_R)=1}}{E} \left[|H^{(x_R \oplus r)} - H^{(r)}| \right] > \underset{x_R, r \in \{0,1\}^{n-1}}{E} \left[|H^{(x_R \oplus r)} - H^{(r)}| \right]$$

and thus

$$\underset{\substack{x_R, r \in \{0,1\}^n \\ g(x_R)=1}}{E} \left[|H^{(x_R \oplus r)} - H^{(r)}| \right] > \underset{\substack{x_R, r \in \{0,1\}^n \\ g(x_R)=0}}{E} \left[|H^{(x_R \oplus r)} - H^{(r)}| \right] .$$

This in turn implies

$$\langle |C^{(x_{R^*})} - C^{(r)}| \rangle_{S_1} \gtrapprox \langle |C^{(x_{R^*})} - C^{(r)}| \rangle_{S_0}$$

and consequently $\Delta^{(2)} \not\approx 0$.

On the contrary, when $k_t = 0$, both sets S_0 and S_1 behave as random (i.e., uncorrelated) sets. Therefore, provided that the number of power consumption traces is sufficiently large, we have

$$\langle |C^{(x_{R^*})} - C^{(r)}| \rangle_{S_1} \approx \langle |C^{(x_{R^*})} - C^{(r)}| \rangle_{S_0}$$

and so $\Delta^{(2)} \approx 0$. $\square$

From

$$\begin{cases} \mathbb{E}_{x_R, r \in \{0,1\}^n}[H^{(x_R \oplus r)} - H^{(r)} \mid g(x_R) = 0] = 0 \\ \mathbb{E}_{x_R, r \in \{0,1\}^n}[H^{(x_R \oplus r)} - H^{(r)} \mid g(x_R) = 1 \text{ and } g(r) = 0] = 1 \\ \mathbb{E}_{x_R, r \in \{0,1\}^n}[H^{(x_R \oplus r)} - H^{(r)} \mid g(x_R) = 1 \text{ and } g(r) = 1] = -1 \end{cases} ,$$

an analysis similar to the one given in [13] would erroneously deduce that $\Delta^{(2)} \approx |\epsilon|$ when $k_t = 1$ and $\Delta^{(2)} \approx 0$ otherwise. The conclusion, however, remains correct.

References

[1] R.M. Avanzi. Countermeasures against differential power analysis for hyperelliptic curve cryptosystems. In C.D. Walter, Ç.K. Koç, and C. Paar, editors, *Cryptographic Hardware and Embedded Systems – CHES 2003*, volume 2779 of *Lecture Notes in Computer Science*, pages 366–381. Springer-Verlag, 2003.

[2] S. Chari, C.S. Jutla, J.R. Rao, and P. Rohatgi. Towards sound approaches to counteract power-analysis attacks. In M. Wiener, editor, *Advances in Cryptology – CRYPTO '99*, volume 1666 of *Lecture Notes in Computer Science*, pages 398–412. Springer-Verlag, 1999.

[3] M. Ciet and M. Joye. (Virtually) free randomization techniques for elliptic curve cryptography. In S. Qing, D. Gollmann, and J. Zhou, editors, *Information and Communications Security (ICICS 2003)*, volume 2836 of *Lecture Notes in Computer Science*, pages 348–359. Springer-Verlag, 2003.

[4] C. Clavier and M. Joye. Universal exponentiation algorithm: A first step towards provable SPA-resistance. In Ç.K. Koç, D. Naccache, and C. Paar, editors, *Cryptographic Hardware and Embedded Systems – CHES 2001*, volume 2162 of *Lecture Notes in Computer Science*, pages 300–308. Springer-Verlag, 2001.

[5] J.-S. Coron. Resistance against differential power analysis for elliptic curve cryptosystems. In Ç.K. Koç and C. Paar, editors, *Cryptographic Hardware and Embedded Systems – CHES '99*, volume 1717 of *Lecture Notes in Computer Science*, pages 292–302. Springer-Verlag, 1999.

[6] E. De Win, S. Mister, B. Preneel, and M. Wiener. On the performance of signature schemes based on elliptic curves. In J.P. Buhler, editor, *ANTS-3: Algorithmic Number Theory*, volume 1423 of *Lecture Notes in Computer Science*, pages 252–266. Springer-Verlag, 1998.

[7] L. Goubin. A refined power analysis attack on elliptic curve cryptosystems. In Y.G. Desmedt, editor, *Public Key Cryptography - PKC 2003*, volume 2567 of *Lecture Notes in Computer Science*, pages 199–211. Springer-Verlag, 2003.

[8] D. Hankerson, J. López Hernandez, and A. Menezes. Software implementation of elliptic curve cryptography over binary fields. In Ç.K. Koç and C. Paar, editors, *Cryptographic Hardware and Embedded Systems - CHES 2000*, volume 1965 of *Lecture Notes in Computer Science*, pages 1–24. Springer-Verlag, 2000.

[9] M.A. Hasan. Power analysis attacks and algorithmic approaches to their counter-measures for Koblitz cryptosystems. In Ç.K. Koç and C. Paar, editors, *Cryptographic Hardware and Embedded Systems - CHES 2000*, volume 1965 of *Lecture Notes in Computer Science*, pages 93–108. Springer-Verlag, 2000.

[10] M. Joye and C. Tymen. Protections against differential analysis for elliptic curve cryptography: An algebraic approach. In Ç.K. Koç, D. Naccache, and C. Paar, editors, *Cryptographic Hardware and Embedded Systems - CHES 2001*, volume 2162 of *Lecture Notes in Computer Science*, pages 377–390. Springer-Verlag, 2001.

[11] P. Kocher. Timing attacks on implementations of Diffie-Hellman, RSA, DSS, and other systems. In N. Koblitz, editor, *Advances in Cryptology - CRYPTO '96*, volume 1109 of *Lecture Notes in Computer Science*, pages 104–113. Springer-Verlag, 1996.

[12] P.C. Kocher, J. Jaffe, and B. Jun. Differential power analysis. In M. Wiener, editor, *Advances in Cryptology - CRYPTO '99*, volume 1666 of *Lecture Notes in Computer Science*, pages 388–397. Springer-Verlag, 1999.

[13] T.S. Messerges. Using second-order power analysis to attack DPA resistant software. In Ç.K. Koç and C. Paar, editors, *Cryptographic Hardware and Embedded Systems - CHES 2000*, volume 1965 of *Lecture Notes in Computer Science*, pages 238–251. Springer-Verlag, 2000.

[14] R. Schroeppel, H. Orman, S. O'Malley, and O. Spatscheck. Fast key exchange with elliptic curve systems. In D. Coppersmith, editor, *Advances in Cryptology - CRYPTO '95*, volume 963 of *Lecture Notes in Computer Science*, pages 43–56. Springer-Verlag, 1995.

[15] J.H. Silverman. *The arithmetic of elliptic curves*, volume 106 of *Graduate Texts in Mathematics*. Springer-Verlag, 1986.

[16] N.P. Smart. An analysis of Goubin's refined power analysis attack. In C.D. Walter, Ç.K. Koç, and C. Paar, editors, *Cryptographic Hardware and Embedded Systems - CHES 2003*, volume 2779 of *Lecture Notes in Computer Science*, pages 281–290. Springer-Verlag, 2003.

[17] E. Trichina and A. Bellezza. Implementation of elliptic curve cryptography with built-in countermeasures against side channel attacks. In B.S. Kaliski Jr., Ç.K. Koç, and C. Paar, editors, *Cryptographic Hardware and Embedded Systems - CHES 2002*, volume 2523 of *Lecture Notes in Computer Science*, pages 98–113. Springer-Verlag, 2003.

DIFFERENTIAL POWER ANALYSIS MODEL AND SOME RESULTS

Sylvain Guilley*, Philippe Hoogvorst* and Renaud Pacalet[†]
GET / Télécom Paris, CNRS LTCI
Département communication et électronique

* *46 rue Barrault, 75634 Paris Cedex 13, France.*

[†] *Institut Eurecom BP 193, 2229 route des Crêtes, 06904 Sophia-Antipolis Cedex, France.*
{sylvain.guilley, philippe.hoogvorst, renaud.pacalet}@enst.fr

Abstract   CMOS gates consume different amounts of power whether their output has a falling or a rising edge. Therefore the overall power consumption of a CMOS circuit leaks information about the activity of every single gate. This explains why, using differential power analysis (DPA), one can infer the value of specific nodes within a chip by monitoring its global power consumption only.

We model the information leakage in the framework used by conventional cryptanalysis. The information an attacker can gain is derived as the autocorrelation of the Hamming weight of the guessed value for the key. This model is validated by an exhaustive electrical simulation.

Our model proves that the DPA signal-to-noise ratio increases when the resistance of the substitution box against linear cryptanalysis increases.

This result shows that the better shielded against linear cryptanalysis a block cipher is, the more vulnerable it is to side-channel attacks such as DPA.

Keywords: Differential power analysis (DPA), DPA model, DPA electrical simulation, substitution box (S-Box), DPA signal-to-noise ratio, cryptanalysis.

Introduction

Power attacks are side-channel attacks on cryptosystems implementing public or private key algorithms. They were first published by Kocher in 1998 [8]. Public key algorithms, like RSA, are vulnerable to simple power analysis (SPA), but can be efficiently secured by algorithmic counter-measures [11], like key and/or data blinding. Secret key algorithms, such as DES or AES, consist in the repetition of several rounds, and are thus threatened by the differential power analysis (DPA).

DPA can attack on either the first or the last round of an algorithm and requires the knowledge of either the cleartext or the ciphertext. The side-channel exploited is the difference between the power consumed by a single gate when its output rises or falls.

Similar attacks take advantage of other types of leakage that disclose information about the internal computation. For instance, CPA [4] monitors the activity of a register: the attack exploits the fact that in CMOS logic, a gate only dissipates energy when it changes states. CPA, unlike DPA, can be modeled with the assumption that the energy dissipation is independent on the gate (either rising or falling) edge. Those attacks can also be conducted by recording a different physical quantity than the power consumption, like the electromagnetic field [6].

The rest of the article is organized as follows: Sec. 1 explains the principle of the DPA attack. In Sec. 2, we present a theoretical model for the DPA. The model is validated against exhaustive electrical simulations in Sec. 3. In Sec. 4, some results prove that the better shielded against linear cryptanalysis a block cipher is, the more vulnerable it is to side-channel attacks such as DPA.

1.   Differential Power Analysis

Measuring the Consumption Bias of a CMOS Inverter

The schematic depicted on Fig. 1(a) has been implemented using discrete transistors to measure the instantaneous current drawn from the power source VDD and sent back to the ground VSS. Fig. 2 shows that the current I(VDD)

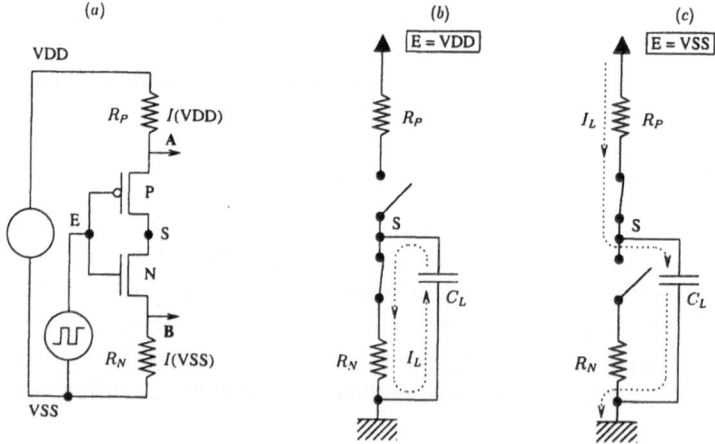

Figure 1.   (a) Experimental setup used to measure the currents I(VDD) and I(VSS) when the output S of a CMOS inverter switches. The currents flows are illustrated in (b) and (c).

flowing throught resistor R_P is the sum of:

- a short-circuit current, I_{short}, whose intensity is independent of the edge of the output S of the inverter and of

- a current I_L, loading a charging capacitance C_L and observed only when S rises from VSS to VDD (Fig. 1(c)), because, otherwise, C_L discharges through R_N only (Fig. 1(b)). The capacitance C_L models both the gate output capacitance, linked to the gate fanout and to the routing wires, as well as the parasitic capacitances.

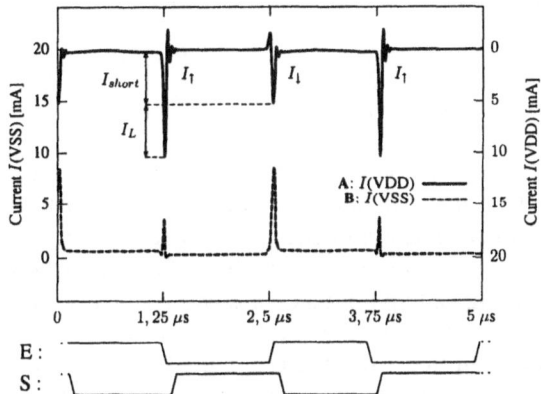

Figure 2. Measures of I(VDD) et I(VSS) of the inverter of Fig. 1 acquired by an oscilloscope.

The current I(VDD) depends on the edge (rising or falling) of S. We denote:

- $I_\downarrow = I_{short}$ the current observed upon a VDD $\rightarrow$ VSS edge and

- $I_\uparrow = I_{short} + I_L$ the one observed upon a VSS $\rightarrow$ VDD edge.

Principle of the DPA Attack

The analysis of the instantaneous power consumption can leak the type of operations being performed. Fig. 3 shows that the power consumption of a DES operator indicates the beginning of every encipherment.

Moreover, a more precise analysis can insulate the activity of a single gate, because:

- the instantaneous consumption of the circuit is the sum of all individual consumptions,

- each gate draws a different intensity ($I_\uparrow$ or $I_\downarrow$) according to its output edge, as shown in the previous example of the CMOS inverter.

The DPA attacks proceed in two phases. First, a large number of power consumption traces for different plaintexts[1] are recorded. Those traces contain the information about the type of edge (via a $I_\uparrow$ or $I_\downarrow$ contribution) of each gate in the design.

The second step consists in extracting this information from the traces $T_x(t)$. In the historical DPA [8], Kocher suggests to partition the traces according to the value of a particular bit i of the algorithm, which (hopefully) corresponds to a particular node in the netlist. One partition, S_0, gathers the traces $T_x(t)$, where $i = 1$, expected to contain an $I_\uparrow$ contribution, whereas the other, S_1, gathers the traces where $i = 0$. Thus the "differential trace", computed as:

$$\frac{1}{\#S_0} \sum_{T_x \in P_0} T_x(t) \;-\; \frac{1}{\#S_1} \sum_{T_x \in P_1} T_x(t)$$

reveals the $I_\uparrow - I_\downarrow$ power consumptions of the target gate i. This *modus operandi* can be used as an oracle to validate or invalidate an assumption. The DPA attack consists in testing whether the differential trace feature a singularity (peak) when analyzing the consumption of a gate i whose unknown state is guessed by making an hypothesis on a secret (typically a part of a round key). When the hypothesis on the key is correct, the differential trace is expected to feature a peak, resulting from the accumulation in a coherent manner of the $I_\uparrow - I_\downarrow$ information extracted from the power traces. More precisely, the peak is expected around the date t_S when the gate switches.

Refinements on this attack have been put forward [10]. The idea is to take into account that, in CMOS technologies, a gate only dissipates power when its output switches. The traces are thus partitioned into three sets. In addition to Kocher S_0 and S_1 sets, the S_2 set contains the traces with no or little dissipated power. Only traces from the sets S_0 and S_1 are used to compute the differential traces. For the sake of clarity, and to prepare for the presentation of our DPA model, we prefer not to present DPA in terms of traces partitioning but rather in terms of traces weighting. This allows us to reformulate the definition of the differential traces as an weighted accumulation of power traces, the weights being +1, −1 and 0 for traces traces belonging to sets S_0, S_1 and S_2.

Ghost peaks in differential traces

It has been reported [4] that "ghost" peaks also appear in differential traces computed with a wrong assumption of the key. We explain in the next section that those secondary peaks can be as high as the peak for the correct key and we provide a theoretical way to compute their relative amplitude.

---

[1]The plaintexts need not be known: the DPA is a ciphertext-only attack.

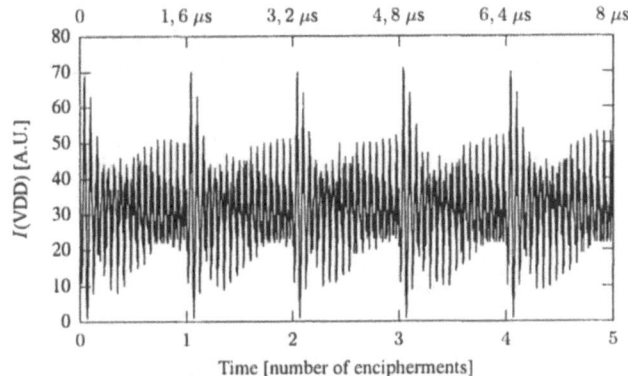

Figure 3.   Power consumed by five DES encipherments (programmed in an FPGA).

2.   DPA Model

Framework for DPA

Model general setup   The DPA model we describe is applicable to hardware implementations of private key product ciphers. The algorithm consists in the repetition of some rounds, the first or the last one being attackable by DPA. Without any loss of generality, we focus on an attack on the last round. Fig. 4 shows the typical dataflow of one round: the "plaintext" corresponds to the intermediate message produced by the penultimate round which is mixed in the last round with the "key" to produce the "ciphertext". The last round features one non-linear function (called S-Box) and one linear function (in our case a bit-to-bit XOR-ing) with some bits of the round key k. Given a known value x and an unknown but constant key k, the value y of all the target bits $i \in [0, q[$ under investigation is derived as:

$$y = F(x \oplus k).\qquad(1)$$

In the original DPA [8], the value of each bit i of y is used to partition the traces so as to build differential traces. In other words, the "selection function" D introduced by Kocher is the projection of Eqn. 1 on i. In our model, the whole value of y is used to weight the traces in a view to obtain one differential trace.

    As explained below, the function of Eqn. 1 applies to both AES and DES.

AES   The schematic of Fig. 4 comes in a direct line from the structure of the last round of AES, with $F = S^{-1} = \texttt{InvSubBytes}$ and $p = q = 8$.

DES   Fig. 5 represents a simplified dataflow of the last round of DES: the permutations and the expansion are left apart since the attacker can work around them. The guess on the bit i (belonging to the right part of the round 15 output) comes down to a guess on a bit of the output y of the S-Box, since C_L is known to the attacker. DPA on DES is therefore a particular case of Fig. 4, where $F = S$ is the direct S-Box: $K^p \to K^q$ with $p = 6$ and $q = 4$ (we denote $K = (\{0, 1\}, \oplus, \cdot)$ the field with two elements).

Fig. 5 schematic actually also applies to any Feistel cipher with constant S-Boxes, in which the attacked bit belongs to the right part of the penultimate round.

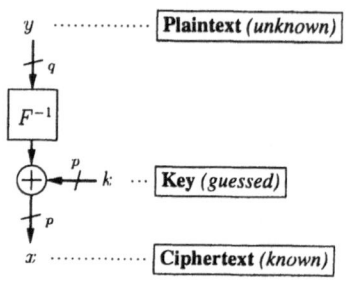 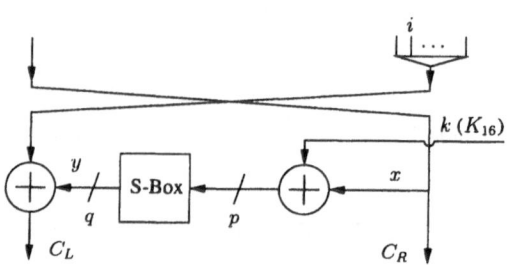

Figure 4.   Schematic DPA setup.

Figure 5.   Simplified DES cipher flow showing a single S-Box out of 8.

Noise Sources Occurring during DPA

There are various sources of noise when doing a DPA:

N1. The activity of the rest of the circuit. This noise can be lowered by the accumulation of many independent traces. Noise spectral power vanishes as the inverse square root of the number of traces recorded.

N2. The jitter on the attacked gate. Depending on the delays in the lines and the type of edges, the switching of a gate output can happen at different dates, which leads to a loss of coherence of the trace accumulation. This is negligible for gates directly fed by registers, as their inputs are perfectly synchronized.

N3. S-Boxes themselves introduce their own bias. Measured traces slightly match the activity deduced from the computation of one plaintext bit $y_i = F_i(k \oplus \cdot)$, as described by Eqn. 1, even if the assumption on k turns out to be wrong. Although substitution box bits are designed to be independent from one another so as to block linear cryptanalysis, DPA *modus operandi* artificially introduces an inter-bit correlation. The plaintext bits y are computed all together which introduces an artificial correlation between them. This notion of S-Box "intrinsic noise" is investigated in the next section. It is also called "ghost peaks" in [4] and "algorithmic noise" in [10].

DPA Intrinsic Noise

The "DPA signal": a model for the differential traces peak amplitude.
In this section we assume that the noise sources N1 and N2 are low enough. Under this condition, the DPA makes it possible to insulate the power consumption ($I_\uparrow$ or $I_\downarrow$) resulting from the activity of one single bit i: this manifests as a peak in the differential trace.

We propose to model the amplitude of the peak observed in the differential trace as a "DPA signal" that is built by an accumulation of scores. Given one ciphertext x, this score is:

+1 if the value $F_i(k \oplus x)$ inferred for the bit i by the selection function (Eqn. 1) is the same as the actual $F_i(k_0 \oplus x)$ for the correct key k_0,

-1 otherwise. As the recomputed bit value is false, the trace is considered to provide a $I_\uparrow$ power consumption contribution whereas the actual contribution is $I_\downarrow$, or vice-versa. Thus, instead of accumulating coherently $I_\uparrow - I_\downarrow$ to the differential traces, the opposite $I_\downarrow - I_\uparrow$ will be added, thus reducing the score coherence.

The scores are accumulated over many encipherments. Asymptotically, the accumulation is done for all the ciphertexts x.

As already mentioned when discussing the noise source N3, all the q bits of $y = F(k \oplus x)$ are guessed at the same time. As they take their values simultaneously, it is impossible to test the $y_i = F_i(k \oplus x)$ independently.

The "DPA signal" is thus the accumulation over all the ciphertexts x of the score obtained by a bit i of the plaintext against the q bits of actual plaintext. As a result, the DPA signal is built from the correlation of plaintext bit i for the trial key k with plaintext bit j for the actual key k_0:

$$\text{Corr}(k_0, k; \mathbb{1}_i, \mathbb{1}_j), \qquad \text{where: } \forall k_0, k \in K^p, \ \forall a, b \in K^q, \qquad (2)$$

$$\text{Corr}(k_0, k; a, b) \stackrel{\text{def}}{=} \frac{1}{2^p} \sum_x (-1)^{\langle a|F(x \oplus k)\rangle \oplus \langle b|F(x \oplus k_0)\rangle}$$

as [11]:

$$\textbf{DPA}(k_0, k; \mathbb{1}_i) = \sum_{j=0}^{q-1} \text{Corr}(k_0, k; \mathbb{1}_i, \mathbb{1}_j). \qquad (3)$$

Moreover, as it is easy to prove that:

$$\text{Corr}(k_0, k; a, b) = \text{Corr}(k_0 \oplus k, 0; a, b),$$

the correlation is independent of the actual key k_0. The relevant parameter is the difference $k_0 \oplus k$ between the actual key and the trial key. We simply denote this difference k, as if the actual key was 0. The correlation (Eqn. 2) is rewritten $\mathbf{Corr}(k; \mathbb{1}_i, \mathbb{1}_j)$. The "DPA signal" is rewritten accordingly: $\mathbf{DPA}(k_0, k; \mathbb{1}_i) = \mathbf{DPA}(k; \mathbb{1}_i)$. The correlation takes its values in $[-1, +1]$ and equals $+1$ if the guess on the key is correct (*i.e.* $k = 0$) and $i = j$.

The "ghost peaks".   When recomputing one bit of the plaintext from the ciphertext and a guessed key, the value can, by chance, match the actual value. If it happens often, the guessed key might be hard to distinguish from the actual key.

     For instance, in the case of DES S-Box #3, there exists one wrong key that leads to a "DPA signal" as high as the one for the correct key: it occurs when the bits 0 or 3 of the S-Box output are guessed.

     Those secondary peaks make it difficult to interpret the differential traces: they make up an artificial noise that was reported as "ghost peaks" [4].

DPA *modus operandi* justification   In this section we assume that the S-Box F is balanced and that the attacker found the correct key (*i.e.* $k = 0$). If the partitioning test is done according to the value of $\langle a|F(x) \rangle$, where a belongs to K^q, (*e.g.* if $a = \mathbb{1}_i$, the sole bit i is used to partition the traces), the attacker computes the following DPA signal:

$$\mathbf{DPA}(0; a) = \sum_{j=0}^{q-1} \frac{1}{2^p} \sum_x (-1)^{\langle \mathbb{1}_j|F(x)\rangle \oplus \langle a|F(x)\rangle} = \frac{1}{2^p} \sum_{j=0}^{q-1} \sum_x (-1)^{\langle \mathbb{1}_j \oplus a|F(x)\rangle}$$

$$= \frac{1}{2^p} \sum_{j=0}^{q-1} 2^p \, \delta(\mathbb{1}_j \oplus a) \qquad \text{(because F is balanced)}$$

$$= \begin{cases} 1 & \text{if there exists one } i \in [0, q[\text{ such as } a = \mathbb{1}_i, \\ 0 & \text{otherwise.} \end{cases}$$

It shows that DPA exhibits a non-zero signal iff the partition is made on one of the q plaintext bits. In this case, the DPA signal is maximum $(+1)$.

A new *modus operandi* for DPA.   The traditional *modus operandi* for the DPA is to compute the differential traces for testing the value of the q bits of F output. However, the q differential traces are not independent, because the each predicted bit i is matched against all the actual plaintext $F(k_0 \oplus \cdot)$. For this reason we consider the sum of the q differential traces. The DPA signal associated is :

$$\mathbf{DPA}(k) \stackrel{\text{def}}{=} \sum_{i=0}^{q-1} \mathbf{DPA}(k; \mathbb{1}_i) = \frac{1}{2^p} \left(\sum_{i=0}^{q-1} (-1)^{F_i} \otimes \sum_{j=0}^{q-1} (-1)^{F_j} \right)(k). \quad (4)$$

As an auto-correlation, the signal $\mathbf{DPA}(k)$ is maximum in absolute value in $k = 0$, *i.e.* when the attacker guess on the key is correct.

The method to compute the differential trace can be reformulated. Let k be the key being evaluated. For every ciphertext x, the power trace is weighted by $W(x, k)$, the centered Hamming weight of the recomputed plaintext (Eqn. 1):

$$W(x, k) \stackrel{\text{def}}{=} \sum_{i=0}^{q-1} F_i(x \oplus k) - q/2. \quad (5)$$

The weighted power traces are accumulated to yield the differential trace.

3.   Electrical Simulation of the DPA

The DPA attack is simulated at the electrical level in order to validate our DPA signal model (Eqn. 4).

We find that, given the long time required by electrical simulations, a 6×4 S-Box like one S-Box of DES cannot be simulated for all the plaintext transitions. Instead of limiting ourselves to a subset of the possible messages, like in [13], we choose to simulate a simpler cryptographic operator. The cipher used is the one shown in Fig. 4, with Serpent [2] S-Box #0 ($p = q = 4$).

The cipher is synthesized using various synthesis constraints into a 130 nm technology. The various logical netlists are translated into SPICE netlists using extracted standard cells in BSIM3V3 model.

The cipher is fed with all the transitions of plaintexts and the currents I(VDD) and I(VSS) are extracted during the simulation.

The exhaustive stimuli space exploration (2^{2q} traces) as well as the accuracy provided by the electrical simulation ensure that the traces we measure and the differential traces we compute emulate a perfectly noise-free DPA attack.

For the cipher described above, the theoretical model (Eqn. 4) predicts a DPA signal whose amplitude is given as an histogram in Fig. 6.

The differential traces depicted on Fig. 7 are computed from the traces acquired during the electrical simulation with the method explained above. The differential traces amplitude for the correct key can reach about 10 mA, which is also more or less the peak amplitude of a typical trace. The differential traces show that the amplitudes of secondary peaks are those predicted by the histogram of Fig. 6 for both side-channel I(VDD) and I(VSS). This conclusion is the same for all the netlists we simulate, which tends to show that DPA does not depend on the implementation.

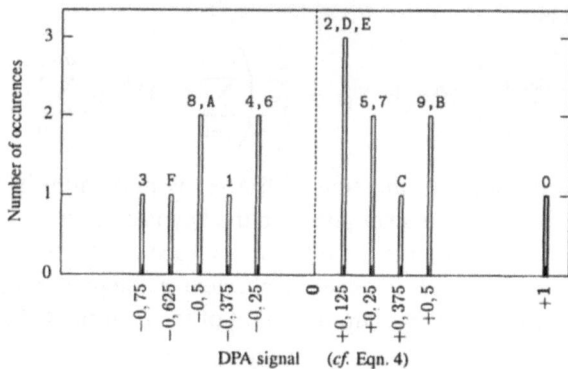

Figure 6.   Theoretical histogram for the DPA signal (Eqn. 4). The hexadecimal values $0, \cdots, F$ on top of the bars are those of $k_0 \oplus k$ (written on $p = 4$ bits). The thick peak for $k_0 \oplus k = 0$ is the DPA peak that betrays the secret key k_0, whereas the others are the "ghost peaks".

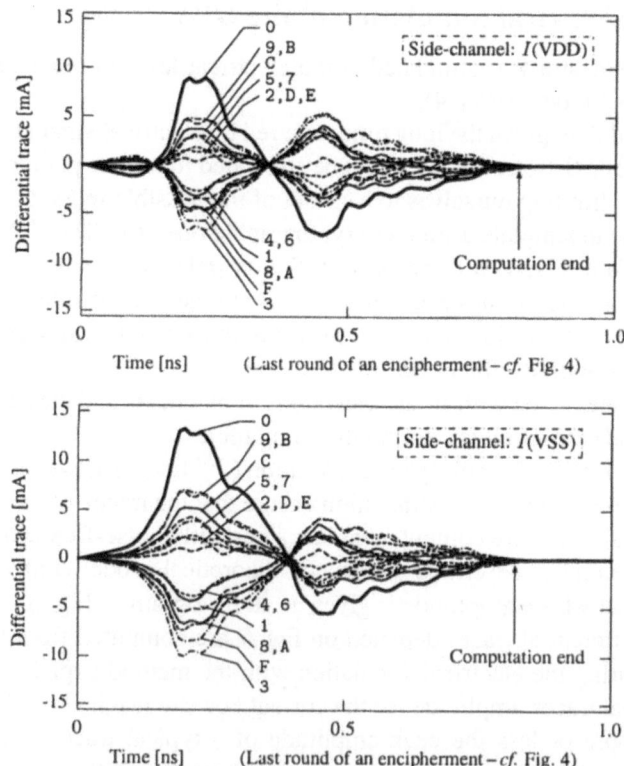

Figure 7.   Electrical simulation of a DPA, using either $I(VDD)$ or $I(VSS)$ as the side-channel. The bold curve is the one computed when the hypothesis on the key is correct ($k_0 \oplus k = 0$).

The DPA signals obtained by simulation match the theory, which justifies the DPA model of Sec. 2 and proves that the difference $I_\uparrow - I_\downarrow$ of power consumption of a single gate can be extracted from the overall power consumed by a cryptographic operator. In addition, the model remains valid during much of the cipher computation time.

4.   Connexions between DPA and Conventional Cryptanalysis

DPA Signal-to-Noise Ratio

DPA work factor is related to actual experiments, where the performance is assessed by a signal-to-noise ratio (SNR). As already mentioned, even if the DPA is not noisy, it does not allow to directly spot the right peak ($k = 0$) because there exists secondary peaks even for wrong keys ($k \neq 0$). Secondary peaks are modeled as noise. DPA quality is thus assessed by the following notion of SNR.

DEFINITION 1 *Signal* **Sig** *SNR.*

$$\mathrm{SNR}(\mathbf{Sig}) \overset{def}{=} \frac{\mathbf{Sig}(k=0) - \overline{\mathbf{Sig}}}{\left(\frac{1}{\#k}\sum_k \left(\mathbf{Sig}(k) - \overline{\mathbf{Sig}}\right)^2\right)^{1/2}}, \textit{where } \overline{\mathbf{Sig}} \textit{ is the signal mean}.$$

As far as the DPA signal (Eqn. 4) is concerned, a balanced S-Box F satisfies:

$$\begin{cases} \mathbf{DPA}(0) &= \sum_{i,j} 2^{-p} \sum_x (-1)^{\langle \mathbb{1}_i \oplus \mathbb{1}_j | F(x)\rangle} \\ &= \sum_{i,j} \delta(\mathbb{1}_i \oplus \mathbb{1}_j) = q, \\ \overline{\mathbf{DPA}} &= 2^{-2p} \sum_{i,j} \sum_x (-1)^{F(x)_i} \left(\sum_k (-1)^{F(x \oplus k)_j}\right) \\ &= 0 \qquad \text{(because } F \text{ is balanced)}. \end{cases} \qquad (6)$$

As a result, DPA SNR is:

$$\mathrm{SNR}(\mathbf{DPA})(F) = q\,2^{2p} \left(\sum_k \left(\sum_{i=0}^{p-1} \widehat{(-1)^{F_i}}(k)\right)^4\right)^{-1/2}, \qquad (7)$$

where $\hat{f}(k) = \sum_x (-1)^{\langle x|k\rangle} f(x)$ is the Hadamard-Walsh transform of the boolean function f.

The DPA SNR expression of Eqn. 7, proper to each S-Box F, fully characterize the DPA discrimination power.

It happens that the value of $\mathbf{SNR}(\mathbf{DPA})(F)$ (Eqn. 7) is significantly lower than $\mathbf{SNR}\left(\sum_{i=0}^{q-1} \mathbf{Corr}(k; \mathbb{1}_i, \mathbb{1}_i)\right)$ (refer to Eqn 2); for instance, those SNR are respectively 9.6 and 15.1 for AES. This proves that it is not realistic to neglect the inter-bit correlations (N3) when doing DPA. The available information in the traces is the sum of the consumptions of the q output bits of the function F and the DPA indeed only reveals the correlation of one bit of the predicted plaintext with the Hamming weight of the full plaintext.

SNR for some typical S-Boxes. A relevant reference for the SNR is the experimental setup where there is no S-Box. Analysis is thus performed behind a set of $q = p$ independent XOR gates. In this case, $\mathbf{DPA}(k) = \sum_i (-1)^{k_i}$ (see Eqn. (8), with $F = I$, the identity matrix) and the SNR is $\sqrt{q}$.

The SNR of the DPA signal is computed using Eqn. 7 for balanced S-Boxes and an *ad hoc* calculus for the bent S-Box. Results are reported in Table 1.

Table 1. Signal-to-noise ratio of DPA signal on some typical S-Boxes.

S-Box	No S-Box ($F=I$)	Linear S-Box	DES S-Box 1	AES	Bent S-Box
p	8	8	6	8	8
q	8	8	4	8	4
DPA SNR	$\sqrt{8} = 2.8$	2.8	3.6	9.6	9.8

DPA SNR for an Affine Balanced S-Box. Let F be an affine balanced S-Box: $F(x) = M \times x \oplus D$.

LEMMA 2

$$\mathbf{Corr}(k; \mathbb{1}_i, \mathbb{1}_j) = (-1)^{\langle \mathbb{1}_j | M \times k \rangle} \delta_{i,j} , \tag{8}$$

thus: $\mathbf{SNR}(\mathbf{DPA})(F) = q^{\frac{1}{2}}$.

DPA SNR for a Bent S-Box. As far as (unbalanced) bent S-Boxes are concerned, DPA expression (Eqn. 4) yields high SNR (cf. Table 1). However, we have not investigated other expressions that could take advantage of the unbalancedness.

Results of Eqn.6 do not apply to an unbalanced S-Box. Instead, if F is bent,

$$\left\{ \begin{array}{rcll} \mathbf{DPA}(0) & = & q + \frac{1}{2^p} \sum_{i \neq j} \widehat{(-1)^{\langle \mathbb{1}_i \oplus \mathbb{1}_j | F \rangle}}(0) , & \text{hence:} \\ |\mathbf{DPA}(0) - q| & \leq & q(q-1)2^{-p/2} \ll q & \text{and} \\ \overline{\mathbf{DPA}} & = & q\,2^{-p} \ll \mathbf{DPA}(0) & \text{if } p, q \to \infty \text{ and } p \geq 2q \text{ [5]} . \end{array} \right.$$

Therefore, at first order in $2^{-p/2}$, the expression of Eqn.7 for DPA SNR still holds. It allows for the derivation of a minoration of $\mathbf{SNR}(\mathbf{DPA})(F)$:

$$\mathbf{SNR}(\mathbf{DPA})(F) \;\gtrsim\; q\,2^{2p} \left(\sum_k \left(\sum_{i=0}^{q-1} 2^{\frac{p}{2}} \right)^4 \right)^{-1/2} = 2^{\frac{p}{2}}/q .$$

DPA SNR Bounds. Let us denote $Y_k = \left(\sum_{i=0}^{q-1} \widehat{(-1)^{F_i}}(k) \right)^2$.

SNR(DPA)(F) is thus rewritten as $q\, 2^{2p} \left(\sum_k Y_k^2 \right)^{-1/2}$. The maximum and the minimum of the SNR correspond to the minimum and the maximum of $\sum_k Y_k^2$. Moreover,

LEMMA 3

- *If F is balanced: $\sum_k Y_k = q\, 2^{2p}$,*

- *otherwise: $\sum_k Y_k \in [0,\, q^2\, 2^{2p}]$.*

DPA SNR maximum bound. The application $C\, :\, (x_0, x_1, \cdots, x_{2^p-1}) \in (\mathbb{R}^+)^{2^p} \rightarrow \sum_{k=0}^{2^p-1} x_k^2 \in \mathbb{R}^+$ is convex. Its minimum is thus reached when C gradient is null, *i.e.* when all the x_k, $k \in [0, 2^p[$, are equal. Given Lem. 3, this value is $q\, 2^p$ if F is balanced and can be as low as 0 otherwise.

Therefore, the maximum SNR of the DPA signal in a balanced S-Box is $2^{p/2}$. We ignore whether there exists S-Boxes that reach this bound. We also ignore whether DPA SNR is maximum bounded if the analyzed S-Box is unbalanced.

DPA SNR minimum bound. As $\sum_k Y_k^2 = \left(\sum_k Y_k \right)^2 - \sum_{k', k''} Y_{k'} Y_{k''}$, DPA SNR is minimum when the sum of positive terms $\sum_{k', k''} Y_{k'} Y_{k''}$ is minimum. It is null (and thus minimum) iff there exists an index k_0 such as all the Y_k, $k \neq k_0$, are null. If F is balanced, $\forall k$, $Y_k = q\, 2^{2p} \delta(k \oplus k_0)$. This lower bound can only be reached provided $q\, 2^{2p}$ (hence q) is a perfect square. If F is unbalanced, Y_k can reach $q^2\, 2^p$. As for all $i \in [0, q[$ and $k \in K^p$, $\widehat{(-1)^{F_i}}(k) \leq 2^p$, Y_k reaches $q^2\, 2^p$ iff for all i and k, $\widehat{(-1)^{F_i}}(k) = 2^p \delta(k \oplus k_0)$. S-Boxes satisfying this constraint are affine S-Boxes whose linear part rank is 1. Their corresponding SNR is $1/q$.

Summary of DPA Range and Typical Values. The results on the SNR of the DPA measured signal obtained in the Sec. 4 are summarized in Tab. 2.

Table 2. DPA signal-to-noise ratio bounds and typical values for different S-Boxes F.

SNR	Bound or typical value / S-Box type
$1/q$	Lower bound for unbalanced S-Boxes, reached only by rank 1 affine S-Boxes
1	Lower bound for balanced S-Boxes. Can only be reached if q is a perfect square
$q^{1/2}$	SNR of rank q affine S-Boxes
$2^{p/2}/q$	Approximative (at first order in $2^{-p/2}$) lower bound for bent S-Boxes
$2^{p/2}$	Upper bound for balanced S-Boxes

Conventional Cryptanalysis evaluators

Algorithmic attacks, like linear [9] or differential [3] cryptanalysis, are measured by a maximum singularity in distributions. For example [5],

$$
\begin{cases}
\Lambda_S \stackrel{\text{def}}{=} \sup_{a \neq 0,\, k} \left| \#\{x \,/\, \langle a|x \rangle \oplus \langle k|S(x) \rangle = 0\} - \frac{2^p}{2} \right| = \sup_{a \neq 0,\, k} \frac{1}{2}\hat{\theta}_S(k,\, a), \\
\Delta_S \stackrel{\text{def}}{=} \sup_{k \neq 0,\, a} \#\{x \,/\, S(x) \oplus S(x \oplus k) = a\} = \sup_{k \neq 0,\, a} (\theta_S \otimes \theta_S)(k,\, a),
\end{cases}
$$

are two parameters that characterize the resistances of an S-Box S against linear and differential cryptanalysis respectively. The lower they are, the more difficult the corresponding attack. We recall that in Eqn. 1, $F = S^{-1}$ for AES and $F = S$ for DES.

Comparing DPA and Conventional Cryptanalysis

The results of the previous section tend to show that the less linear a S-Box (and thus the higher its cryptographic quality), the higher the DPA SNR. The histograms of the occurrences of the SNR signal amplitudes are shown for some S-Boxes in appendix.

On the other hand, linear S-Boxes, the poorest protection against cryptanalysis, are the most difficult to attack by DPA.

DPA SNR Connexion with Conventional Cryptanalysis. The SNR of the DPA signal (Eqn. 7) is related to the two quantities Λ_S and Δ_S that characterize linear and differential cryptanalysis on S-Box F by:

$$
\mathbf{SNR(DPA)}(F) \;\geq\; \frac{2^{\frac{3p}{2}-2}}{q\,\Lambda_S^2} = \mathcal{O}\left(\frac{1}{\Lambda_S^2}\right), \tag{9}
$$

$$
\mathbf{SNR(DPA)}(F) \;\geq\; \frac{2^p}{\Delta_S} = \mathcal{O}\left(\frac{1}{\Delta_S}\right). \tag{10}
$$

The best shielded against linear or differential cryptanalysis (Λ_S or Δ_S low), the more vulnerable to DPA attack ($\mathbf{SNR(DPA)}(S)$ high).

5.   Conclusion

The overall power consumption of a circuit leaks the activity of every single gate. The DPA attack exploits this side-channel to retrieve one secret kept within the circuit. The signal that an attacker computes to perform a DPA can be modeled as the auto-correlation of the Hamming weight of a given temporary variable used in the cryptographic algorithm. This auto-correlation function is maximum when the attacker key guess is correct. We validate this model against an electrical simulation of a block cipher. The SNR of the DPA signal increases when the resistance against linear or differential cryptanalysis increases. The SNR is bounded, the lower bound being reached by the poorest cryptographic S-Boxes, namely affine S-Boxes. High quality cryptographic S-Boxes (AES, bent S-Boxes) feature high SNR, close to the maximum bound. As a consequence, DPA is fostered on devices implementing a high cryptographic quality private key algorithm.

Special care is thus needed while designing cryptoprocessors. As no trade-off is possible as for resistance against cryptanalysis, specific counter-measures must be devised. A possible counter-measure is to use secured logic gates [13]. However, those gates leak information because of parasitic effects: algorithmic counter-measures can thus be an adequate solution. For instance, the combination of a high SNR followed by a low SNR (in terms of *DPA* SNR) cipher on the same chip could provide a protection against both DPA and conventional cryptanalysis. Masking [1] and duplication [7] method are other counter-measures that require to re-design the ciphers.

References

[1] M. Akkar and C. Giraud. An Implementation of DES and AES secure against Some Attacks. *Proc. of CHES'01*, (2162):309–318, 2001.

[2] Ross J. Anderson. Serpent website (former candidate to the AES), 1999. http://www.cl.cam.ac.uk/~rja14/serpent.html.

[3] E. Biham and A. Shamir. Differential cryptanalysis of DES-like cryptosystems. *Journal of Cryptology*, 4(1):3–72, 1991.

[4] Eric Brier, Christophe Clavier, and Francis Olivier. Optimal statistical power analysis. 2003. http://eprint.iacr.org/.

[5] Florent Chabaud and Serge Vaudenay. Links between Differential and Linear Cryptanalysis. *Proc. of Eurocrypt'94*, 950:356–365, 1995.

[6] K. Gandolfi, C. Mourtel, and F. Olivier. Electromagnetic Analysis: Concrete Results. *Proc. of CHES'01*, 2162:251–261, 2001.

[7] L. Goubin and J. Patarin. DES and Differential Power Analysis: The Duplication Method. *Proc. of CHES'99*, (1717):158–172, 1999.

[8] Paul Kocher, Joshua Jaffe, and Benjamin Jun. Differential Power Analysis: Leaking Secrets. *Proc. of CRYPTO'99*, 1666:388–397, 1999.

[9] M. Matsui. Linear cryptanalysis method for DES cipher. *Proc. of Eurocrypt'93*, (765):386–397, 1994.

[10] Thomas S. Messerges, Ezzy A. Dabbish, and Robert H. Sloan. Investigations of Power Analysis Attacks on Smartcards. *USENIX Workshop on Smartcard Technology*, pages 151–162, May 1999.

[11] Elisabeth Oswald. *On Side-Channel Attacks and the Application of Algorithmic Countermeasures*. PhD thesis, may 2003. http://www.iaik.tu-graz.ac.at/aboutus/people/oswald/papers/PhD.pdf.

[12] Takashi Satoh, Tetsu Iwata, and Kaoru Kurosawa. On Cryptographically Secure Vectorial Boolean Functions. *Proc. of Asiacrypt'99*, 1716:20–28, 1999.

[13] K. Tiri and I. Verbauwhede. Securing Encryption Algorithms against DPA at the Logic Level: Next Generation Smart Card Technology. *Proc. of CHES'03*, 2779:126–136, 2003.

Appendix: Illustration of DPA Signal-to-Noise Ratio on Histograms

The figures of this appendix show the histograms of occurrence of a given DPA signal amplitude. The actual signal is the peak of amplitude q (4 or 8), whereas the other peaks make up the S-Box intrinsic noise. It clearly appears in Fig. A.1(a) that a linear S-Box has a weak SNR (namely $\sqrt{q}$). Usual cryptosystems DES (Fig. A.1(b)) and AES (Fig. A.1(c)) have a better SNR. The SNR is still better for a bent S-Box of Maionara-McFarland type [12] (Fig. A.1(d)).

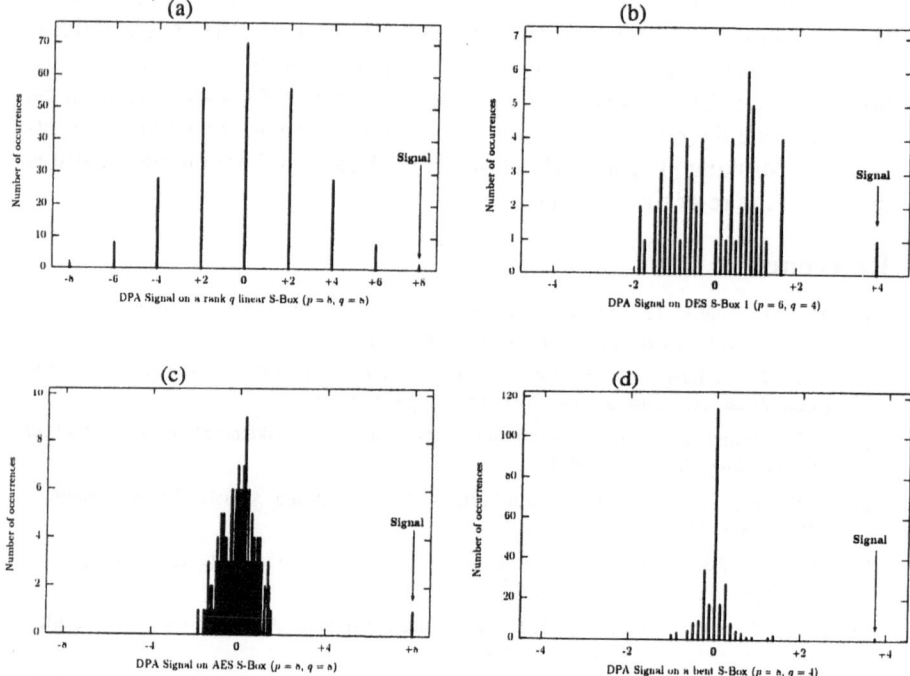

Figure A.1.   Histogram of occurrences of the DPA signal measured on:

(a) a linear S-Box. $\mathbf{SNR(DPA)}(F) = \sqrt{8} \sim 2.8$ (Eqn. 7),

(b) DES S-Box 1. $\mathbf{SNR(DPA)}(F) = 3.6$,

(c) AES. $\mathbf{SNR(DPA)}(F) = 9.6$,

(d) a bent S-Box. $\mathbf{SNR(DPA)}(F) = 9.8$.

PLACE AND ROUTE FOR SECURE STANDARD CELL DESIGN

Kris Tiri and Ingrid Verbauwhede
UC Los Angeles

Abstract: Side channel attacks can be effectively addressed at the circuit level by using dynamic differential logic styles. A key problem is to guarantee a balanced capacitive load at the differential outputs of the logic gates. The main contribution to this load is the capacitance associated with the routing between cells. This paper describes a novel design methodology to route a design in which multiple differential pairs are present. The methodology is able to route 20K+ differential routes. The differential routes are always routed in adjacent tracks and the parasitic effects between the two wires of each differential pair are balanced. The methodology is developed on top of a commercially available EDA tool. It has been developed as part of a secure digital design flow to protect security applications against Differential Power Analysis attacks. Experimental results indicate that a perfect protection is attainable with the aid of the proposed differential routing strategy.

Key words: Place & Route, Differential Pair, Differential Power Analysis, Side Channel Attacks

1. INTRODUCTION

Much design effort is spent in developing secure protocols and selecting strong encryption algorithms to achieve the security level envisioned in the specifications of the smart card application. Any security application however, is only as safe as its weakest link. Information related with the physical implementation of the device, such as variations in time delay and power consumption, has been used repeatedly to find the secret key in so-called Side Channel Attacks [1]. Especially the Differential Power Analysis (DPA)

[2] is of great concern as it is very effective in finding the secret key and can be mounted quickly with off-the-shelf devices. The attack is based on the fact that logic operations have power characteristics that depend on the input data. It relies on statistical analysis to retrieve the information from the power consumption that is correlated to the secret key.

At first, DPA has been thwarted with ad hoc countermeasures, which essentially concealed the supply current variations. Examples are for instance the addition of random power consuming operations or a current sink. Yet, the attacks have evolved and become more and more sophisticated. To address the problem, countermeasures need to be provided at different design abstraction levels. At the algorithmic level, an example is masking [3]. This technique prevents intermediate variables to depend on an easily accessible subset of the secret key.

Only recently, at the circuit level, dedicated hardware techniques have been proposed [4],[5]. Instead of concealing or decorrelating the side channel information, these techniques aim at not *creating* any side channel information. Goal of these countermeasures is to balance the power consumption of the logic gates. When the power consumption of the smallest building block is a constant and independent of the signal activity, no information is leaked through the power supply and a DPA is impossible.

Both in the synchronous [4] and in the asynchronous [5] approach, dynamic differential logic is employed. In this logic, every signal transition, also e.g. a degenerated 0 to 0 transition, is represented with an actual switching event, in which the logic gate charges a capacitance. Besides a 100% switching factor, it is essential in order to achieve constant power consumption that a fixed amount of charge is used per transition. This means that the load capacitances at the differential output should be matched. The load capacitance has 3 components: the intrinsic output capacitance, the interconnect capacitance and the intrinsic input capacitance of the load. Through a careful layout of the standard cells, the intrinsic input capacitances of a gate can be matched, as well as the intrinsic output capacitances. Yet with shrinking channel-length of the transistors, the share of the interconnect capacitance in the total load capacitance increases and the interconnect capacitances will become the dominant capacitance. Hence, the issue of matching the interconnect capacitances of the signal wires is crucial for the countermeasures to succeed [4],[5]. To our knowledge, this publication is the first to address this problem.

As we will derive in section 3, the best strategy to achieve matched interconnect capacitances is to route the output signals differentially. It is important to note that in this manuscript, differential routing denotes that the 2 output signals are at all times routed in adjacent tracks. Yet, this is the very opposite of current commercial cross-talk aware routers, which precisely avoid running wires in parallel. As a response, we have elaborated a technique to force commercial EDA tools to route multiple differential pairs. In this technique, each output pair is routed as 1 'fat' wire, which has among other characteristics the width of 2 parallel wires. Afterwards, the fat wires are split into the 2 differential lines.

Differential pair and shielded routing has been available through shape-based routers whose antecedents are in the PCB domain, where electrical constraints are historically more dominant. However, router performance and completion rate degrade rapidly with increasing number of such constraints. Gridded routers, with both routing resource representation and heuristic search optimized for speed and capacity, are very difficult to adapt for connection of 'wide wires' or 'co-constrainted' wires [6]. Recently, shielded routing capability has been migrated to gridded routers. We cannot directly use this approach for differential pair routing because (1) the two parallel wires are not VDD or VSS line (that are typically used for shielding); (2) the spacing between two differential wires will be larger due to the shielded signal wire in the middle; and (3) the two differential wires cannot be guaranteed to have similar length. Therefore, we present a way to work around tool limitations in section 3.

The remainder of this paper is organized as follows. In section 2, a place & route approach is developed in order to thwart DPA. Section 3 discusses the differential routing technique. In section 4, an experiment is setup in which the technique is applied to route the DES encryption algorithm and results of a DPA are provided. Finally a conclusion will be formulated.

2. BALANCING INTERCONNECT LOADS

2.1 Given Information

A standard cell of a dynamic differential cell library has 2 differential outputs A and A', which connect to k differential input pairs $\{(I_1, I_1'), ... (I_k, I_k')\}$. Each standard cell has a balanced design. This means that (1) the intrinsic output capacitances $C_{o,A}$ and $C_{o,A'}$ seen at outputs A and A' are equal; that (2) the intrinsic input capacitances $C_{i,Ij}$ and $C_{i,Ij'}$ seen at inputs I_j and I_j' are equal; and that (3) the drive strengths at output A and A' are equal. The standard cell has exactly 1 switching event per cycle. This event is always the same and consists of 2 transitions: (1) in the evaluation phase: both outputs are at 1 and 1 output discharges to 0; and (2) in the precharge phase: 1 output is at 1; the other output is at 0 and is charged to 1.

2.2   Place & Route Constraints

2.2.1   Match Load Capacitance

In addition to engaging in 1 charging event per clock cycle, it is mandatory for input independent power consumption that the load capacitance is a
constant. The differential standard cell has a load capacitance at each output.
Since only 1 output undergoes a transition per switching event, the total load
at output A should match the total load at output A': $C_A = C_{A'}$. This, as can
be seen in Figure 1, can be restated as: $C_{o,A} + C_{w,A} + C_{i,I1} + \dots C_{i,Ik} = C_{o,A'} +
C_{w,A'} + C_{i,I1'} + \dots C_{i,Ik'}$, which can be reduced to $C_{w,A} = C_{w,A'}$ because of the
balanced standard cell design. This means that the Steiner routing tree over
$\{A, I_1, \dots I_k\}$ must have the same total capacitance as the Steiner routing tree
over $\{A', I_1', \dots I_k'\}$.

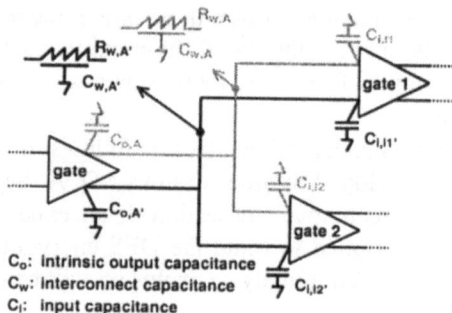

C_o: intrinsic output capacitance
C_w: interconnect capacitance
C_i: input capacitance

Figure 1. Load capacitance decomposition.

It is not necessary to balance the routing tree over $\{A, I_1, \dots I_k\}$ with the
routing tree over $\{B, I_m, \dots I_s\}$ (where A and B are the outputs of two different gates). For the total encryption module to have constant power consumption, it is sufficient that the power consumption of each building block is
input independent. There is no need for mutually matching the routing trees.

Cross-talk, which is the phenomenon of noise induced on one wire by a
signal switching on a neighboring wire, influences not only the delay, but
also the power consumption. Therefore the pair of interconnects need to be
routed with the same capacitance and with control over any cross-talk.

2.2.2   Match Source-Sink Delays

An attacker will analyze all information available from the power consumption. He/she will not settle for the total charge per switching event, but
will trace the instantaneous power consumption. To assure that only minor

instantaneous current variations exist between different switching events, each pair of differential routes should have a constant source-sink delay: the delay from A to I_i must be equal to the delay from A' to I_i' for each $i = 1, ...$ k. It is not necessary that the delay from A to I_i is equal to the delay from A to I_j (where A is connected to the inputs I_i of gate B and I_j of gate C), nor that the delay from A to I_i is equal to the delay from B to I_j, (where the gates A and B are connected to the inputs I_i and I_j of gate C).

2.2.3   Miscellaneous Constraints

Implementations of encryption algorithm generally require 20K+ gates. This means that 20K+ differential routes must be balanced. It might be possible to only implement the most sensitive parts of the encryption module and reduce this number. Yet, the size of the problem will not be reduced with an order of magnitude.

The differential routing procedure must complete the missing part of a secure digital design flow [7]. This design flow does not restrict the fanout, which is the number of gates a gate connects to.

2.3   Place & Route Approach

If two gates are connected with parallel routes that are at all times in adjacent tracks and on the same layers, then since independent of the placement, the geometric distances are balanced, the two routes have to a first order the same capacitances and the same delays. Yet, this is not completely true because both nets may have different parasitics and cross-talk effects.

Parasitic effects are caused by the distributed resistance of the interconnect and by the distributed capacitance of the interconnect to the substrate and to neighboring wires in other metal layers. Though aside from process variations, these effects are equal for both nets. The resistance is the same since both interconnects have the same number of vias and have the same length in each metal layer. The capacitance to the other layers is ideally the same since in general the length of the differential route is orders of magnitude larger then the pitch between the 2 differential routes and one can therefore argue that both nets travel in the same environment. Making every other metal layer a ground plane would completely control the capacitance to other layers. Yet this would not only reduce the solution space but also increase the total capacitance.

Cross-talk effects are caused by the distributed capacitance to adjacent wires in the same metal layer. Routing the two output nets in parallel removes the uncertainty of one neighbor: during a switching event only one output line switches, the other output line remains quiet. All uncertainty can be removed by shielding the differential routes on either side with a VDD or VSS line. Besides the loss in routing tracks, it will also be hard to get multiple sets of 4 routes into and out a standard cell. Yet, reserving 1 grid line out of 3 upfront for a power line reduces the problem again to routing 2 differential lines. Note that the approach of alternating signal lines and quiet power lines has been shown to produce predictable interconnect parasitics [8]. Alternatively, the cross-talk effects can be controlled by merely increasing the distance between different differential routes. This can easily be done with the differential routing methodology we will present now.

3.   DIFFERENTIAL ROUTING

The methodology that we propose is to abstract the differential pair as a single fat wire. The differential design is routed with the fat wire and at the end the fat wire is decomposed into the differential wire.

3.1   Basic Ideas

3.1.1   Fat Wire Definition

The fat wire covers the two differential wires. The centerline of the fat wire is the centerline between the two differential wires. The width of the fat wire W_f is set by the summation of the pitch P_n of the normal wires and 2 times half the width of the normal wire W_n: $W_f = P_n + 2W_n/2$. The pitch, which is the distance between the centerline of two adjacent wires, P_f of the fat wires is set by the summation of 2 times half the width of the fat wire and the desirable distance Δ between the fat wires: $P_f = 2W_f/2 + \Delta$. The distance Δ can be made large to reduce cross-talk effects. The minimum spacing rules do not change.

3.1.2   Transformation

After place & route with the fat wire, the resulting design must be transformed into the final differential design. The transformation consists of 2 translations of the fat wire and a width reduction to the normal width.

Since the centerline between two normal wires is the centerline of the fat wire, a translation of the fat wire in the positive direction will result in one differential line and a negative translation in the in the other line. The translation must occur both in the horizontal and the vertical direction. As shown in Figure 2, a consistent shift of all segments of the fat wire with a ΔX in the X direction and a ΔY in the Y direction will result in one wire; a shift with a

-ΔX and a -ΔY in the other wire. The shifts ΔX and ΔY are half the pitch lengths of the normal wires in the X and Y direction.

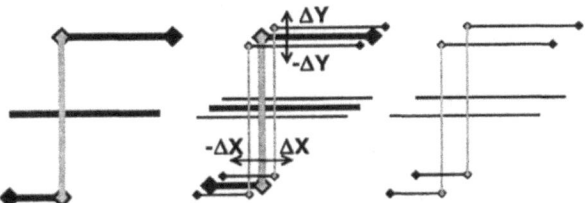

Figure 2. Fat routes (left); translation operation (middle); and differential routes (right).

The resulting differential wires have the same number of vias and segments. Each segment has the same length in both wires and is routed over the same number of wires in the other metal layers. As a result, both lines have the same distributed resistances and parasitic capacitances to the substrate and to the routes in the other metal layers.

3.2 Practical Aspects

This section discusses some of the practical issues we have come across. While these issues are independent of the tool, the guidelines presented are for Silicon Ensemble [9], which we have used as the place & route tool.

3.2.1 Restrictions on Differential Standard Cell

As can be seen in Figure 2, the vias are all aligned on a positive tilted diagonal. The in- and output pins of the standard cells must also be aligned likewise and with the same offsets. The upper pin is the pin associated with the true net, the lower with the false net. Only then the translation can be done in a consistent way.

3.2.2 Fat Wire Definition

Silicon Ensemble extends routes with at least half the width at their endpoints. Because of the increased width, the fat wire extends too far at its endpoints and covers an area where there is no actual normal route. As a result, spacing errors are generated for certain patterns of wires, which are only virtual errors. To address this problem, the original normal wire is routed on a large grid that has been defined such that there will be no spacing violations after splitting. Doubling the original grid pitches results in such a grid. Now, the .lef library database [10], which contains all the information that is relevant for the router tool, such as routing layers, via rules, grid definition, spacing rules and abstract views of the standard cells, can be left unchanged except for a new grid definition and the abstract views of the cells with fat pin information. Note that from now on, we will route a normal wire, but we will continue to refer to this wire as the fat wire.

3.2.3   Grid Definition

To facilitate the placement, the height and width of a standard cell should be a multiple of the horizontal and vertical pitch respectively. In addition, since we can only route on the grid the pins should be situated on the grid crossings. The most straightforward is to take the pitch of the fat design a multiple of the original grid. The minimum pitch between the fat wires is 2 times the pitch between the normal wires.

We defined the grid and the standard cells as follows: (1) the horizontal and vertical pitches of the fat grid are double the ones of the normal grid; and (2) the normal and fat grids have an offset of half their pitch length in both the horizontal and vertical direction. With this definition all requirements previously derived are fulfilled: (1) the standard cell dimensions are multiples of the horizontal and vertical pitch of the fat and the normal grid; (2) the fat pins are situated on the crossings of the fat grid, the differential pins on the crossings of the normal one; and (3) the differential pins can obtained by shifting the fat pin with half a pitch length of the normal grid in both the horizontal and vertical direction.

3.2.4   Non-preferred Routing Direction

If the fat wire takes a turn in a metal layer, the wires of a differential route may cross in the same metal layer and result in an electric short between both wires. This however, can only happen if a metal layer is used in the vertical and the horizontal direction. Even though each metal layer has a preferred routing direction, this does not guarantee that the routing layer is only used in that direction. This required us to force Silicon Ensemble to only route in the preferred direction.

3.2.5   Transformation Procedure

The transformation procedure consists of two parts: (1) parsing the placed and routed fat design to reflect the differential design and (2) reading in the differential library database. The differential 'diff.lef' library database contains the normal grid definition, normal wire definition, normal via definition and the differential gates with differential pin information.

The wires in the routed 'fat.def' design file are described as lines between 2 points and vias are assigned as points. The wire width and via characteristics are defined in the .lef library database. As a result the parser only needs to translate the (X,Y) coordinates of the end points without to worry about the wire characteristics. The translation is done by (1) repeating each statement that defines a net; (2) attaching the first statement to the positive pins and translating it in a positive $(\Delta X, \Delta Y)$ direction; and (3) attaching the second statement to the negative pins and translate it in a negative $(\Delta X, \Delta Y)$ direction. Recall that ΔX and ΔY are half the pitch lengths of the normal wires in the X and Y direction. Besides the translation of the nets, each fat gate in the 'fat.def' file is substituted by its corresponding differential gate.

Figure 3 summarizes the differential routing methodology.

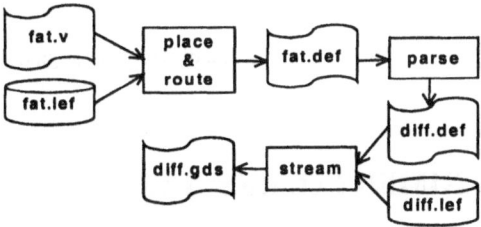

Figure 3. Differential routing methodology.

3.2.6 Differential and Single Ended Routing

Up to now, we only presented a methodology to route a design of which all wires are differential. It is however possible to combine single ended routing and differential routing. There are 3 options. The design can be routed in 2 stages. First the differential lines are routed, and subsequently with a new library database the single ended lines are routed, or visa versa. The design can also be routed concurrently by defining the fat routes or the single ended routes as nondefault routing rules. Or, one could route every wire as a fat wire and subsequently transform the single ended signals into a single line and the differential signals into 2 lines. The last option is not preferred if the single ended routes are in the majority and area constraints are tight.

3.3 Design Example

Figure 4 demonstrates the differential routing technique. The figure shows an arbitrary placed and routed design consisting of 7 differential gates. At the left, the result is shown of the fat routing. At the right, the result after transformation is shown. Each fat wire is replaced by 2 normal wires.

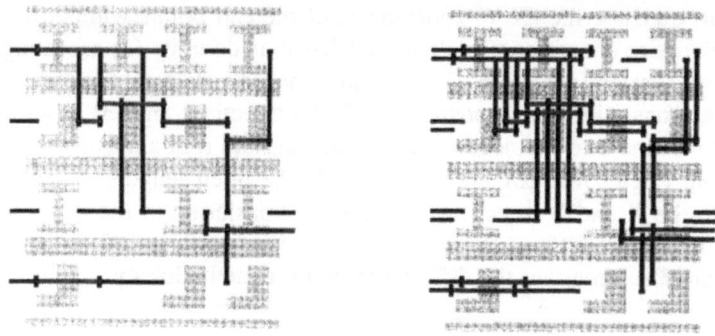

Figure 4. Example: fat (left); and differential design (right).

4.   EXPERIMENTAL RESULTS

In this section, we compare two dynamic differential implementations of the DES algorithm [11]. One implementation uses the differential routing technique described above. For the other implementation, we use the default standard cell route as generated by the EDA router. We mounted a DPA on both differential implementations. We have chosen DES, because much research has focused on both the implementation and prevention of DPA on DES. The first subsection presents the basic procedures of the DPA. Then the experimental setup is discussed and implementation details are given. The last subsection covers the experimental results. Please note that a DPA attack on a regular (non differential) standard cell implementation is described in [4].

4.1   Differential Power Analysis

A DPA attack is carried out in several stages. First, the power consumption is recorded for a large number of encryptions. This is done by measuring the instantaneous supply current. Next, the measurements are partitioned over 2 sets based on a so-called selection function, which uses a guess on a subset of the secret key to make its decision. At the end, the difference is calculated between the typical supply currents of the 2 sets. The difference is referred to as Differential Trace (DT). If the DT has noticeable peaks, the guess on the secret key was correct.

The selection function is a behavioral model of the encryption module and predicts 1 state bit of the module. If the secret key has been guessed correctly, the outcome of the selection function is always equal to the actual state bit and is therefore correlated with the power consumption of the logic operations that are affected by the state bit. Measurement errors and the power consumption of the other logic operations are uncorrelated. As a result, the DT will approach the effect of the state bit on the power consumption. If on the other hand the guess on the secret key was incorrect, the result of the selection function is uncorrelated with the state bit and the DT will approach 0.

4.2   Experimental Setup

The experimental setup is depicted in Figure 5 [4]. The module forms part of the DES encryption algorithm. Based on 4 bits of the left plaintext P_L, 6 bits of the right plaintext P_R and 6 bits of the secret key K, it calculates 4 bits of the left ciphertext C_L. The selection function D(K,C), which is used in the DPA, predicts the first bit of the register that stores the left plaintext P_L. In the selection function, only K is guessed, C_L and C_R are known.

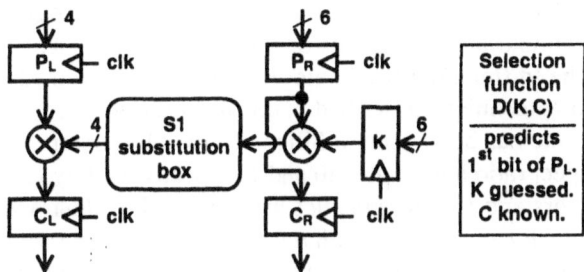

Figure 5. Experimental setup: DPA on submodule of the last round in the DES-algorithm [4].

The experimental setup is a necessary and sufficient subset of the DES encryption algorithm on which a DPA can be mounted [4]. The algorithm has been reduced to this setup such that it becomes computationally feasible to simulate it with Hspice. High-level power estimators, such as cycle accurate simulators, cannot be used. A DPA attack uses statistical methods that can detect very small power variations, which are plainly neglected by high-level power estimators. Furthermore, in order to retrieve as much hidden information as possible, an attacker will sample several times per clock cycle.

4.3   Implementation Details

We have implemented the module in Wave Dynamic Differential Logic (WDDL) [7]. WDDL is a logic style that has dynamic differential behavior. Yet, it is implemented with static complementary CMOS logic, which is the default logic style used in standard cell libraries. In WDDL, static CMOS standard cells are combined to form secure compound standard cells, which have a reduced power signature. It has the advantage that it can readily be implemented without the investment in a custom designed standard cell library. We used a commercially available standard cell library developed for a 0.18μm, 1.8V CMOS technology. A WDDL gate actually is a double-height cell that consists of 2 abutted static CMOS standard cells, which are extended with filler cells in which the pin placement requirements are fulfilled. A WDDL gate does not have balanced input capacitances, nor balanced output capacitances. The filler cells could also incorporate additional capacitances such that the intrinsic capacitances become balanced. The implementation consists of 194 WDDL gates, which are 440 actual static CMOS standard cells.

Two different layouts have been created. Both designs have exact the same floor plan and cell placement. The difference is in the routing procedure. One implementation, which we will call the 'regular route'-design, has been routed without any constraints or special techniques. The other, which we will call the 'differential pair route'-design, has been routed with the differential routing technique described in this manuscript.

Place & route has been done in Silicon Ensemble version 5.3. Row utilization and aspect ratio are set at 0.80 and 1 respectively. Layout-to-netlist, in which transistors and layout parasitics are extracted, is done with Virtuoso. Simulations are done in Hspice, with the transient increment set at 10ps. The clock frequency of the circuit is chosen at 125MHz, and therefore 800 'measurement' samples are made per clock cycle. The clock and input signals are driven by cascaded inverters in order to provide realistic data and clock signals. The power consumption of the additional input circuitry is excluded from the measurements. In total, 2000 clock cycles have been simulated with a random input at the plaintext P_L and P_R, and with a fixed secret key K, equal to 46.

4.4   Experimental Results

Silicon Ensemble required 8 and 3 CPU seconds on a SUN ULTRA 5 to route without any violations the regular route and the differential pair route respectively. It took 0.85 CPU seconds to parse and create the final differential netlist. Note that many floorplan settings can be evaluated; only the final design should be parsed. To have a comparison, we have also made an attempt to use Cadence Chip Assembly Router version 11.0.06 [12] to route the placed design. This is one of the commercially available tools that has the capability of routing differential pairs. Only the 442 internal nets have been defined as 221 differential pairs. In total, 100 iterations have been performed. This required 7 hrs 56 min and 33 sec in CPU time without generating a completely routed result. It still had 972 conflicts and 125 unconnected nets.

Figure 6 and Figure 7 show 2 histograms in which the internal interconnect capacitances of the regular route and the differential pair route are compared. The capacitance per net was reported directly from Silicon Ensemble using Simcap. Figure 6 depicts the distribution of the ratio between the capacitance at the true signal net and the capacitance at the corresponding false signal net. The variation between the capacitances at the differential nets is up to a factor 4 for the regular route procedure. On the other hand, the differential pair route procedure does not show any variation. In fact, the tool always returned exact the same values.

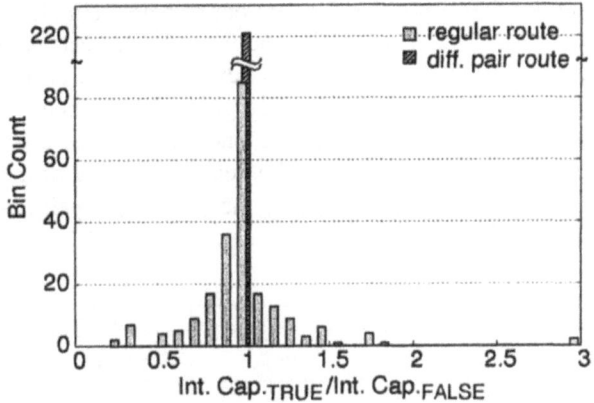

Figure 6. Ratio between interconnect capacitances at true and false nets.

The distributions of the absolute values of the capacitances, which are shown in Figure 7 are very much similar between the 2 routing procedures. This indicates that the mean power consumption and the time delay of the 2 implementations will be alike.

In the transient simulation, the mean energy consumption per clock cycle is 42.72pJ and 44.21pJ for the regular route and the differential pair route respectively. The normalized energy deviation, which specifies the absolute range of the variation on the energy consumption per cycle, is 1% for the regular route and 0.7% for the differential pair route. The normalized standard deviation is 0.2% and 0.1% respectively.

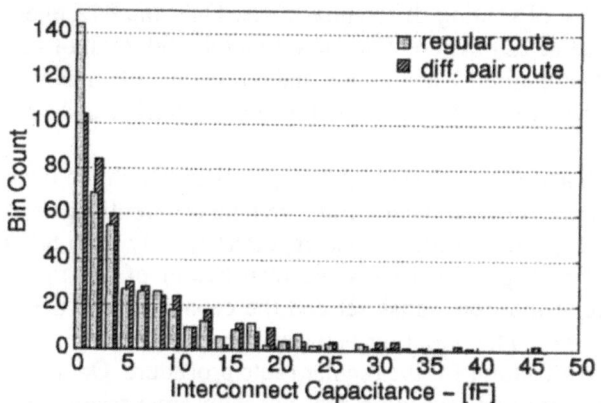

Figure 7. Absolute interconnects capacitances.

Figure 8 shows DTs from the DPA on our transient simulation. For each implementation, 2 DTs are shown: the one from the correct secret key guess and one from an arbitrary incorrect secret key guess. For transparency of the figure, the DTs of the other incorrect key guesses have been omitted. The omitted DTs are in accordance with the one shown. The differential pair routing has been effective in reducing the peaks in the DT of the correct secret key. Compared to the regular routing, the differential pair route achieves a complete reduction.

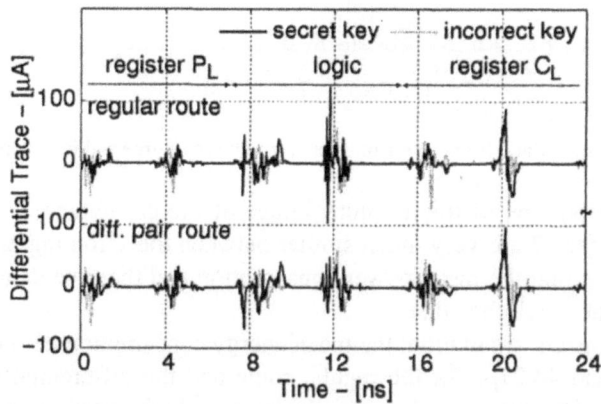

Figure 8. Differential trace for correct and arbitrary incorrect key guess.

Figure 9 shows the peak-to-peak value of the differential traces of the secret key guesses. For the regular routed design, the DT of the secret key stands out. On the other hand, for the differential pair routed design, the DT does not reveal the secret key.

Several approaches, which we have mentioned throughout the paper, such as for instance shielded lines or a larger pitch, balanced intrinsic capacitances and ground planes, are still available to improve the results. These options all have in common that they are not for free: one requires more design time, the other more area, etc. This is typically for security applications, where the higher the level of security is aspired, the more expensive the implementation will be. Yet, note also that we have performed a perfect attack with perfect measurement results. In real life, many factors, such as other circuitry, noise, decoupling, sample frequency, jitter, etc. will influence final result. The more the attacker is willing to invest in his measurement setup, the better his results may be.

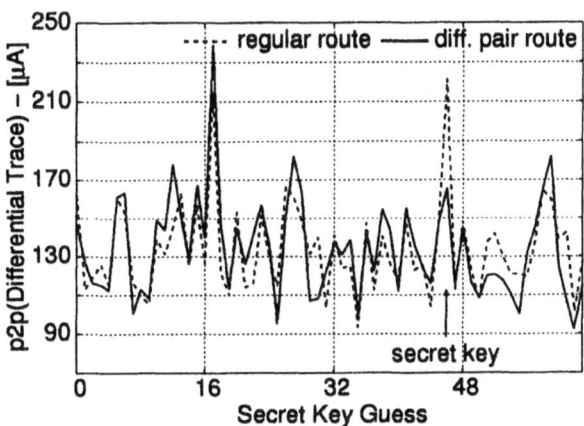

Figure 9. Peak-to-peak value of differential traces.

It does not come as a surprise that the secret key is still detectable in the regular routed design, even though the cycle-to-cycle variation on the power consumption is a mere 1%. DPA is a very powerful attack that, as we mentioned before, will reveal apparent insignificant power differences. Therefore it is also necessary to proof the resistance of a technique against DPA with the results of an actual DPA and not to rely on any form of visual inspection of the power consumption behavior.

5.   CONCLUSIONS

We have presented a methodology to route multiple differential wires between secure dynamic differential standard cells. In the methodology, a wire code is assigned that abstracts the differential pair as a single fat wire. Subsequently, the router is run using this fat wire and at the end the fat wire is decomposed back into the 2 wires. Experimental results have shown that a differential design is routed 2.6 times faster with this methodology compared with the case that the same differential design is routed without any constraints. The differential routing effectively helps in controlling the parasitic effects between the two wires of each differential pair. Experimental results show that the differential pair routing is an essential technique to successfully thwart the DPA.

ACKNOWLEDGEMENTS

The authors would like to acknowledge the support of the National Science Foundation, grant NSF – CCR 0098361.

6.   REFERENCES

1. E. Hess, N. Janssen, B. Meyer and T. Schuetze. "Information Leakage Attacks Against Smart Card Implementations of Cryptographic Algorithms and Countermeasures – a Survey," Proc. of Eurosmart Security Conference pp. 55–64, June 2000.
2. P. Kocher, J. Jaffe, B. Jun, "Differential Power Analysis," Proc. of CRYPTO'99, LNCS 1666, pp. 388–397, Jan. 1999.
3. S. Chari, C. S. Jutla, J. R. Rao and P. Rohatgi, "Towards Sound Approaches to Counteract Power-Analysis Attacks," Proc. of CRYPTO'99, LNCS 1666, pp. 398-412, Jan. 1999.
4. K. Tiri and I. Verbauwhede, "Securing Encryption Algorithms against DPA at the Logic Level: Next Generation Smart Card Technology," Proc. of CHES 2003, LNCS 2779, pp. 125–136, Sept. 2003.
5. J. Fournier, S. Moore, H. Li, R. Mullins and G. Taylor, "Security Evaluation of Asynchronous Circuits," Proc. of CHES 2003, LNCS 2779, pp. 137–151, Sept. 2003.
6. R. Brashears, and A. Kahng, "Advanced routing for deep submicron technologies," vlsi-cad.ucsd.edu/Presentations/talk/supporting.html, May 1997.
7. K. Tiri, I. Verbauwhede, "A Logic Level design methodology for a secure DPA resistant ASIC or FPGA implementation," Proc. of DATE 2004, pp. 246-251, Feb. 2004.
8. S. Khatri et al. "A novel VLSI layout fabric for deep sub-micron applications," Proc. of DAC 1999, pp. 491-496, June 1999.
9. Silicon Ensemble, www.cadence.com/products/digital_ic/sepks
10. LEF/DEF Language Reference 5.5, Jan. 2003, www.openeda.org
11. NIST FIPS PUB 46-2 "Data Encryption Standard," Dec. 1993. www.itl.nist.gov/fipspubs/fip46-2.htm
12. Cadence Chip Assembly Router, www.cadence.com/products/custom_ic/chip_assembly

A SURVEY ON FAULT ATTACKS

Christophe Giraud
Oberthur Card Systems
25, rue Auguste Blanche
92 800 Puteaux, France.
c.giraud@oberthurcs.com

Hugues Thiebeauld*
Thales Microelectronics
CNES LAB. BPI 1414
18, Avenue Edouard Belin
31 401 Toulouse Cedex 9, France.
hugues.thiebeauld@cnes.fr

Abstract   Fault attacks described in cryptographic papers mostly apply to cryptographic algorithms, yet such attacks may have an impact on the whole system in a smart card. In this paper, we describe what can be achieved nowadays by using fault attacks in a smart card environment. After studying several ways of inducing faults, we describe attacks on the most popular cryptosystems and we discuss the problem of induced perturbations in the smart card environment. Finally we discuss how to find appropriate software countermeasures.

Keywords: Fault Attack, Differential Fault Attack, Side-Channel Attack, Tamper Resistance, Smart Card.

1.   Introduction

In the smart card industry, fault attacks have been increasingly studied since the publication of Boneh, DeMillo and Lipton's paper [10] in 1996. At that time, the attack described in the Bellcore paper was merely theoretical, but since then improvements in technology have allowed attackers to put fault attacks into practice. In 2001, Skorobogatov and Anderson presented a practical attack [27] requiring very cheap equip-

---

*Research done while at Oberthur Card Systems.

ment: by using a photoflash lamp, they succeeded in flipping a bit in a memory cell. Despite the fact that the component was an old unprotected chip, their paper introduced a means of putting into practice theoretical fault attacks on cryptographic algorithms and they also showed that every software application dealing with security can be threatened by fault attacks. Nowadays, faults attacks are very powerful and may allow an attacker to break an unprotected system faster than any other kind of side-channel attacks, such as SPA, DPA or EMA.

In this paper, we discuss what is now possible concerning fault attacks from both a theoretical and a practical point of view. We deal with fault attacks applied to cryptographic algorithms but also with fault attacks applied to the whole system used in a smart card environment.

Firstly, we describe different ways to set up a fault attack. After studying the different kinds of induced faults and their implications during the execution of a software, we describe how to use these faults to break the most well-known cryptographic algorithms such as DES, AES, RSA, DSA and ECDSA. Finally, we discuss the security of software applications including cryptographic algorithms.

2.   How to perform a fault attack ?

There are many ways of performing a fault attack. We briefly describe the three which are most often published in cryptographic papers. These techniques are all moderately difficult to carry out and require relatively simple equipment. Moreover, they are all non-invasive or semi-invasive, which means that they do not require any physical tampering with the chip, although the packaging may be destroyed in some cases. Finally, we discuss the importance of synchronization and localization to successfully perform such attacks.

2.1   Glitch attacks

Historically the first way to induce a faulty behaviour in a smart card was to induce a glitch of power on one of the contacts. Good results were obtained with a power glitch applied to Vcc (power supply), GND (ground) or to the clock. A very powerful and brief addition of power can perturb the component, and hence generates a fault. However, the additional signal has to be short enough to avoid its detection or the destruction of the card. In order to achieve a successful attack, one can only set the timing of the glitch, its amplitude and its duration.

The obvious advantage is that the attack is very easy to set up, because it is truly non invasive and the equipment needed is easily available and cheap. The main drawback is that it is impossible to perturb a specific part of the chip, since the perturbation has an impact on the whole smart card.

Moreover, nowadays smart cards resist this kind of attack : thanks to filters, DC converters or power sensors, the card can either detect or cancel an out of specifications peak of power that might induce faulty behaviour. The hardware reacts accordingly by, for example, forcing a reset.

2.2   Light attacks

Light attacks were introduced in 2000 by Skorobogatov and Anderson [27].

Nowadays this is the most common way to induce faults on smart cards. The idea behind this attack is to use the energy of a light emission to perturb the silicon of a chip. For this kind of attack the preparation of the chip is more difficult than for a glitch attack because it is semi-invasive : the plastic layer on the component has to be removed in order to reach the silicon with the light beam. Such a preparation is described in [1] and [5].

Physical Effects.   Why is the component perturbed ? With a penetration depth dependant on the light wavelength, the energy carried by the light beam is absorbed by the silicon's electrons, which allows them to jump from the valence band to the conductance band. Each new electron in the conductance band creates an electron-hole pair and the set of all of these pairs forms a local photoelectric current. The logical gates or the memory cells of the component, depending on their technology, are sensitive to such a current and behave accordingly. This can therefore be exploited for an attack

Parameters.   To succeed in generating a fault, we have to take into account the following parameters : the energy of the light, its wavelength, the duration of the emission, the location and the extent of the impact zone. Let us detail each of them :

- The energy of the light beam must be considered as an energy per surface on the chip. The amount of energy per square inch must be enough to allow the electrons to jump to the conductance band, but low enough not to destroy the chip.

- The penetration depth of the energy depends on the wavelength of the light. For CMOS technology, if we succeed in creating electron-hole pairs close to the N-channel or the P-channel, the memory cells become very sensitive ; the wavelength has to be chosen accordingly.

- The longer the light emission, the larger the electron-hole pairs that are created. If the duration of the emission can be adjusted, the duration of the local current generated can be controlled.

- One of the advantages of light attacks compared to glitch attacks is that one can choose the location of the attack on the chip. The thinner the source of light, the better the precision of the impact on the chip.

Material.   In view of the previous parameters, the source of light has to be well chosen. The characteristics of the two most common sources for such an application are described here.

From [27], a simple camera flash installed on a microscope allows the execution of a light attack. The spectrum of a flash is composed of all visible wavelengths (white light) and a great part of infra red. The main advantage of this equipment is that it is very cheap and easy to install. Many attacks can be performed, but the attacks remain very limited. For example, with current smart cards, it is very difficult to change the value of a memory cell or to perturb the execution of the code. With a camera flash the different parameters described above (duration, power and localization of the emission) are difficult to control.

By using laser equipment one can significantly enhance the accuracy of the attack. The main characteristics of a laser are a discrete wavelength emission, a very thin beam and an emission which is easy to control in terms of duration and power. As previously explained, the depth penetration of the energy is in direct relation with the light wavelength, so the spectrum of the laser has to be chosen accordingly. The maximal power emission also has to be taken into account. The single drawback of using a laser is the price of the material, clearly much more expensive than a camera flash.

Chips have been equipped for some time with hardware countermeasures that render simple flash attacks quite inoperative. However, an attacker, with experience in the field and more enhanced equipment, can still find ways to perform light attacks.

2.3   Magnetic attacks

Introduced in [26], another way to induce transient faults is to emit a powerful magnetic pulse near the silicon. The magnetic field creates local currents on the surface of the component, which can generate a fault. From a practical point of view, the attack is performed with very cheap material : a wire is wound around a very sharp needle, a current flowing through the wire creates an oriented magnetic field. For better results, the attack has to be semi-invasive : the needle must be as close as possible to the silicon, so the plastic layer has to be removed.

Depending on the equipment used and its characteristics (size of the needle, number of loops around the needle) the following parameters have to be taken into account : the power of the magnetic field, its duration and localization on the chip. This technique seems difficult to set up practically, because the generation of a high magnetic field requires a high current, which would destroy the wire. Moreover, even if the needle concentrates the magnetic field, the laser still focuses the perturbation on a very small part of the chip much more accurately than the needle.

2.4   Synchronization issues

In order to set up a successful attack, the attacker must understand which errors have to be introduced from a theoretical point of view. He must then try to put these errors into practice. According to the theoretical analysis, induced faults can be more or less coarse (change any or several bits versus changing one given bit), which explains why the synchronization has to be more or less precise. For example, a DFA attack on RSA only needs a faulty computation on one of the CRT components; as the computations are quite long, a precise synchronization is not necessary. On the other hand, a DFA attack on a hardware DES (cf. section 4.1) requires the uttermost precision in synchronization due to the very fast execution of the computation we want to perturb. As the attack needs to be accurate in time, a good synchronization is critical.

There are many ways of synchronizing an attack. It is possible to use side channel information, such as monitoring the power consumption or the magnetic radiations resulting from the activity of the chip. Events such as writing in EEPROM can thus be recognized allowing us to determine when the attack must be performed. Signal processing can enhance the quality of the signal and its interpretation. All these techniques can be combined to obtain complementary information to help target the attack more accurately.

2.5   Localization issues

As we have already pointed out, a glitch attack perturbs every part of the chip. On the other hand, light or magnetic attacks allow a more accurate localization. It is easy to understand that to perturb only a specific part of the chip, for example a DES module, a light or magnetic emission has to be focused on this specific part. Sometimes, in order to induce desired faults, the targeted area has to be accurately circumscribed and exposed. An optical circuit is thus necessary in order to perform a light attack with the required precision.

3.   Kind of faults and consequences

As soon as the attacker is able to perform a perturbation on the chip, many errors are observed. Two kinds of faults, permanent or transient, can be generated.

Permanent faults.   Permanent faults occur when the value of a cell is definitively changed. It can concern either data (contained in EEPROM or in RAM) or code (contained in EEPROM). Modifying data definitively could be very powerful, particularly when the data is related to sensitive objects of the smart card, like a PIN or a key. Nevertheless, perturbing many successive logical cells or choosing the value of the perturbed cell is very difficult when the data is ciphered or the memory is scrambled.

Transient faults.   Transient faults are the most common ; they occur when a code execution or a computation is perturbed. An error in a code execution can be caused when the logical parts, including CPU and registers (data and control) are affected, or when reading code or data is perturbed. On the other hand, a computation error occurs when CPU or peripherals (such as cryptographic modules) are affected.

Consequences on the execution.   A single modified byte can severely perturb the execution flow of a code. At the lowest level of the micro-controller, each operation is encoded over several bytes : some for the instruction and others for the parameters. So assuming the attacker modifies one byte of code during its execution, the fault thus generated can affect either an instruction or a parameter.

The following cases can then occur:

- An instruction byte is modified : a different operation is executed. First consequence, the expected instruction is not executed : a call to a subroutine can be skipped, a test can be avoided, etc. Second consequence, one or several different instructions are executed, which can result in strange behaviour.

- A parameter byte is modified : a different address or value is considered. In the former case the target address is not modified but another is, or a wrong value is fetched from the memory, or the program counter is indirectly modified in a case of a JUMP instruction. In the latter case the operation is performed with a different operand, with the consequence that the result is wrong.

Given dedicated tools, techniques and some knowledge of the topology of the chip, most parts of a micro-controller can be attacked to cause transient faults.

4. Algorithms attacks

Fault attacks on cryptographic algorithms have been studied since 1996 and since then, nearly all the cryptographic algorithms have been broken by using such attacks. In this section we describe fault attacks on the best-known cryptosystems. We begin with symmetric algorithms such as DES and AES and then we focus on asymmetric algorithms such as RSA and DSA.

4.1 Symmetric algorithms

In this section we focus on the two best-known block-ciphers: the DES and the AES. Firstly we describe DFA attacks on these two block-ciphers and then we describe an efficient countermeasure against these attacks.

DES. The main paper about DFA on DES was written in 1996 by Biham and Shamir [7]. In this paper, they explain how to obtain the secret key by using between 50 and 200 faulty ciphertexts. Let us describe how it works.

We suppose that a single-bit fault is induced on the right part of the temporary result at the beginning of the last round. We denote respectively by C and C^* the correct and the faulty ciphertexts of a message M. We also denote respectively by T_{0-31} and T_{32-63} the left and the right parts of a 64-bit value T.

If we compare the left parts of the correct and the faulty temporary results at the end of the 16th round (i.e. $IP(C)_{32-63}$ and $IP(C^*)_{32-63}$), there is only one bit which differs. This bit corresponds to the induced fault so we obtain the position of the faulty bit and we deduce which S-boxes have been affected. By computing $EP(IP(C)_{32-63}) \oplus EP(IP(C^*)_{32-63})$ where EP is the DES expansive permutation, we obtain the difference Δ_{inputs} between the correct and faulty inputs of the S-Box in the last round.

Then we look at the difference $\Delta_{outputs}$ between the correct and faulty outputs of the S-Box in the last round, which is equal to $P^{-1}(IP(C)_{0-31})$

$\oplus P^{-1}(IP(C^*)_{0-31})$ where P is the permutation used at the end of each round.

Δ_{inputs} (respectively $\Delta_{outputs}$) contains non-zero bits only in input bits (respectively outputs bits) of the one (or two) S-box(es) which has been affected by the fault.

In the following part, we suppose that the faulty bit affected only one S-box, let us say $Sbox_k$. In order to find the 6-bit input of this S-box, we sort the possible values out by using the following procedure:

1 List all the 6-byte couples (B_i, B_j) such as $B_i \oplus B_j = \Delta_{inputs}$

2 For each couple in this list, if $Sbox_k(B_i) \oplus Sbox_k(B_j) \neq \Delta_{outputs}$ then eliminate the couple (B_i, B_j).

By using several faulty ciphertexts, we recover the 6-bit input of the kth S-box.

By iterating this attack we know the whole 48-bit input $Input\text{-}SB_{16}$ of the S-Box in the last round. Then we can compute the 16th subkey which is equal to $EP(IP(C)_{32-63}) \oplus Input\text{-}SB_{16}$. From this subkey we find the whole DES key by a very fast exhaustive search on the last unknown 8 bits.

We can extend this attack with a fault induced at the beginning of the 11th, 12th, 13th, 14th or 15th round. But in these cases we use a counting method: instead of eliminating the couple (B_i, B_j) which does not verify the relation $Sbox_k(B_i) \oplus Sbox_k(B_j) = \Delta_{outputs}$, we increase the counter by one of any pair which verifies the previous relation, the right value is expected to be counted more frequently than any wrong value.

This attack can also be extended if we induce a fault on a whole (or even several) byte(s) or if we disturb the key scheduling.

Such an attack has been successfully achieved on an unprotected software DES. By using a camera flash, as described in section 2.2, we succeeded in recovering the secret key by using only 2 ciphertexts. This could be achieved thanks to the fact that in practice we disturb one or several bytes compared to the single-bit fault model used by Biham and Shamir. Thus one faulty ciphertext provides information on several bytes of the 16th subkey.

AES. On the 2nd October 2000, the AES was chosen to be the successor of the DES. Since then, many papers have been published about DFA attacks on the AES [9, 12, 16, 17, 25]. Here we focus on the most powerful attack which was published recently [25].

For the sake of simplicity, we suppose the attack is done on an AES-128. This attack is based on the following observation: the MixColumns function operates on its input 4 bytes by 4 bytes, so if we induce a fault on one byte of one of this 4-byte block, the number of possible differences at the input of the MixColumns transformation is 255*4. Due to the linearity of the MixColumns function, we have 255*4 possible differences at its output.

If we suppose that a fault on one byte has been induced on the input of the last MixColumns and if we denote by M the plaintext, C and C^* the corresponding correct and faulty ciphertexts, the attack can be mounted as described hereafter:

1. Compute the 255*4 possible differences at the output of the MixColumns function and store them in a list $\mathcal{D}$

2. C and C^* differ only on 4 bytes, let us say bytes 0, 13, 10 and 7 (it corresponds to a fault on one of the first four bytes of the input of the last MixColumns). Take a guess on the 4 bytes in the same position of the last round key (i.e. $K^{10}_{0,13,10,7}$).

3. Compute $invSB(invSR(C_{0,13,10,7} \oplus K^{10}_{0,13,10,7})) \oplus invSB(invSR(C^*_{0,13,10,7} \oplus K^{10}_{0,13,10,7}))$ and check if this value is in $\mathcal{D}$. If so, add the round key to the list $\mathcal{L}$ of possible candidates.

4. Go back to step 2 by using another correct/faulty ciphertexts pair (D, D^*) (which could be obtained from another plaintext) which differs on the same bytes as C and C^*, and by choosing $K^{10}_{0,13,10,7}$ from the list $\mathcal{L}$. Repeat until there remains only one candidate in $\mathcal{L}$.

After this attack, 4 bytes of the last round key are known and we re-iterate three times this attack with pairs (C, C^*) which differ respectively on bytes (1, 14, 11, 4), (2, 15, 8, 5) and (3, 12, 9, 6).

This attack implies a guess on 4 bytes which is not very practical. In [25], an ingenious implementation of this attack is described by guessing only 2 bytes of the last round key at each iteration.

Moreover Piret and Quisquater remark that if we induce a fault on one byte between the two calls to the MixColumns function of the 7th and the 8th round, we obtain a fault on one byte for each 4-byte block of the input of the MixColumns function of the 9th round. So we obtain information on 16 bytes of the last round key instead of information on only 4 bytes. By using this attack the key could be retrieved by using only 2 faulty ciphertexts.

Countermeasures.   If we suppose that an attacker cannot induce
the same fault twice, one of the best countermeasures to protect the
DES and the AES is to compute the last rounds twice (including the
key scheduling). For the DES, doubling the last 8 rounds is efficient
and for the AES, doubling the last 4 rounds prevents nearly all known
DFA attacks[1]. Of course this kind of countermeasure must be adapted
for each symmetric cryptosystem, but doubling the whole or a part of
symmetric algorithms is generally an effective countermeasure against
fault attacks.

4.2   Public Key Algorithms

In this section we firstly describe fault attacks applied to the RSA, before looking at fault attacks on some signature schemes such as DSA and
EC-DSA. Finally, we present fault attacks on the scalar multiplication
used in Elliptic Curve Cryptography (ECC).

RSA.   The first published fault attack was applied to the RSA [10]
and was improved shortly after by Lenstra [22]. This attack on the
RSA-CRT is very simple and efficient in practice.

Firstly, we describe how to sign a message by using the RSA-CRT and
how to find the secret key by using the DFA attack described in [10].
We then describe how Lenstra improved this attack.

Let $N = pq$ be a product of two large prime integers, e chosen such
as $gcd(e, (p-1)(q-1)) = 1$ and $d = e^{-1}$ mod $(p-1)(q-1)$, (d, p, k) is
the secret key and (e, N) is the public key.

To sign a message $m < N$, the signer computes $s = m^d$ mod N. To
improve the time calculation, this signature is performed by using the
Chinese Reminder Theorem:

1 Compute $d_p = d$ mod $p - 1$ and $d_q = d$ mod $q - 1$

2 Compute $s_p = m^{d_p}$ mod p and $s_q = m^{d_q}$ mod q

3 Compute $s = ((((s_q - s_p)$ mod $q) * (p^{-1}$ mod $q)) * p + s_p)$ mod N

We can find two integers a and b such as $s = as_p + bs_q$ where:

$$\begin{cases} a \equiv 1 \text{ mod } p \\ a \equiv 0 \text{ mod } q \end{cases} \text{ and } \begin{cases} b \equiv 0 \text{ mod } p \\ b \equiv 1 \text{ mod } q \end{cases} \tag{1}$$

because $s_p = s$ mod p and $s_q = s$ mod q.

If the signer signs the same message twice, a fault attack can then
be performed: during the second signature, a fault is induced during

the computation of s_p (this attack works in the same way if a fault is induced during the computation of s_q). So we obtain a faulty signature $\tilde{s}$:

$$\begin{cases} s & = & as_p + bs_q \bmod N \\ \tilde{s} & = & a\tilde{s}_p + bs_q \bmod N \end{cases} \tag{2}$$

We remark that:

$$s - \tilde{s} \equiv a(s_p - \tilde{s}_p) \bmod N \tag{3}$$

Moreover $a \equiv 0 \bmod q$, so if p does not divide $s_p - \tilde{s}_p$ then:

$$\gcd(s - \tilde{s}, \ N) = \gcd(a(s_p - \tilde{s}_p), \ N) = q \tag{4}$$

Then we can easily find the other part p of the secret key by dividing the modulus N with q.

An improvement on this attack was found by Lenstra [22]: by using the same fault as above (i.e. fault induced during the computation of s_p), we have $s \equiv \tilde{s} \bmod q$ and $s \not\equiv \tilde{s} \bmod p$ so:

$$\gcd(m - \tilde{s}^e, N) = q \tag{5}$$

where e is the public exponent used to verify the signature. By using this attack, we only need one faulty signature of a known message to find the secret key.

An efficient countermeasure against this attack is to verify the signature by using the public key. As the public key is often very short ($e = 2^{16} + 1$ for example), the verification by using the RSA-SFM is very fast. A faster countermeasure was described by Blömer *et al.* in [8]: instead of verifying the signature, they rewrote the CRT recombination in such a way that if a fault is induced on a CRT component (s_p or s_q), the error also affects the other CRT component (respectively s_q or s_p).

DSA. Let us briefly describe the signature of the DSA: firstly the signer chooses a 160-bit prime number q and a 1024-bit prime p such as q divides $p - 1$. Then he chooses a positive integer g less than $p - 1$ of order q modulo p. Finally he chooses a positive integer a less than $q - 1$ and computes $A = g^a \bmod p$. His public key is $(p, \ q, \ g, \ A)$ and his secret key is a.

To sign a message m, the signer chooses a non-null random k less than $q - 1$ and computes

$$\begin{cases} r & = & (g^k \bmod p) \bmod q \\ s & = & k^{-1}(h + a.r) \bmod q \end{cases} \tag{6}$$

where h is the hash of m obtained by using the SHA-1 algorithm. The signature of the message m is the couple $(r, \ s)$.

The only existing DFA attack on the DSA was published in [4]. Thanks to this attack we can find a bit of the secret key each time we succeed in flipping a bit of the secret key during the computation of the second part of the signature.

Let us describe this attack: if an attacker succeeds in inducing a fault on only one bit of the secret key a, he obtains a faulty signature $(r, \tilde{s})$ where $\tilde{s} = k^{-1}(h + \tilde{a}.r) \bmod q$. The attacker can compute

$$
\begin{aligned}
T &= g^{\tilde{u}.h \bmod q}.A^{\tilde{u}.r \bmod q} \bmod p \bmod q \\
  &= g^{\tilde{u}(h+a.r) \bmod q} \bmod p \bmod q
\end{aligned}
\tag{7}
$$

where $\tilde{u} = \tilde{s}^{-1} \bmod q$. Let

$$
R_i = g^{\tilde{u}.r.2^i \bmod q} \bmod p \bmod q \text{ for } i = 0, 1, ..., 159
\tag{8}
$$

then he obtains from (7) and (8):

$$
TR_i = g^{\tilde{u}(h+r(a+2^i)) \bmod q} \bmod p \bmod q
\tag{9}
$$

and

$$
T/R_i = g^{\tilde{u}(h+r(a-2^i)) \bmod q} \bmod p \bmod q
\tag{10}
$$

So he obtains

$$
\begin{cases}
TR_i &= r \bmod p \bmod q \text{ if the } i^{\text{th}} \text{ bit of } a \text{ is } 0, \\
T/R_i &= r \bmod p \bmod q \text{ if the } i^{\text{th}} \text{ bit of } a \text{ is } 1.
\end{cases}
\tag{11}
$$

By iterating i from 0 to 159 and by comparing r with TR_i and T/R_i, the attacker can discover the value of a bit of the secret key. The complete secret key can be recovered by performing 161 signatures of the same (unknown) message: one correct and 160 faulty.

ECDSA. With the same approach described previously, we can attack the ECDSA. This attack was firstly described in [14].

We denote by (q, a, b, r, G, W) the signer's public key and by s his private key where q is a 160-bit prime, a and $b \in GF(q)$ the coefficients defining the elliptic curve, G a curve point generating the subgroup of order r and $W = sG$.

To sign a message m the signer generates a random u and computes the point $V = uG = (x_V, y_V)$. He then converts x_V into an integer k. Finally he computes the two parts of the signature:

$$
\begin{cases}
c &= k \bmod r, \\
d &= u^{-1}(f + s.c) \bmod r.
\end{cases}
\tag{12}
$$

where $f = SHA - 1(m)$.

If an attacker succeeds in inducing a fault on only one bit of the secret key s during the computation of d, he obtains a faulty signature $(c, \tilde{d})$ where $\tilde{d} = u^{-1}(f + \tilde{s}c) \bmod r$. The attacker can compute

$$\left\{ \begin{array}{rcl} h_{1,i,+} & = & \tilde{d}^{-1}(f + 2^i c) \bmod r \\ h_2 & = & c\tilde{d}^{-1} \bmod r. \end{array} \right. \tag{13}$$

He then computes the point

$$P_{i,+} = h_{1,i,+}G + h_2 W = \left(u\frac{f + (2^i + s)c}{f + \tilde{s}c} \right)G = (x_{P_{i,+}}, \, y_{P_{i,+}}) \tag{14}$$

He then converts $x_{P_{i,+}}$ into an integer $k_{P_{i,+}}$.

Now, we have two cases:

1. If the fault has flipped the ith bit of s from 1 to 0, then $\tilde{s} = 2^i + s$ so from (14) $P_{i,+} = uG$. By iterating the value of i from 0 to 159 and by testing if $k_{P_{i,+}} \bmod r = c$, the attacker can discover the value of a bit of the secret key.

2. But if the fault has flipped the ith bit of s from 0 to 1 then $\tilde{s} = s - 2^i$. The attacker can also discover the value of this bit by applying the same method as described above with $P_{i,-}$ instead of $P_{i,+}$ where $P_{i,-} = h_{1,i,-}G + h_2 W$ and $h_{1,i,-} = \tilde{d}^{-1}(f - 2^i c) \bmod r$.

So by using several faulty signatures (at least 160), an attacker can recover the whole value of the secret key a.

Remark.   The fault attacks on the DSA and the ECDSA described in this paper use the fact that the value of the secret key is disturbed during the computation of the second part of the signature. To protect such signature schemes against this kind of attack, we can implement one of the following countermeasures:

- we can check the integrity of the secret key at the end of the signature computation, for example by adding a CRC to the secret key: we check if the CRC of the secret value used during the signature computation is the same as the CRC of the secret key stored in non-volatile memory,

- we can also verify the signature by using the public key before sending the signature out. But this countermeasure is very costly in terms of execution time because the verification takes nearly twice as long as the signature computation.

Summary table.

	Fault model	Fault location	Minimum number of required faulty results
DES	Byte *(could be more)*	Anywhere among the last 6 rounds	2
AES	Byte	Anywhere between the MixColumns of the 7th and 8th round	2
RSA-CRT	Size of the modulus	Anywhere during the computation of one of the CRT components	1
DSA	Bit	Anywhere among 20 bytes	160
ECDSA	Bit	Anywhere among 20 bytes	160

ECC.   Only two papers dealing with elliptic curve scalar multiplication in the presence of faults have been published. The first fault attacks on scalar multiplication were published at Crypto 2000 by Biehl, Meyer and Müller [6]. In 2003, Ciet and Joye relaxed assumptions on these attacks [13].

In the sequel, we suppose that a scalar multiplication is performed with a scalar d and a point P which lays on the elliptic curve E: $y^2 + a_1xy + a_3y = x^3 + a_2x^2 + a_4x + a_6$ where the a_i's belong to a finite field $\mathcal{K}$.

Let us describe the fault attacks of Biehl *et al.* [6]: by observing that the elliptic curve parameter a_6 is not used in the addition and the doubling formulas on elliptic curves, they remark that addition and doubling operations with a random point P' lying on a curve E': $y^2 + a_1xy + a_3y = x^3 + a_2x^2 + a_4x + a_6'$ are the same as addition and doubling operations with a point P lying on the elliptic curve E. By supposing the Elliptic Curve Discrete Logarithm Problem (ECDLP) could be solved on E', if the smart card executes the scalar multiplication with P' and sends out the result $d.P'$, we could then obtain the secret scalar d.

As a result of this observation, they developed two attacks. For the first one, they induce a bit-fault on one of the coordinates of the point $P \in E$. They obtain a point P' which lies on an elliptic curve E' of the form $y^2 + a_1xy + a_3y = x^3 + a_2x^2 + a_4x + a_6'$. After the scalar multiplication, they obtain a point $d.P'$ which lies on the same curve as P'. From dP' they can compute a_6' and so define the elliptic curve E'. By trying all the different candidates P' which only differ from the known point P by one bit, they check if this candidate is a point on E' and if so, they try to solve the ECDLP on E'.

For the second attack, a fault on one bit is induced on a temporary point $Q^{(i)}$ during the scalar multiplication. We make a supposition on which $Q^{(i)}$ the fault is induced and we guess the latest $n - i$ bits of the scalar. By using these $n - i$ bits, we compute $\tilde{Q}^{(i)}$ from $\tilde{Q}^{(n)}$. By flipping one bit of $\tilde{Q}^{(i)}$, we obtain $\hat{Q}^{(i)}$ and we then compute $\hat{Q}^{(n)}$. If $\hat{Q}^{(n)} = Q^{(n)}$, the right value for the latest $n-i$ bits of the secret scalar has been guessed. Otherwise they continue the attack by changing another bit of $\tilde{Q}^{(i)}$. If the attack is not successful after trying each possibility they make another supposition about which $Q^{(i)}$ was perturbed.

By exploiting random faults induced in either coordinates of P, in the elliptic curve parameters or in the field representation, Ciet and Joye showed in [13] that it is possible to recover a part of the secret scalar.

All of these attacks can be avoided by checking that $dP = (x, y)$ satisfies the relation $y^2 + a_1 xy + a_3 y = x^3 + a_2 x^2 + a_4 x + a_6$.

5.　About security

5.1　The importance of securing hardware and software

As attacks are developed and improved, hardware evolves and gets more secure. However, protecting all the silicon surface against each and every type of attack is a difficult and very costly process. On the other hand, securing an unprotected micro-controller by only adding software countermeasures is also extremely costly in terms of execution time and memory space.

It appears that following a careful analysis of the attacks described above, a combination of hardware and software countermeasures yields a very good security/cost ratio.

Moreover, it is extremely important to consider the security of an embedded platform as a whole, no level being excluded from the analysis, whether it is the cryptographic functionalities, the low level operating system functions or the virtual machine and the high level applications.

For example, having a very secure PIN check is pointless if it is easy to bypass the high level call to this function and continue as if it had been satisfactorily executed.

5.2　How can we determine appropriate countermeasures?

Firstly one has to fix what an attacker is able to do : for example is he able to induce the same fault twice in the same code execution ? Is he able to choose a modified value ? ...

All these assumptions allow a developer to have a framework, which helps him to choose the most efficient countermeasures.

Next one must decide what has to be secured : which objects and what kind of processing are sensitive. From there on, one can go up in the software architecture and decide up to which layers the security level has to be enforced. Finally one must add the security features which are needed : redundancy on the objects, ensuring that operations have been well executed (for example by doubling them), securing the cryptographic algorithms, ...

One very helpful way to make it difficult to perform a fault attack efficiently is to desynchronize as much as possible the execution of the sensitive parts, by using for example random waits, dummy instructions, jitter on clocks, etc.

As it appears clearly in this paper, making an embedded platform secure demands a thorough understanding of the attacks, which comes only through practical experiments, through a good knowledge of making codes secure and through fully exploiting all the hardware features.

6.   Conclusion

Although the equipment to set up fault attacks appears to be quite common, we have seen that putting such attacks into practice requires technical experience. Moreover, it is increasingly difficult to perturb the latest micro-controllers, which are designed to resist fault attacks. Nonetheless, the danger exists and the risk has to be seriously considered. As we have seen, protecting algorithms and software implemented on smart cards or any tamper-proof device is best achieved by combining both software and hardware countermeasures. Efficient countermeasures are well-known, but they have to be implemented with care and sparingly in view of the cost of security in terms of time and memory space, especially in the restrictive smart card environment.

Notes

1. Blömer and Seifert presented in [9] a bit-fault attack on the AES with a very restrictive fault model: they suppose it is possible to set to zero one *chosen* bit. With such an attack, doubling the last 4 rounds is pointless, but in any case their fault model is too restrictive to be put into practice. Indeed, a lot of improvements in perturbation techniques would have to made for their attack to be effective from a practical point of view.

References

[1] R. Anderson and M. Kuhn. Tamper Resistance - a Cautionary Note. In *Proceedings of the 2nd USENIX Workshop on Electronic Commerce*, pages 1–11, 1996.

[2] R. Anderson and M. Kuhn. Low cost attacks on tamper resistant devices. In B. Christianson, B. Crispo, T. Mark, A. Lomas, and M. Roe, editors, 5th *Security Protocols Workshop*, volume 1361 of *LNCS*, pages 125–136. Springer, 1997.

[3] C. Aumüller, P. Bier, W. Fischer, P. Hofreiter, and J.-P. Seifert. Fault attacks on RSA with CRT: Concrete Results and Practical Countermeasures. In B. Kaliski Jr., Ç.K. Koç, and C. Paar, editors, *Cryptographic Hardware and Embedded Systems - CHES 2002*, volume 2523 of *LNCS*, pages 260–275. Springer, 2002.

[4] F. Bao, R. Deng, Y. Han, A. Jeng, A. D. Narasimhalu, and T.-H. Ngair. Breaking Public Key Cryptosystems an Tamper Resistance Devices in the Presence of Transient Fault. In 5th *Security Protocols WorkShop*, volume 1361 of *LNCS*, pages 115–124. Springer-Verlag, 1997.

[5] F. Beck. *Integrated Circuit Failure Analysis - A Guide to Preparation Techniques*. Wiley, 1998.

[6] I. Biehl, B. Meyer, and V. Müller. Differential Fault Analysis on Elliptic Curve Cryptosystems. In M. Bellare, editor, *Advances in Cryptology - CRYPTO 2000*, volume 1880 of *LNCS*, pages 131–146. Springer-Verlag, 2000.

[7] E. Biham and A. Shamir. Differential Fault Analysis of Secret Key Cryptosystem. In B.S. Kalisky Jr., editor, *Advances in Cryptology - CRYPTO '97*, volume 1294 of *LNCS*, pages 513–525. Springer-Verlag, 1997.

[8] J. Blömer, M. Otto, and J.-P. Seifert. A New RSA-CRT Algorithm Secure Against Bellcore Attacks. In *ACM-CCS'03*. ACM Press, 2003.

[9] J. Blömer and J.-P. Seifert. Fault based cryptanalysis of the Advanced Encryption Standard. In R.N. Wright, editor, *Financial Cryptography - FC 2003*, volume 2742 of *LNCS*. Springer-Verlag, 2003.

[10] D. Boneh, R.A. DeMillo, and R.J. Lipton. On the Importance of Checking Cryptographic Protocols for Faults. In W. Fumy, editor, *Advances in Cryptology - EUROCRYPT '97*, volume 1233 of *LNCS*, pages 37–51. Springer-Verlag, 1997.

[11] D. Boneh, R.A. DeMillo, and R.J. Lipton. On the Importance of Eliminating Errors in Cryptographic Computations. *Journal of Cryptology*, 14(2):101–119, 2001. An earlier version was published at EUROCRYPT'97 [10].

[12] C.-N. Chen and S.-M. Yen. Differential Fault Analysis on AES Key Schedule and Some Countermeasures. In R. Safavi-Naini and J. Seberry, editors, *Information Security and Privacy - 8th Australasian Conference - ACISP 2003*, volume 2727 of *LNCS*, pages 118–129. Springer-Verlag, 2003.

[13] M. Ciet and M. Joye. Elliptic Curve Cryptosystems in the Presence of Permanent and Transient Faults. In *Designs, Codes and Cryptography*, 2004. To appear.

[14] E. Dottax. Fault Attacks on NESSIE Signature and Identification Schemes. Technical report, NESSIE, Available from https://www.cosic.esat.kuleuven. ac.be/nessie/reports/phase2/SideChan_1.pdf, October 2002.

[15] E. Dottax. Fault and chosen modulus attacks on some NESSIE asymetrique Primitives. Technical report, NESSIE, Available from https://www.cosic. esat.kuleuven.ac.be/nessie/reports/phase2/ChosenModAtt2.pdf, February 2003.

[16] P. Dusart, G. Letourneux, and O. Vivolo. Differential Fault Analysis on A.E.S. Cryptology ePrint Archive, Report 2003/010, 2003. http://eprint.iacr.org/.

[17] C. Giraud. DFA on AES. Cryptology ePrint Archive, Report 2003/008, 2003. http://eprint.iacr.org/.

[18] M. Joye, A.K. Lenstra, and J.-J. Quisquater. Chinese Remaindering Based Cryptosystems in the Presence of Faults. *Journal of Cryptology*, 12(4):241–246, 1999.

[19] M. Joye, J.-J. Quisquater, F. Bao, and R.H. Deng. RSA-type Signatures in the Presence of Transient Faults. In M. Darnell, editor, *Cryptography and Coding*, volume 1355 of *LNCS*, pages 155–160. Springer-Verlag, 1997.

[20] M. Joye, J.-J. Quisquater, S.-M. Yen, and M. Yung. Observability Analysis - Detecting When Improved Cryptosystems Fail -. In B. Preneel, editor, *Topics in Cryptology - CT-RSA 2002*, volume 2271 of *LNCS*, pages 17–29. Springer-Verlag, 2002.

[21] V. Klíma and T. Rosa. Further Results and Considerations on Side Channel Attacks on RSA. In B. Kaliski Jr., Ç.K. Koç, and C. Paar, editors, *Cryptographic Hardware and Embedded Systems - CHES 2002*, volume 2523 of *LNCS*, pages 244–259. Springer-Verlag, 2002.

[22] A.K. Lenstra. Memo on RSA Signature Generation in the Presence of Faults. Manuscript, 1996. Available from the author at arjen.lenstra@citicorp.com.

[23] F. Koeune M. Joye and J.-J. Quisquater. Further results on Chinese remaindering. Technical Report CG-1997/1, UCL, 1997. Available from http://www.dice.ucl.ac.be/crypto/techreports.html.

[24] D.P. Maher. Fault Induction Attacks, Tamper Resistance, and Hostile Reverse Engineering in Perspective. In R. Hirschfeld, editor, *Financial Cryptography - FC '97*, volume 1318 of *LNCS*, pages 109–121. Springer-Verlag, 1997.

[25] G. Piret and J.-J. Quisquater. A Differential Fault Attack Technique Against SPN Structures, with Application to the AES and KHAZAD. In C.D. Walter, Ç.K. Koç, and C. Paar, editors, *Cryptographic Hardware and Embedded Systems - CHES 2003*, volume 2779 of *LNCS*, pages 77–88. Springer-Verlag, 2003.

[26] D. Samyde, S. Skorobogatov, R. Anderson, and J.-J. Quisquater. On a New Way to Read Data from Memory. In *First International IEEE Security in Storage Workshop*, pages 65–69. IEEE Computer Society, 2002.

[27] S. Skorobogatov and R. Anderson. Optical Fault Induction Attack. In B. Kaliski Jr., Ç.K. Koç, and C. Paar, editors, *Cryptographic Hardware and Embedded Systems - CHES 2002*, volume 2523 of *LNCS*, pages 2–12. Springer, 2002.

[28] S.-M. Yen and J.Z. Chen. A DFA on Rijndael. In A.H. Chan and V. Gligor, editors, *Information Security - ISC 2002*, volume 2433 of *LNCS*. Springer, 2002.

[29] S.-M. Yen and M. Joye. Checking before output may not be enough against fault-based cryptanalysis. *IEEE Trans. on Computers*, 49(9):967–970, 2000.

[30] S.-M. Yen, S.-J. Kim, S.-G. Lim, and S.-J. Moon. A Countermeasure against one Physical Cryptanalysis May Benefit Another Attack. In K. Kim, editor, *Information Security and Cryptology - ICISC 2001*, volume 2288 of *LNCS*, pages 414–427. Springer-Verlag, 2001.

[31] S.-M. Yen, S.J. Moon, and J.-C. Ha. Permanent Fault Attack on RSA with CRT. In R. Safavi-Naini and J. Seberry, editors, *Information Security and Privacy - 8th Australasian Conference - ACISP 2003*, volume 2727 of *LNCS*, pages 285–296. Springer-Verlag, 2003.

DIFFERENTIAL FAULT ANALYSIS ATTACK RESISTANT ARCHITECTURES FOR THE ADVANCED ENCRYPTION STANDARD[*]

Mark Karpovsky, Konrad J. Kulikowski, Alexander Taubin
Reliable Computing Laboratory, Department of Electrical and Computer Engineering, Boston University, 8 Saint Mary's Street, Boston, MA 02215 {markkar, konkul, taubin}@bu.edu

Abstract: We present two architectures for protecting a hardware implementation of AES against side-channel attacks known as Differential Fault Analysis attacks. The first architecture, which is efficient for faults of higher multiplicity, partitions the design into linear (XOR gates only) and nonlinear blocks and uses different protection schemes for these blocks. We protect the linear blocks with linear codes and the nonlinear with a complimentary nonlinear operation resulting in robust protection. The second architecture uses systematic nonlinear (cubic) robust error detecting codes and provides for high fault detection for faults of low and high multiplicities but has higher hardware overhead.

Key words: Advanced Encryption Standard; Differential Fault Analysis

1. INTRODUCTION

Cryptographic algorithms are designed so that by observing only the inputs and outputs of the algorithm it is computationally infeasible to break the cipher, or equivalently determine the secret key used in encryption and decryption. Thus, the algorithm itself does not leak enough useful information during its operation to compromise its security. However, when a physical implementation of the algorithm is considered, additional information like power consumption, behavior as a result of internal faults, and timing of the circuit implementing the algorithm can provide enough

[*] This work was supported by the Academy of Finland, Project No 44876 (Finish Center of Excellence Program (2000-2005))

information to compromise the security of the system. Attacks based on the use of this implementation specific information are known as Side Channel Attacks (SCA) [1,2].

In this paper we focus on the SCA's known as Differential Fault Analysis (DFA) attacks against the Advanced Encryption Standard (AES) [3]. DFA attacks are based on deriving information about the secret key by examining the differences between ciphers resulting from correct operations and faulty operations. DFA attacks have been applied to AES in [4-8]. Several methods for protection of AES have been developed. However, the current methods do not provide an adequate solution since they either require duplication in time or space [9], or they are not effective for all fault attacks [10,11].

We propose two methods for the protection of one round of AES. The first is a hybrid method which partitions AES into linear (XOR only) and nonlinear blocks and uses different protection techniques for the two different types of circuits. We protect the nonlinear blocks by performing nonlinear complimentary operation with respect to the function of the original block. The linear block is protected with linear codes. Using this hybrid partitioning method allowed us save on redundant hardware.

The second method uses systematic nonlinear robust codes. For the robust codes used in the design the probability of error detection depends not only on the error pattern (as in the case for linear codes) but also on the data itself. If all the data vectors and error patterns are equiprobable, then the probability of injecting an undetectable error if the device is protected by our robust codes is 2^{-2r} versus 2^{-r} if the device is protected by any linear code with the same r (r is a number of redundant bits which are added for data protection).

The error detection procedures of both designs can be used to detect a DFA attack and disable the card preventing further analysis.

2.   DFA ATTACK FAULT MODELS

We refer to a *fault* as a physical malfunction of a part of a circuit. An *error* is a manifestation of fault at the output of the device. An error is the difference (componentwise XOR) of the correct and distorted outputs of the device.

In this paper, we consider protection against a probabilistic attack. This attack does not necessitate chip depackaging nor expensive probing equipment and is therefore one of the more accessible attacks. In this model the attacker subjects the device to abnormal conditions which will generate faults in the circuit (radiation, high temperature, etc). We consider that under these conditions the locations of faults is uniformly distributed throughout the circuit and that the probability that a fault will occur in any wire is characterized by the wire distortion rate p which is a characteristic of the attack performed. Thus the number of actual faults injected into a circuit is dependent on the size N of the circuit and the expected number of faults (multiplicity of faults) is pN where N is the number of gates in the circuit.

We present two architectures for the protection of a round of AES from probabilistic attacks. The first method, based on partitioning, is an efficient and effective method under an assumption that probabilistic attacks have a high wire distortion rate and therefore result in the injection of many faults at a time. For the cases where no assumptions can be made about the wire distortion rates, we propose an architecture based on robust codes which is effective for all fault multiplicities, but has a higher hardware overhead than the first.

3. PROTECTION OF ONE ROUND OF AES BY HYBRID PARTITIONING

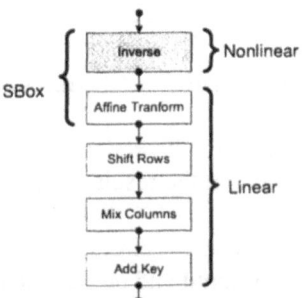

Fig. 1. Transformations involved in one typical round of encryption of AES

Encryption in AES-128 (AES with a 128-bit key) involves performing 10 rounds of transformations on a block of 128 bits with the last tenth round having one less transformation and with the first round being preceded by a round key addition. (The complete AES specification can be found in [3]) In each of the nine typical rounds there are four transformations: SBox, Shift Rows, Mix Columns, and Add Round Key. The last round differs from the rest in that it does not contain the Mix Columns transformation. The SBox transformation actually involves two operations: inversion in $GF(2^8)$ followed by an affine transform which involves a matrix multiplication M over $GF(2)$, followed by addition of a constant vector τ. With the exception of inversion, all other transformations and operations are linear (Fig. 1). That is, they can all be implemented using XOR gates only.

When considering only one round, the 128-bit data path can be divided into four identical independent 32-bit sections. Furthermore, in each of the four partitions the nonlinear inversion is performed on 8-bit data block. Thus, the nonlinear section is composed of 16 disjoint blocks and the linear portion composed of four identical disjoint blocks (Fig. 2).

Based on this partitioning, we designed redundant protection hardware for each of the two types of blocks in the design. The details of each block's method of protection are discussed in the next section.

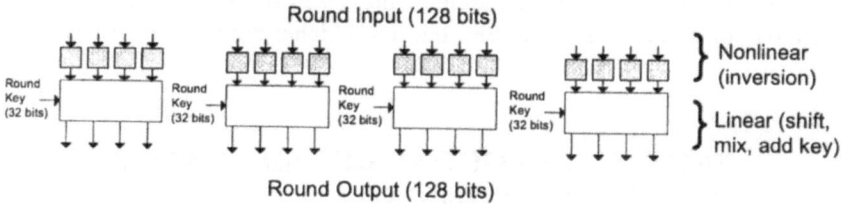

Fig 2. The nonlinear portion of one round can be separated into 16 identical independent blocks. The linear portion can be separated into 4 identical independent blocks.

3.1   Protection of Nonlinear Blocks

The nonlinear block performs inversion in $GF(2^8)$. Since zero does not have an inverse it is defined that the result of the inverse operation on zero is zero.

Our proposed fault detection circuitry for inverters is based on multiplication in $GF(2^8)$ of input and output vectors to verify the condition

$$X * X^{-1} = \begin{cases} 00000001 = I_8, \text{if } X \neq 0 \\ 00000000, \text{if } X = 0 \end{cases}.$$

Instead of computing the whole eight bit product, we compute only the least $r=2$ bits of the product.

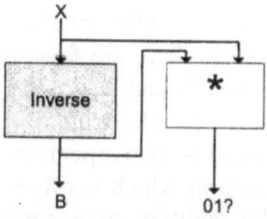

Fig 3. Architecture for protection of nonlinear block. The redundant portion performs partial multiplication in $GF(2^8)$

3.2 Analysis of Error Detecting Probabilities for Nonlinear Blocks

The number of bits, r, ($r<9$) in the signature (the number of bits resulting from partial multiplication), directly translates into the error detection capability of the protection scheme.

The probability that an error in the inverter will be missed is equal to the probability that two uniformly distributed random 8-bit vectors multiplied together will produce the expected r-bit constant I_r.

This protection scheme also has the advantage of being *robust* with respect to the $(8+r)$-bit output of the protected inverter (The probability of missing an error in the inverter depends not just only on the error itself but also on the input X). An error $e = (e_B, e_R)$ of $(8+r)$-bits, where e_B is an error at the output of the inverter and e_R is an error at the output of the redundant portion, is missed iff

$$[X^{-1} \oplus e_B] * [X] = I_r \oplus e_R$$

where $e_R \in GF(2^r), e_B \in GF(2^8)$ and $\oplus$ is bitwise XOR, or iff

$X * e_B = e_R$ where $X * e_B$ denotes r least significant bits of the product between X and e_I in $GF(2^8)$.

Thus, with the exception of an input X of all zeros, all error patterns e are detectable with probability of $1 - 2^{-r}$ for any given input X. Also, since error detection is dependant on the data X, the probability that an error will be missed after m random inputs is 2^{-rm}.

In one round of encryption of AES there are T=16 disjoint inverters, each with its own independent error detection. While for a single inverter the probability of missing an error is constant for all fault multiplicities that is not the case when multiple inverters are considered together. The probability that a fault will not be detected if it affects t inverters is q^t where q is the probability of missing a fault in one inverter.

Assuming that the distribution of faults is uniform, the probability that a fault of multiplicity l will affect t out of T inverters can be determined as:

$$P_T(t,l) = \frac{N_T(t,l)}{2^l} \quad \text{where} \quad N_T(t,l) = \binom{T}{t}[t^l - \sum_{j=1}^{t-1} N_t(j,l)] \ .$$

Thus, for AES and its T=16 inverters the probability of missing a fault of multiplicity l in the whole nonlinear portion of encryption of one round is

$$Q_T(l) = \sum_{i=1}^{\min(T,l)} q^i (P_T(i,l))$$

The detection probabilities for the sixteen inverters of AES with two bit signatures (r=2) were simulated using C++. A two input gate level C++ model of the circuit was built with the ability to induce faults at the output of each gate. The model was simulated for different multiplicities of faults. Two types of simulations were performed. The first considered each of the inverters to have an independent error signal. That is, it was assumed that the circuitry which checked the 32 bits of the total error signature for 16 invertors was fault free. Another simulation was performed with each of the error outputs of the nonlinear block being combined together to produce only two error signals for the whole nonlinear portion. In this simulation, the error signals which were expected to have a value of one (the least significant bit) were ANDed together while the other bit, which was expected to be zero, was ORed together. In this simulation this circuitry was **not** assumed to be fault free. In both of the simulation types, a XOR type of fault was induced (the output of the faulty gate was flipped from its correct value).

As Fig.4a shows, the computed and the experimental miss rates for the case of independent errors in the inverters are quite similar but are not exactly equal. Their difference can be accounted to the approximation in the calculated value of q, the miss rate for one inverter. In the calculation it was assumed to be constant for all multiplicities l. This approximation was not completely correct. As Fig.4b shows there are variations in this probability for small fault multiplicities, but a constant value of $2^{-r} = 0.25$ was used in the calculations.

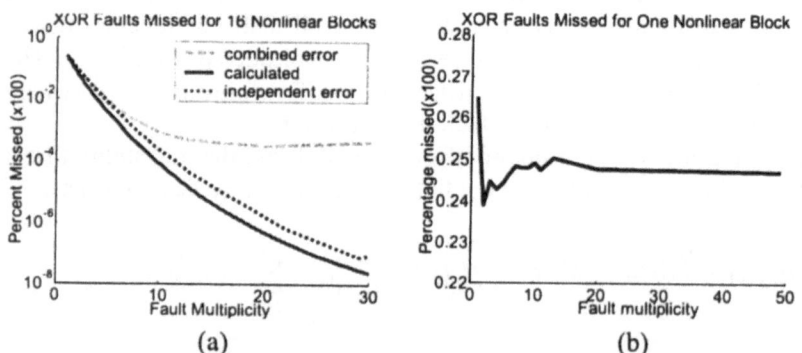

Fig 4. a. Theoretical and experimental miss rates for 16 inverters with independent error signals (dotted line) and with combined signals (dashed line) for r=2 bits in inverter's signatures for 1 input text. b. Experimental miss rate for one nonlinear block as a function of multiplicity of faults.

The simulation results in which the error outputs were combined together are significantly worse (dashed line). For fault multiplicities of ten and higher the miss rate reached a constant of about 0.1 % for one input text.

3.3 Protection of Linear Blocks

Each on of the four linear blocks has 64 bits of input (32 bits from the nonlinear portion and 32 bit of round key) and a 32 bits of output. Due to its large number of inputs and outputs and a relatively small gate count, a linear code proved to be the most cost efficient in terms of its hardware overhead to error detection ratio. The linear block performs three transformations: affine transform, mix columns, and add RoundKey.

The outputs Y can be written in terms of the inputs B in the following way:

$$Y1 = 02 \bullet (M(B1) \oplus \tau) \oplus 03 \bullet (M(B2) \oplus \tau) \oplus$$
$$M(B3) \oplus \tau \oplus M(B4) \oplus \tau \oplus RK1,$$
$$Y2 = M(B1) \oplus \tau \oplus 02 \bullet (M(B2) \oplus \tau) \oplus$$
$$03 \bullet (M(B3) \oplus \tau) \oplus M(B4) \oplus \tau \oplus RK2,$$
$$Y3 = M(B1) \oplus \tau \oplus M(B2) \oplus \tau \oplus$$
$$02 \bullet (M(B3) \oplus \tau) \oplus 03 \bullet (M(B4) \oplus \tau) \oplus RK3,$$
$$Y4 = 03 \bullet (M(B1) \oplus \tau) \oplus M(B2) \oplus \tau \oplus$$
$$M(B3) \oplus \tau \oplus 02 \bullet (M(B4) \oplus \tau) \oplus RK4,$$

where $\bullet$ is multiplication in $GF(2^8)$, M is the binary *(8x8)* matrix , $M(Bi)$ is multiplication over GF(2), τ is a constant as defined in AES and RKi are round keys[6].

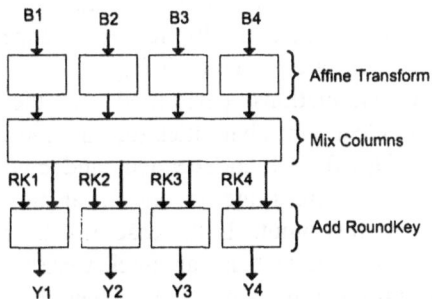

Fig 5. Transformations performed in one linear block.

The design of the linear code for the block was based on the observation that an implementation of the sum $Y1 \oplus Y2 \oplus Y3 \oplus Y4$ is much simpler than of the original block. Indeed:

$$S = Y1 \oplus Y2 \oplus Y3 \oplus Y4$$
$$= M(B1 \oplus B2 \oplus B3 \oplus B4) \oplus RK1 \oplus RK2 \oplus RK3 \oplus RK4.$$

This function S is computed by the linear predictor and used as an eight-bit redundant signature for the original linear block (see Fig.6). Under fault-free operation, the output of the linear predictor should be equal to the sum of the output of the original linear block. The Error Detecting Network (EDN) sums the output (block P in Fig 6) and compares it to the expected value.

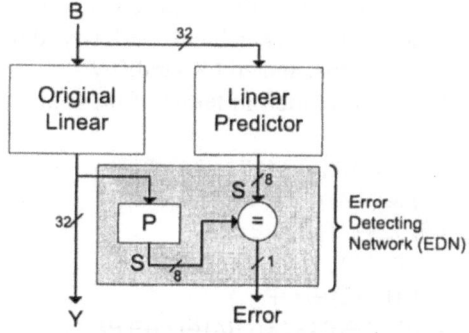

Fig 6. Architecture for protection of linear block.

3.4   Analysis of Error Detection Probabilities for Linear Blocks

A gate-level model of the linear block was built and simulated in C++. Like in the simulations for nonlinear blocks, faults were injected randomly (with equal probabilities of a fault at outputs of the gate) into the circuit with random and uniformly distributed multiplicities in range from 1 to 50. The results of these simulations are presented in Fig.7.

Similarly to the simulations performed on the nonlinear blocks, simulations for fault detection probabilities for one linear block (Fig.7b) and four linear blocks (Fig.7a) with independent and combined error signals were performed. The fault miss rate for one linear block resulted in a miss rate of $5 \pm 6\%$ for one text input. In the case of 4 linear blocks the design where the error signals from each linear block were not fault-free (Fig.7a, dashed line) were significantly worse than when the errors signals were independent for each block (Fig.7a, dotted line)

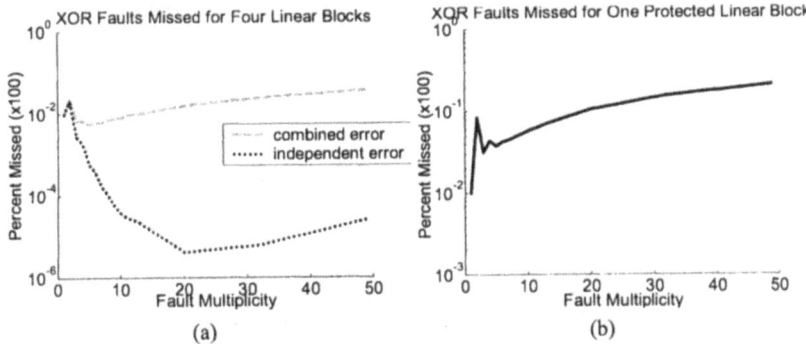

Fig 7. a. Experimental miss rates for 4 linear blocks with independent error signals (dotted line) and with combined signals (dashed line) for one input text. b. Experimental miss rate for one linear block as a function of multiplicity of faults.

3.5   Complete Round of Encryption

One typical round of encryption was constructed from the protected linear and nonlinear blocks. Fig.8 shows one forth of one round of the encryptor. The complete round is composed of four identical blocks arranged in parallel. The error signals of the nonlinear portion are chained together to output only a 2 bit signature for all of the 16 inverters. Likewise, the error signal from the four linear blocks is chained together to produce 1 error signal for the whole linear portion of the round. Thus there are 3 error outputs for the whole round. Under fault free operation the nonlinear error outputs should have a value of 01 (excluding an input of all zeros in the input) and the linear error output should have a value of 0.

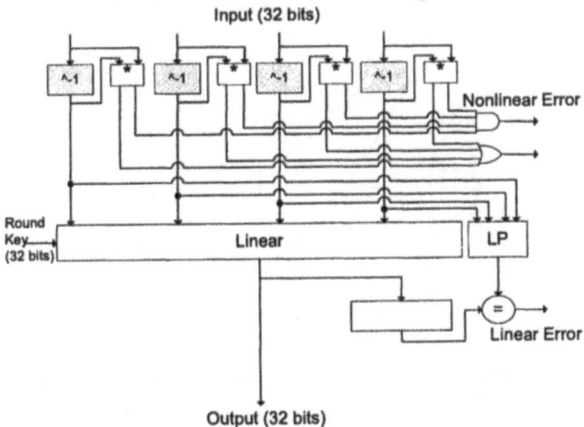

Fig 8. One fourth of a typical round of encryption.

The complete protected round has a total hardware overhead (in terms of 2-input gates) of 35%. Table 1 summarizes the overheads for each type of block.

A C++ model of one complete protected round of encryption was built. For random and uniformly distributed texts and round keys, faults of different multiplicities were injected into the circuit. The results for stuck-at-one, stuck-at-zero and XOR fault simulation are presented in Fig.9.

Table 1. Sizes of Components of One Complete Round of Encryption in Terms Two-Input Gates

Component	Gate Count for Original AES	Gate Count for the Redundant Portions	Overhead
Linear Portion (4 blocks)	896	460	51.3%
Nonlinear Portion (16 blocks)	2800	800	28.5%
Error Chaining	0	33	-
Total	3696	1293	35%

When fault detection is considered for one input text, as in Fig.9, the experimental miss rate for stuck-at-one (sa1) and stuck-at-zero (sa0) faults was higher than that for XOR faults. For the unidirectional faults (only sa1 or only sa0), not every injected fault will manifest itself in the circuit. On average, only about 50% of the injected unidirectional faults will manifest themselves at an output of a gate. For XOR faults, since the fault involved flipping a value, 100% of the faults are manifested. Thus, the miss rate curves of the unidirectional faults presented in Fig.9 should be shifted to more precisely reflect the fault multiplicity.

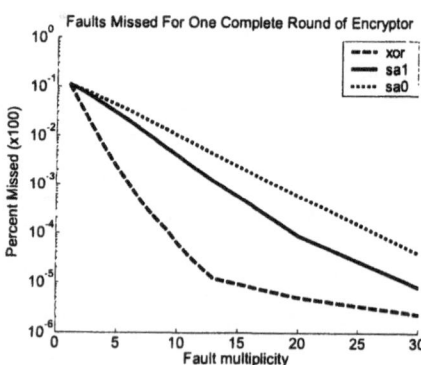

Fig 9. Fault Miss Rates for one complete round of encryption for one input text.

As mentioned in Section 3.2, the design of the nonlinear block resulted in robust protection with respect to its output. Since the nonlinear blocks account for a large portion of the total hardware, it was expected that the whole round will exhibit partial robust behavior. That is, as long as fault affects the nonlinear blocks, it is expected that the detection of that fault is dependent on the input text. Thus, fault miss rate should decrease when multiple random text inputs are considered for the same fault. Simulation results for multiple text inputs for unidirectional and XOR faults are presented in Fig.10.

The simulation results in Fig.10 show considerable improvement for unidirectional faults when multiple text inputs are considered. XOR faults showed limited improvement. The manifestation of stuck-at-faults is different depending on the data, resulting in a different error distribution for each input. That is not the case for XOR faults. XOR fault error manifestation is much less dependant on the data.

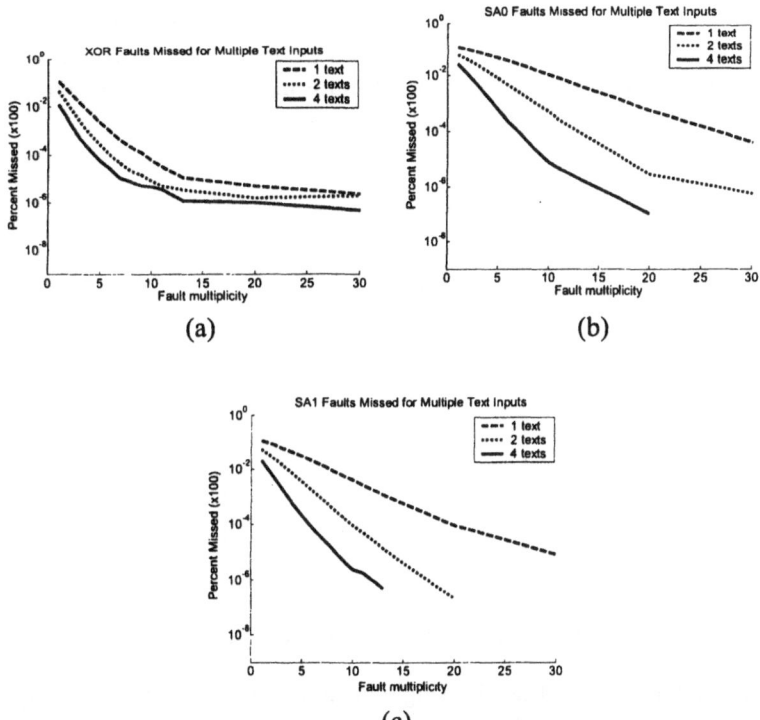

Fig 10. Simulated miss rates for one round of encryptor for multiple texts with a. XOR faults b. Stuck-at-zero faults c. Stuck-at-one faults

The simulation results show that the miss rate improves as fault multiplicity increases. With 4 random text inputs the miss rate drops to 0.0001% for stuck at one faults with a fault multiplicity of 14 (see Fig.10c).

The detection for small fault multiplicities is considerably worse since they will affect a small number of blocks. When only one block is affected the detection is only as good the detection in one block.

The design and simulations were only performed for one block of encryption.

4. PROTECTION OF AES BY NONLINEAR SYSTEMATIC ROBUST CODES

4.1   General Robust Architecture

Robust codes [12] can be used to extend the error coverage of linear prediction schemes for AES. Only two extra cubic networks computing $y(x) = x^3$ in $GF(2^8)$ are needed, one in the extended device, and one in the Error Detection Network. The architecture of one round AES encryption with robust protection is presented in Fig.11.

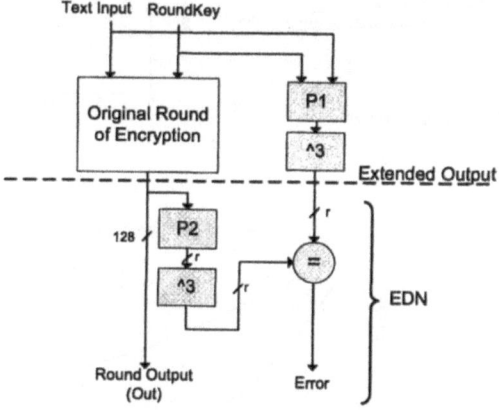

Fig 11. General architecture which uses systematic nonlinear robust protection.

In the architecture in Fig.11 a single linear predictor (block P1) is required for the encryptor. (Note that in this context a linear predictor is such that it generates a signature, which is a linear combination of the outputs of the round. It does not mean that the predictor contains only linear elements.) The r-bit signature of the linear predictor, is cubed in $GF(2^r)$ to produce an r-bit output signature (block ^{3} in Fig.11), which is nonlinear with respect to the output of the round.

For the robust architecture we have designed a linear predictor which can be used to generate a r=32-bit signature. The predictor P1 is designed in a similar fashion to the linear predictor for the linear block presented in the previous section. For four bytes of the output, the predictor predicts one byte, $L'(j)$ $(j = 0,1,2,3)$. For encryption this simplifies to eliminating the mix columns transformation in the predictor P1.

The output of the linear predictor, $L'(j)$, is a 4-byte word which is linearly related to the output of one round of AES. The function of $L'(j)$ with respect to *Out(i,j)* can be written as:

$$L'(j) = \bigoplus_{i=0}^{3} Out(i, j) \text{ where } j \in \{0,1,2,3\}.$$

Thus, the following expressions are valid for AES:

$$
\begin{aligned}
L'(0) = {} & 01 \bullet Sub(In(0,0)) \oplus 03 \bullet Sub(In(1,0)) \oplus Sub(In(2,0)) \oplus Sub(In(3,0)) \oplus \\
& Sub(In(0,0)) \oplus 02 \bullet Sub(In(1,0)) \oplus 03 \bullet Sub(In(2,0)) \oplus Sub(In(3,0)) \oplus \\
& Sub(In(0,0)) \oplus Sub(In(1,0)) \oplus 02 \bullet Sub(In(2,0)) \oplus 03 \bullet Sub(In(3,0)) \oplus \\
& 03 \bullet Sub(In(0,0)) \oplus Sub(In(1,0)) \oplus Sub(In(2,0)) \oplus 02 \bullet Sub(In(3,0)) \\
& \oplus RK(0,0) \oplus RK(1,0) \oplus RK(2,0) \oplus RK(3,0)
\end{aligned}
$$

$$
\begin{aligned}
= {} & Sub(In(0,0)) \oplus Sub(In(1,0)) \oplus Sub(In(2,0)) \oplus Sub(In(3,0)) \\
& \oplus RK(0,0) \oplus RK(1,0) \oplus RK(2,0) \oplus RK(3,0),
\end{aligned}
$$

where $\bullet$ is multiplication in $GF(2^8)$, $In(i, j)$ is a text input byte to the round, $RK(i, j)$ is one byte round key, and $Sub(In(i, j))$ is the SubBytes transformation on the byte $In(i, j)$ as defined in the AES standard [6].

Since $\quad Sub(In(i, j)) = M(In(i, j)^{-1}) \oplus \tau$,

We have $L'(0)$:

$$
\begin{aligned}
L'(0) = {} & M(In(0,0)^{-1} \oplus In(1,1)^{-1} \oplus In(2,2)^{-1} \oplus In(3,3)^{-1}) \\
& \oplus RK(0,0) \oplus RK(1,0) \oplus RK(2,0) \oplus RK(3,0).
\end{aligned}
$$

Extending the procedure to the rest of the bytes of encryption yields:

$$
\begin{aligned}
L'(1) = {} & M(In(0,1)^{-1} \oplus In(1,2)^{-1} \oplus In(2,3)^{-1} \oplus In(3,0)^{-1}) \\
& \oplus RK(0,1) \oplus RK(1,1) \oplus RK(2,1) \oplus RK(3,1), \\
L'(2) = {} & M(In(0,2)^{-1} \oplus In(1,3)^{-1} \oplus In(2,0)^{-1} \oplus In(3,1)^{-1}) \\
& \oplus RK(0,2) \oplus RK(1,2) \oplus RK(2,2) \oplus RK(3,2), \\
L'(3) = {} & M(In(0,3)^{-1} \oplus In(1,0)^{-1} \oplus In(2,1)^{-1} \oplus In(3,2)^{-1}) \\
& \oplus RK(0,3) \oplus RK(1,3) \oplus RK(2,3) \oplus RK(3,3).
\end{aligned}
$$

In the Error Detecting Network (EDN) the block P2 compresses the 128 bits into $r=32$ by bitwise XOR to match the output of the predictor P1. The output of the block P2 is also cubed in $GF(2^r)$. It is this cubed compressed output which is compared to the cubed output of the predictor P1. Under correct operation these two outputs should be equal. It was shown in [12] that the introduction of the nonlinear cubic operation resulted in the reduction of the fraction of undetectable errors at the extended output from 2^{-r} to 2^{-2r} without increasing the redundancy r of the original linear code.

4.2   Analysis for Error Detecting Probabilities for Robust Architecture

A gate-level model of this design was simulated using C++. Two types of simulations were performed. In the first faults were injected into all parts of the circuits. The other assumed that the error detecting network (block P2, the second cubic network and the comparator) were fault-free. The results of these gate level simulations are presented in Fig.12.

To explain the results of simulations of Fig.12 we note that the EDN in simulated architecture was unprotected. As a result, it had a significant impact on the overall fault detection for the proposed design. As Fig.12 shows, when faults were included in the EDN (dashed and dotted lines for stuck-at-zero and stuck-at-one respectively), the fault detection probabilities remained almost constant for higher fault multiplicities. In contrast, when the EDN was considered to be fault free, the fault miss rate quickly dropped as fault multiplicity decreased. The EDN accounts for about 25% of the complete protected round.

The detection performance differed substantially for this approach than that of the first. In this design, even for low fault multiplicities the miss rate remained at a low 0.1%. This improvement came at a price. When a single round of encryption is protected using this approach the overhead of the protection exceeds 150% in terms of the gate count. This high overhead is a result of the high cost of the cubic networks. The overhead can be decreased when the complete AES is protected using this method, including decryption and key expansion. Protecting a larger design offsets the large cost of the cubic networks.

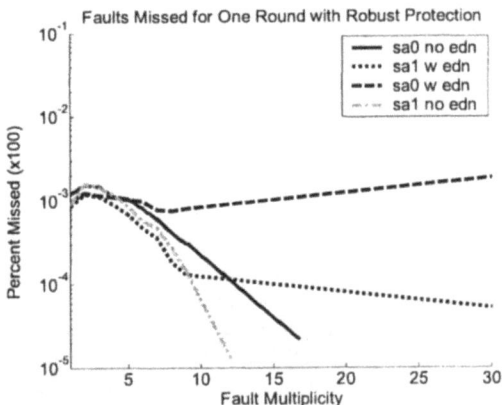

Fig12. Simulation results for one round of encryption with robust protection for one input text.

We note also that this architecture provides for robust detection of XOR faults such that the corresponding errors do not depend on input texts (these may be faults in the linear parts or in the output register of the round). For these faults the probability of missing a fault for one input text is $2^{-r+1} = 2^{-31}$ and probability of missing a fault for all texts is $2^{-2r} = 2^{-64}$.

5. CONCLUSIONS

We presented two methods for protecting the Advanced Encryption Standard against Differential Fault Attacks. The two methods had different overheads and different fault detection probabilities characteristics. We presented also gate-level simulation results for one round of encryption for both architectures.

The first method, which is useful for attacks with high wire distortion rate and based on a hybrid partitioning, had a low hardware overhead of 35%. This method was able to achieve a fault miss rate of 0.01% for one stuck-at-fault with multiplicity of 30 for one text input. For faults of small multiplicity the method's detection rate was substantially worse.

The second method, which is efficient for all wire distortion rates and is based on systematic robust codes, had a high hardware overhead of 150%. However, this method had a much lower miss rate for faults of small multiplicities. Even for faults of multiplicity of one, the miss rate was only about 0.1%. This method is also very efficient for XOR faults resulting in errors which do not depend on input texts.

References

[1] P. Kocher, "Timing Attacks on Implementations of Diffie-Hellman, RSA, DSS, and Other Systems," Crypto 96, Proceedings, Lecture Notes In Computer Science ol. 1109, N. Koblitzed., Springer-Verlag, 1996.

[2] P. Kocher, J. Jaffe, B. Jun, "Differential Power Analysis," Advances in Cryptology - Crypto 99 Proceedings, Lecture Notes In Computer Science Vol. 1666, M. Wiener ed., Springer-Verlag, 1999. [2] J. Kelsey, B. Schneier, D. Wagner, and C. Hall, *Side Channel Cryptanalysis of Product Ciphers*, ESORICS '98 Proceedings, 1998, pp. 97-110.

[3] FIPS PUB 197: *Advanced Encryption Standard,* http://csrc.nist.gov/publications/fips/fips197/fips-197.pdf

[4] C.N. Chen and S.-M.Yen, *Differential Fault Analysis on AES Key Schedule and Some Countermeasures*, ACISP 2003, LNCS 2727, pp.118-129, 2003

[5] P. Dusart, G. Letourneux, O. Vivolo, *Differential Fault Analysis on AES*, Cryptology ePrint Archive, Report 2003/010. Available: http://eprint.iacr.org/2003/010.pdf

[6] C. Giraud. *DFA on AES*. Cryptology ePrint Archive, Report 2003/008. Available: http://eprint.iacr.org and http://citeseer.nj.nec.com/558158.html

[7] Johannes Blömer, Jean-Pierre Seifert: Fault Based Cryptanalysis of the Advanced Encryption Standard (AES). Financial Cryptography 2003: pp. 162-181

[8] Jean-Jacques Quisquater, Gilles Piret, "A Differential Fault Attack Technique Against SPN Structures, with Application to the AES and KHAZAD", (CHES 2003), Volume 2779 of Lecture Notes in Computer Science, pages 77-88, Springer-Verlag, September 2003

[9] Ramesh Karri, Kaijie Wu, Piyush Mishra, Yongkook Kim, Concurrent Error Detection of Fault Based Side-Channel Cryptanalysis of 128-Bit Symmetric Block Ciphers. *IEEE Transactions on Computer-Aided Design of Integrated Circuits and Systems,* Vol.21, No.12, pp. 1509-1517, 2002

[10] G. Bertoni, L. Breveglieri, I. Koren, P. Maistri and V. Piuri, *Error Analysis and Detection Procedures for a Hardware Implementation of the Advanced Encryption Standard,* IEEE Transactions on Computers, VOL. 52, NO. 4, 2003

[11] Ramesh Karri, Grigori Kuznetsov, Michael Gössel: Parity-Based Concurrent Error Detection of Substitution-Permutation Network Block Ciphers. CHES 2003. pp.113-124

[12] M.G.Karpovsky and A. Taubin, "A New Class of Nonlinear Systematic Error Detecting Codes", *to be published in IEEE Info Theory, 2004*

SECURE NETWORK CARD
Implementation of a Standard Network Stack in a Smart Card

Michael Montgomery, Asad Ali, and Karen Lu
Axalto, 8311 North FM 620 Road, Austin, Texas, 78726, USA

Abstract: This paper covers the philosophy and techniques used for implementation of a standard networking stack, including the hardware interface, PPP, TCP, IP, SSL/TLS, HTTP, and applications within the resource constraints of a smart card. This implementation enables a smart card to establish secure TCP/IP connections using SSL/TLS protocols to any client or server on the Internet, using only standard networking protocols, and requiring no host middleware to be installed. A standard (unmodified) client or server anywhere on the network can securely communicate directly with this card; as far as the remote computer can tell, the smart card is just another computer on the Internet. No smart card specific software is required on the host or any remote computer.

Key words: Internet; smart card; network; SSL; TLS; TCP/IP; PPP; resource constraints.

1. INTRODUCTION

Smart cards have been in use for more than two decades now, but they have yet to achieve wide acceptance in mainstream computing. Smart card advantages such as security, portability, wallet compatible form factor, and tamper resistance make smart cards potentially useful for a wide variety of applications. However, these applications are hindered because of the mismatch between smart card communication standards and the communication standards for mainstream computing and networking.

For years, efforts have been underway to connect smart cards to the Internet. Early pioneers include the University of Michigan Webcard [1], Bull iSimplify [2], Guthery's GSM Web server [3], and the Gemplus prototype [4]. This paper expands on the years of work in this area, detailing some of the drawbacks of the approaches to date, and how many of these drawbacks can be overcome by using a standard network stack. Finally, techniques are presented for implementing a standard network stack within the resource constraints of a smart card.

2.   MOTIVATION

The current network smart cards have some key areas of weakness. By analyzing these weaknesses, we can better understand how to achieve our vision of widespread acceptance of smart cards into the network computing mainstream.

2.1   Security

Security is the most important aspect of most smart card applications. Much effort has gone into improving overall smart card tamper resistance and defenses against specific attacks. APDU communication can be protected by encryption. Challenge/response can verify trusted terminals.

Networking adds an extra security challenge [5]. There are currently two primary techniques for establishing network security with smart cards.

One technique is to write remote applications that encrypt data within the remote application as shown in Figure 1. The remote applications do not depend upon the network layer for security, but use an encryption scheme that is shared with the card applications. The drawback of this approach is that smart cards can only interact securely with remote applications that have been written or modified with smart cards in mind.

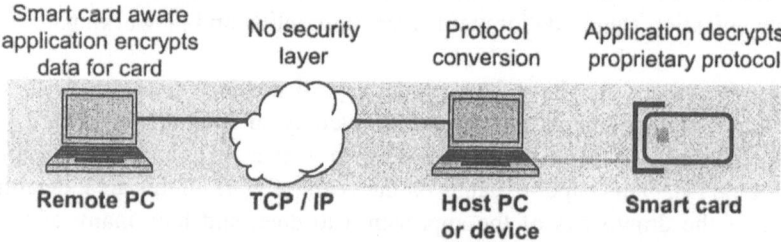

| Smart card aware application encrypts data for card | No security layer | Protocol conversion | Application decrypts proprietary protocol |

| **Remote PC** | **TCP / IP** | **Host PC or device** | **Smart card** |

Figure 1. End-to-end security using card specific security protocols.

A second technique avoids this drawback, and allows the smart card to interact with applications that are not smart card aware. With this technique, the host computer establishes a secure network connection with the remote application using standard SSL or TLS protocols, as illustrated in Figure 2. When the remote application sends data, the network layer automatically encrypts it before it is transmitted. The host then decrypts the data, packages it in APDUs, and can encrypt the APDUs before sending to the card if desired. The problem with this approach is that when the data in the host is decrypted, it is vulnerable to interception within the host computer. Since host computers are typically much more vulnerable to attack than smart cards, this becomes the weak link in the security chain.

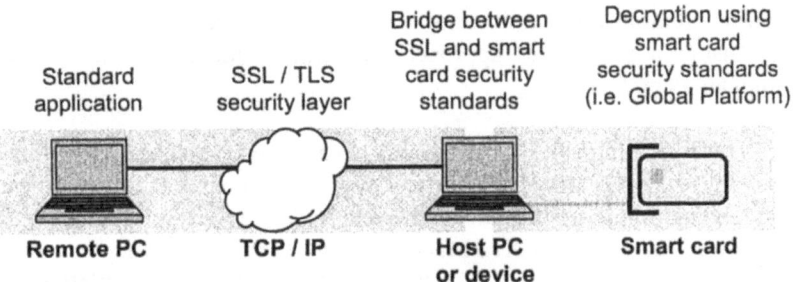

Figure 2. Bridging standard security layers in the host.

To avoid the problems associated with these two techniques, a first key requirement for a secure network card is to implement the TCP/IP network stack and SSL/TLS security layer inside the card, as shown in Figure 3. This allows the card to establish end-to-end security with a remote application that is not smart card aware, so that the communications are protected from all computers along the communications path, including the host computer. The security of the host is no longer an issue, allowing even an untrusted kiosk computer to be potentially used safely as a host.

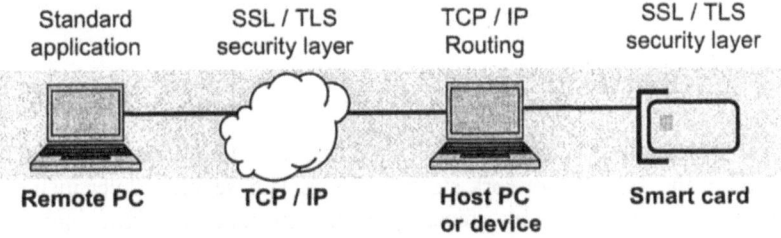

Figure 3. End-to-end security using SSL/TLS security within smart card.

2.2   Middleware

Thus far, all cards have required some middleware [5,6] to be installed on the host computer to facilitate the connection of the cards to the Internet. Typically, the middleware resides on the host computer, acting as a proxy for the card for establishing network communication, handling the network protocols, and repackaging the data to send APDUs to the smart card and receive the card responses. Sometimes the middleware handles the security layer as well, though this can lead to drawbacks as shown earlier. In addition, remote applications for accessing a smart card were sometimes built using a proxy/stub approach, requiring middleware on the remote computer as well.

Requiring middleware has several negative consequences. Middleware must be developed and tested for any platform to enable the use of the card. Middleware for various platforms such as Windows 98, Windows 2000, Windows NT, Windows XP, MacOS X, and Linux is expensive to develop and test, and even more expensive to maintain. Often platforms are excluded because of the high cost of middleware. Users dislike middleware; this is often one of the greatest barriers to user acceptance of a product. Users are now accustomed to plug-and-play operation, and reject products that do not use standard drivers built into the operating system.

Therefore, a second key requirement for a secure network smart card is to use standard interfaces and drivers that are built into most operating systems, so that no middleware is required. This allows a network smart card to freely move between computers, without having to install middleware every time a new host computer is used. This potentially enables host computers such as kiosks or corporate computers, where users may be forbidden to install middleware.

3.   ARCHITECTURE

An architecture was selected based on our two key requirements: implement the TCP/IP network stack and SSL/TLS security layer inside the card; and use standard interfaces and drivers that are built into most operating systems, so that no middleware is required. This architecture is shown in Figure 4.

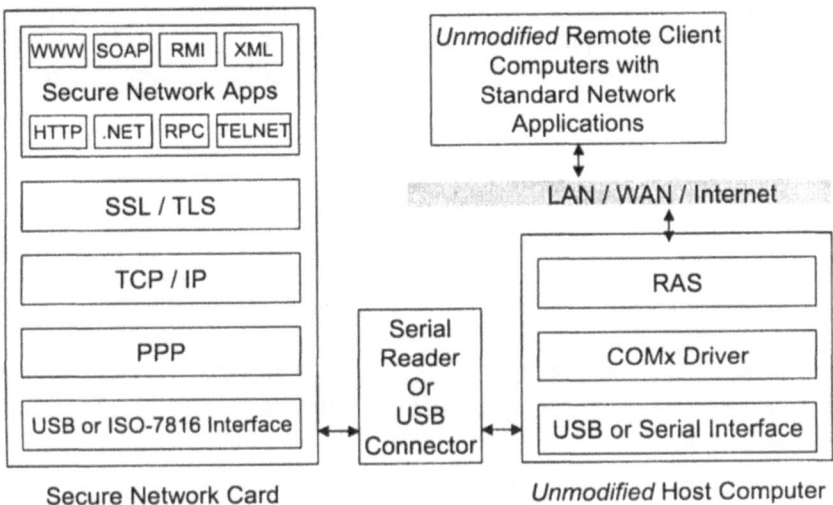

Figure 4. Secure network card architecture.

The secure network card contains a USB or ISO 7816 physical layer, a complete network stack consisting of PPP, TCP/IP, and SSL/TLS, and various network applications. If a USB interface is used, then a USB connector is used to connect to the host computer. If an ISO 7816 interface is used, a specialized smart card reader is used to convert this to full duplex serial or USB, and is connected to a serial or USB interface on the host computer.

The host computer can be any platform that is configured to permit network access from a serial or USB port. This includes most workstation, desktop, and laptop platforms including Windows, MacOS X, Linux, and Unix platforms, as well as some mobile palmtop and handset devices. In the case of Windows platforms, configuration is a simple task requiring less than 10 seconds using the New Connection Wizard (a standard utility that comes with all Windows operating systems) to specify a direct connection to another computer. The host is unaware that the computer being connected is a smart card; it treats the smart card as any other computer requesting a connection. Middleware or other smart card specific software is not required for any platform.

The host computer functions simply as a router to connect the smart card to the network, where it may be accessed by various remote computers. These remote computers are also unmodified, with no middleware or added software. A remote client or server anywhere on the network can securely communicate directly with this card using standard network applications; as far as the remote computer can tell, the smart card is just another standard computer on the Internet. No smart card specific applications are required.

4.   HOST OPTIONS

The main issue with the host was how to get from a physical layer compatible with a smart card to the TCP/IP layer. Once the TCP/IP layer was reached, standard network services were available.

The host could potentially use any physical interface. In practice, a physical interface using standard serial or USB protocols was needed, since these are the interfaces currently available on smart cards and readers.

There are several possible ways to get from serial or USB physical interfaces to the networking stack, while adhering to the requirement of no middleware or smart card specific software. Here are some options that were considered:

1. Serial → PPP (Point-to-point protocol) → RAS (Remote Access Server)
2. USB (Universal serial Bus) → encapsulated Serial → PPP → RAS
3. USB → RNDIS (Remote Network Device Interface Specification)
4. USB → Ethernet over USB → Ethernet drivers

Option 1 is standard for dial-up networking. Option 2 is the same as option 1, except that it uses an encapsulation protocol to create a virtual serial port. There are many drivers that support serial-over-USB encapsulation[1]. Option 3 using RNDIS suffers from two major drawbacks: there are no standard non-Windows implementations, and it has a heavy device footprint. Option 4 would likely be ideal, but at this time, the standard is still under development by the USB communications working group.

Options 1 and 2 were used. This required no software for the host computer. The computer need only be set up to accept an incoming connection over the appropriate serial port, a very simple task for Windows, Macintosh, Linux, and most other computing platforms.

---

[1] The FTDI driver (www.ftdichip.com), which is built into Windows, is one of many drivers that support serial encapsulation over USB.

5.   PROTOCOL STACK: PPP, TCP/IP

The secure network card implements a TCP/IP protocol [7,8,9,10] stack in order to be a standalone Internet node. Depending on the I/O characteristics of the physical connection with the host device, the card may have various link layer protocols.

As listed in section 4, Option 1, one approach to connect the card to the Internet is to connect a smart card reader to the serial port of a PC with Internet connectivity. All Windows platforms define a direct serial cable connection as one of the modem types. RAS, which is a standard part of Windows platforms, provides services that enable a dial-up device, in this case the smart card, to connect to the Internet. RAS uses PPP [11,12] to communicate with the dial-in device that acts as a client and initiates the PPP negotiation. RAS can acquire an IP address on behalf of this client via DHCP. After the PPP connection is established with the client, RAS merely acts as a router between the Internet and the client. With this approach, the smart card has its own IP address, and can act as an autonomous node on the Internet.

With a standard full-duplex serial I/O, a device can connect to the serial port (COM port) of a PC and establish connection with RAS to gain Internet access without loading any additional software on the PC. Since RAS speaks PPP, for a smart card to connect to PC via serial connection, the card needs to implement PPP in addition to TCP/IP. However, smart card communication standards (ISO 7816, or ISO 14443) specify half-duplex I/O, while PPP presumes a full duplex channel. (Current USB smart cards are also based on ISO 7816 APDUs and use a half-duplex logical protocol.)

To bridge the current gap between available half-duplex smart cards, and the required full-duplex behavior for supporting PPP, a token passing protocol was devised. This APDU-based protocol allows peer-to-peer I/O in a logical full-duplex mode, where either the card or host can initiate I/O.

A smart card reader was developed which implemented this token passing protocol with both serial and USB host interfaces. Thus, the encapsulated PPP serial data was recovered from the APDUs and presented to the host over a standard serial interface (option 1) or encapsulated over a USB interface (option 2). Ideally, the USB interface should be built into the card, so that neither the reader nor PPP nor APDUs are required (option 4). Once the Ethernet over USB standard is established, this would become the preferred standard to implement in a USB smart card.

6.   PROTOCOL STACK OPTIMIZATIONS

Smart cards have limited RAM and non-volatile memory compared to most computers running network stacks. Current cards may have up to 8K bytes of RAM, and up to 512K bytes of non-volatile memory. To fit the standard protocol stack of PPP and TCP/IP in the limited resources of a smart card, the following optimizations were done. These optimizations affect both the design and implementation of these protocols in a secure network card.

6.1   Protocol Feature Subset

To conserve memory resources, only those features of PPP and TCP/IP were implemented that are essential for making a smart card an independent node on the Internet. These features are listed below:

- The PPP layer supports dynamic IP addressing and AHDLC processing. It has LCP and IPCP finite state machines for link/network layer negotiation. However, it does not support all PPP options.

- The IP layer processes basic IP datagrams, but currently does not support fragmentation.

- The TCP layer provides reliable transmission for multiple connections. It supports PUSH and delayed ACK for interactive data flow, timeout, round trip time (RTT) measurement, and retransmission time out (RTO) computation using Jacobsen's algorithm.

6.2   Buffer Management

Prudent buffer management is an integral part of the design and implementation of standard protocol stacks like PPP and TCP/IP. Since I/O buffers pose heavy RAM requirements, it is even more critical to have an optimized design for resource-constrained devices like smart cards. Some of the key techniques used in the secure network card to optimize use of the limited RAM resources are listed below:

Chained Buffer: To allow flexibility in use of small as well as large data, a chained-buffer mechanism is used to store and process data. Similar mechanisms have been used in various BSD-style TCP/IP implementations and some embedded developments [13,14,15]. The details of chained buffer data structures vary from one approach to another, but the basic mechanisms are similar. The design used in the secure network card is based on the Packet Buffers (*pbufs*) defined in lwIP [14]. Figure 5 shows the chaining of *pbufs*.

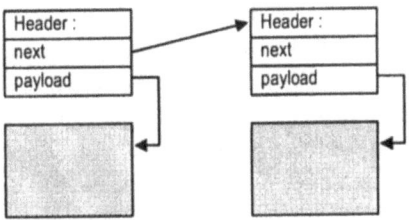

Figure 5. A pbuf chain with two pbufs.

The advantage of a chained buffer approach over fixed length buffer is conservation of memory. With a fixed buffer, memory is wasted by upfront allocation of the largest possible buffer. Most of this allocated memory lies unused during normal processing. With a chained buffer, a new buffer is allocated from a pool on an as-needed basis and then "chained" to existing buffer(s) to give a logically contiguous data array.

In a secure network card, all upper layer protocol modules, including PPP, IP, and TCP, share the *pbuf* chain allocated by the AHDLC module for input processing. During PPP negotiation, the PPP module uses *pbufs* for input and output. A *pbuf* is dynamically allocated from a pool of buffers; it is released after the input packet is processed or the output packet is sent.

AHDLC Processing: To optimize AHDLC processing, a technique of in-place handling of incoming frames is used. Escape characters in each frame are handled without allocating a separate buffer. Similarly, data can flow from the smart card hardware interface directly into application buffers without any additional copy. Figure 6 shows a typical processing of an input AHDLC frame containing a PPP LCP configuration request with no options.

Initial Input buffer

```
7E EF 7D 23 C0 21 7D 21 7D 21 7D 20 7D 24 D1 B5 7E
```

Input buffer after in-place AHDLC processing

```
EF 03 C0 21 01 01 00 04 D1 B5
```

Figure 6. Typical in-place processing of an AHDLC frame data.

Socket Interface: A traditional BSD socket interface has copy semantics. This is because the application and the system usually reside in different protection domains. Since data must be copied during a call across such domains, a socket APIs effectively doubles the memory requirement per packet. Several options have been proposed for zero-copy I/O mechanism [16,17,18,19]. We follow a similar approach where the data buffer, instead of the data itself, is transferred between the communication layer and the application layer. This design has the advantage of reduced memory usage and increased I/O performance.

7.   SSL/TLS LAYER

Secure Sockets Layer (SSL) and its successor Transport Layer Security (TLS) are the de facto standards for securing Internet web communication. Implementing SSL in a smart card can improve the card to support end-to-end network security with any unmodified client on the Internet. However, current smart cards do not support SSL implementations. This is partly due to the absence of an underlying reliable bi-directional transport layer, and partly due to the heavy memory demands imposed by the SSL protocol stack and the cryptographic computations that are required by the protocol. Description of the SSL/TLS protocol stack is outside the scope of this paper. SSL/TLS are open specifications and can be found in various books and RFCs [20,21,22,23].

The secure network card overcomes these challenges by various optimization techniques that reduce the RAM utilization of the SSL layer to less than 1.5 Kbytes. These design optimizations can be broadly divided into four categories.

7.1   SSL Feature Subset

Due to the limited resources of smart cards, the first challenge was to select a minimal feature subset from the SSL/TLS protocol specification without compromising either the specification or compatibility with existing standard clients, the mainstream web browsers. Three browsers were considered for gauging this compatibility: IE 6.0, Netscape 7.0, and Mozilla 1.5. A close examination of the SSL/TLS protocol and the selected browsers [24] led to the following decisions regarding feature subset:

*   There was little value in mixing multiple protocol versions - SSL 2.0, SSL 3.0, and TLS 1.0 - in the same implementation. Instead two separate

implementations were completed: one using TLS 1.0 that can be used for all future work, and one using SSL 2.0 that can be used for extremely low end devices without cryptographic accelerators.

- Instead of supporting multiple cipher suites, the design focused on a single one, TLS_RSA_WITH_DES_CBC_SHA that was available on all mainstream browsers. It uses RSA for authentication and key exchange, DES for encryption, and SHA-1 for digest. This design allowed a fast streamlined implementation of a single cipher suite on the smart card while still providing hooks to add additional cipher suites the in future.

7.2   Stack vs. Heap

To better manage the limited RAM resources, stack size and depth were kept to a minimum. Instead, all memory required for maintaining TLS context state and for performing cryptographic operations was allocated on the heap. This was done by a customized heap management subsystem so that buffers could be dynamically allocated and de-allocated as needed. The TLS implementation requested buffers from this heap and then freed them once the task was complete. The same RAM space could then be used for other operations. Since the TLS state machine knew exactly when it was safe to free a buffer, premature and accidental buffer release was not an issue.

7.3   Buffer Reuse

As an additional optimization of dynamic heap management, an allocated buffer is used in more than one context within the same allocation-release cycle. This eliminates the overhead of releasing a buffer and then allocating another one from the heap. The TLS implementation in the secure network card carefully uses this technique. Some examples are listed below:

- During the full-handshake phase of TLS negotiation [21], the pre-master secret and master secret values are stored in a single common buffer. Although both values are critical during the handshake, they are not used concurrently.

- While processing the Client-key-exchange message during the TLS handshake, the value of the encrypted pre-master secret is not copied to a separate buffer. Instead, it is kept in the same I/O buffer used for processing all incoming TLS records.

- When performing DES encryption and decryption, the same buffer is used for input as well as output.

While the buffer reuse technique reduces the RAM footprint in most cases, it is not viable in all scenarios. For example, during the TLS handshake process a lot more information needs to be kept in memory than the allocated RAM pool will allow. In these situations, we follow an approach that is unique to smart card development. The unused data is swapped to non- volatile memory (NVM), which is much more abundant in the smart card. The RAM buffer is reassigned to hold some other data and perform a different computation. Once this computation is complete, the saved data is reloaded from NVM and the RAM context is restored to its original state. One example of this approach occurs when performing RSA decryption during TLS handshake.

7.4   Application Interface

Following the completion of the TLS handshake, the smart card has established a set of session keys that can be used to encrypt and decrypt application data. It can now begin to securely exchange data with an unmodified client on the Internet. For a resource-constrained device, this secure application data exchange presents a unique challenge. This is due to the limited size of the receive buffer into which TLS records are written after being read from an underlying socket layer. Figure 7 explains this application level call to the TLS layer.

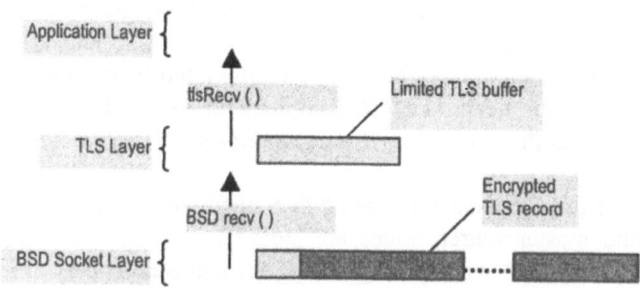

Figure 7. Reading a larger TLS record using a small TLS buffer

In a secure network card, the TLS receive buffer is set to 200 bytes. However, browsers can send TLS records of much larger sizes. Even the simplest of HTTP GET requests from a browser can be a single TLS record of 500 bytes or more. Since symmetric encryption and MAC are applied over the complete TLS record, the challenge is to use a 200-byte TLS receive buffer to process TLS records of much larger sizes. This processing involves decryption as well as verification of the MAC.

We solve this problem by using two distinct approaches, each with its own advantages.

The first approach is a *performance-critical* approach. In this approach, the TLS record is read in blocks of 200 bytes or less. As each block is read, incoming data is decrypted, and the corresponding plain text data is passed to the application layer. In addition, the MAC context and initialization vector are also updated. When the final block of data in the TLS record is read, we perform an additional step of verifying the MAC over the complete TLS record. If this MAC fails, an error is flagged. With this approach, the application layer gets data as soon as it is read, without having to pay the penalty of larger RAM buffers. However, since MAC verification is not possible until the entire TLS record has been processed, any errors in secure transmission are not flagged until we read the entire record. In several applications, this slight delay in receiving a MAC error is acceptable, particularly if this behavior is requested to improve performance.

The second approach is the *error-critical* approach and is illustrated in Figure 8. In this approach, an application can mandate that no application level data be passed to it unless the MAC has been verified. To achieve this behavior we successively read the TLS record, in chunks of 200 bytes, and write it to NVM. Once the complete TLS record is read, we use the NVM buffer to decrypt data and then verify the MAC. If the MAC succeeds, the requested amount of data is passed to the application. On subsequent read calls, the remaining data in NVM is passed directly to the application without any need for decryption or MAC verification.

In this approach, the first application level read call is slow, but subsequent calls for data in the same TLS record are much faster. Overall, this provides a more secure application interface.

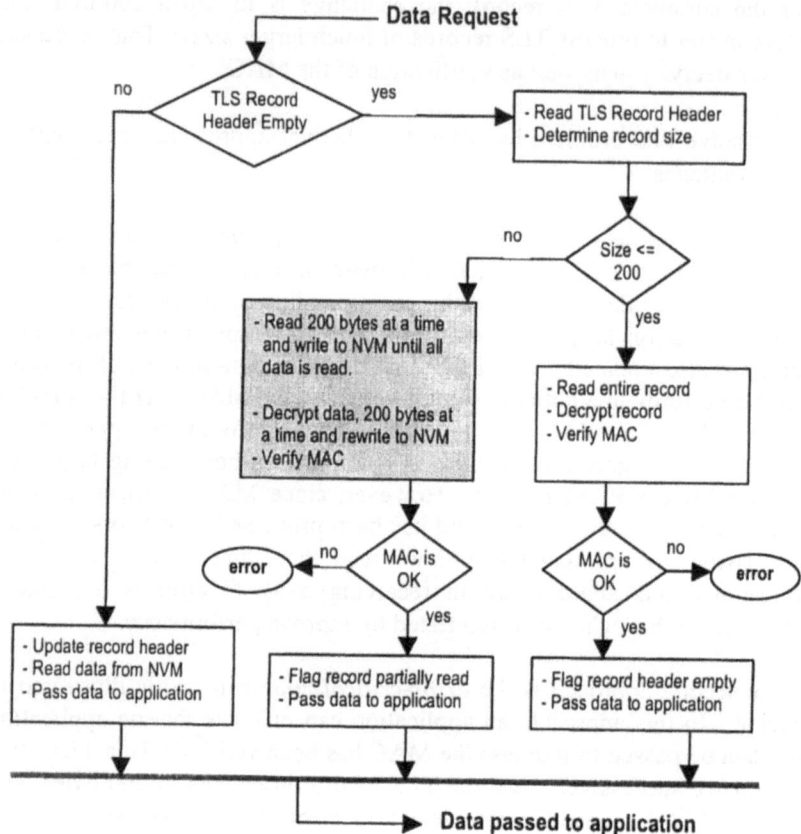

Figure 8. Error-critical design of reading TLS Record with limited I/O buffer.

8. PROTOTYPE AND APPLICATIONS

This project was originally implemented as a Windows simulation and an ARM smart card simulation. SSL/TLS and several applications were developed and simulated. For the prototype card, Samsung Jumbo was chosen to avoid masking. This had the drawback of having no crypto processor; establishing a TLS connection on the prototype took about 15 seconds. So SSL 2.0 with a shorter key length was implemented, to reduce the connection time to less than 1 second. With either TLS or SSL, once the connection was established, the performance was acceptable: pages would load across the network without significant lag time.

Performance varied significantly between the various browsers tested: Internet Explorer, Netscape, and Mozilla. Mozilla offered the best overall performance. After a production card is completed, detailed performance benchmarks will be run and published.

Applications loaded in the prototype cards included a web server/agent, Telnet server, and a mini-shell. Many services were loaded into the card that were accessible via HTTP, including a secure stock trading service, a secure email/encryption service, a secure e-commerce service, and a secure ticket service. Total NVM footprint was 76K Code and 34K Data. This left 412K available for other applications and services to be loaded. The RAM footprint was carefully optimized to just under the 6K available. These applications/services were demonstrated at Cartes 2003 and CT/ST 2004.

The smart card SSL/TLS library provides a simple application level interface that can be used by on-card applications to establish secure end-to-end network connections with any remote unmodified clients on the Internet. New application frameworks such as .NET, SOAP, and RMI can be easily added to enrich the application versatility.

The secure web server implemented in the prototype card can serve both static and dynamic HTML content. Static content is supported by reading the requested file from the card file system. Dynamic content is supported by invoking the requested application through a CGI interface and redirecting the results to the browser. The mini-shell is accessible through Telnet or HTTP, providing a very powerful way of interacting with the card.

9. CONCLUSIONS

The technology presented in this paper enables smart cards to participate as first-class citizens on the Internet within the established infrastructure, with no host or remote application changes required to accommodate smart cards. With this technology, smart cards are like any other computer on the Internet, while providing portability, enhanced security, and tamper resistance. Internet applications or services can be migrated to a smart card, increasing the security of critical information such as certificates, private keys, and passwords, while maintaining compatibility with existing clients. With smart card deployment unshackled from the infrastructure, the barrier to entry for smart cards in mainstream applications is removed. We foresee this technology triggering an unprecedented growth in deployment of smart card applications for the Internet.

REFERENCES

1. Rees, J., and Honeyman, P. "Webcard: a Java Card web server," Proc. IFIP CARDIS 2000, Bristol, UK, September 2000.
2. Urien, P. "Internet Card, a smart card as a true Internet node," Computer Communication, volume 23, issue 17, October 2000.
3. Guthery, S., Kehr, R., and Posegga, J. "How to turn a GSM SIM into a web server," Proc. IFIP CARDIS 2000, Bristol, UK, September 2000.
4. Muller, C. and Deschamps, E. "Smart cards as first-class network citizens," 4th Gemplus Developer Conference, Singapore, November 2002.
5. Itoi, N., Fukuzawa, T., and Honeyman, P. "Secure Internet Smartcards," Proc. Java on Smart Cards: Programming and Security, Cannes, France, September 2000.
6. Urien, P. "Internet smartcard benefits for Internet security issues," Campus-Wide Information Systems, Volume 20, Number 3, 2003, pp. 105-114.
7. Postel, J. "Internet Protocol," RFC 791, September 1981.
8. Postel, J. "Transmission Control Protocol," RFC 793, September 1981.
9. Socolofsky, T. "A TCP/IP Tutorial," RFC 1180, January 1991.
10. Almquist, P. "Type of Service in the Internet Protocol Suite," RFC 1349, July 1992.
11. Simpson, W. "The Point-to-Point" Protocol (PPP)," RFC 1661, July 1994.
12. Carlson, J. "PPP Design, Implementation, and Debugging," second edition, Addison-Wesley, 2000.
13. Wright, G.R. and Stevens, W.R. "TCP/IP Illustrated, Volume 2," Addison-Wesley professional Computing Series, 1995.
14. Dunkels, A. "lwIP – A Lightweight TCP/IP Stack." More details are available at http://www.sics.se/~adam/lwip/ .
15. Lancaster, G., et al. uC/IP (pronounced as meu-kip) is an open source project to develop TCP/IP protocol stack for microcontroller. It is based on BSD code. For details, see http://ucip.sourceforge.net/ .
16. Chihaia, I. "Message Passing for Gigabite/s Networks with Zero-Copy under Linux," Diploma Thesis Summer 1999, ETH Zurich.
17. Pai, V.S. and Druschel, P. and Zwaenepoel, W. "IO-Lite: A Unified I/O Buffering and Caching System," Rice University.
18. Thadani, M. N. and Khalidi, Y.A. "An Efficient Zero-Copy I/O Framework for Unix," SMLI TR-95-39.
19. Abbott, M., and Peterson, L. "Increasing network throughput by integrating protocol layers," IEEE/ACM Transactions on Networking, 1(5):600-610, October 1993.
20. Freier, Alan O., et al. "The SSL Protocol, Version 3.0," Internet Draft, November 18, 1996. Also see the following Netscape URL: http://wp.netscape.com/eng/ssl3/ .
21. Dierks, T., Allen, C., "The TLS Protocol, Version 1.0," IETF Network Working Group. RFC 2246. See http://www.ietf.org/rfc/rfc2246.txt .
22. Elgamal, et al. August 12, 1997, "Secure socket layer application program apparatus and method." United States Patent 5,657,390.
23. Rescorla, E., SSL and TLS, "Designing and Building Secure Systems," 2001 Addison-Wesley. ISBN 0-201-61598-3.
24. Goldberg, I., and Wagner D., "Randomness and the Netscape Browser," Dr. Dobbs Journal, January 1996.

A PATTERN ORIENTED LIGHTWEIGHT MIDDLEWARE FOR SMARTCARDS

J-M. Douin
CEDRIC-CNAM
292 rue St Martin
75141 Paris cedex 03 France
douin@cnam.fr

J-M. Gilliot
GET - ENST Bretagne
Technopole Brest-Iroise - CS 83818
29238 BREST cedex 3 France
jm.gilliot@enst-bretagne.fr

Abstract   Smartcards are a very interesting means to include our own datas and code in a distributed system, during our interaction with it. To achieve this, smartcards integration must be ensured. A transparent usage of card services is necessary to a more wide-spread use. This usage should be available remotely within a distributed environment. Additionally other features such as possible upgrades of code or addition of new services, notification of connections and disconnections, structuration of numerous smartcards are key requirements.

Our middleware is described in three steps. First, we describe mechanims to turn Java Card Applets into webservices and show how to implement them in a lightweight infrastructure. Secondly, a mechanism, based on a collaboration pattern to become JavaCards active and to permit spontaneous discovery of services. Thirdly, we define a structuration of numerous JavaCards and Java Card Applets, as services dynamically available, to give opportunity to access and manage them easily.

1.   Introduction

Smartcards are a very interesting means to include our own datas and code in a distributed system, during our interaction with it. To achieve this, Smartcards integration must be ensured. As noticed in many papers (see for example (Kehr et al., 2000a)), the basic use of Java Card suffers from its specificities to access applets. A transparent usage of card services is needed for a more widespread use. This usage should be available remotely within a distributed environment and with minimal effort. Additionally other features such as possible upgrades of code or addition of new services, notification of connections and disconnections, structuration of numerous smartcards are key requirements. We propose in this paper an approach to develop infrastructure or middleware able to undertake those requirements. Flexibility is a central point with smartcards, new services may be added or removed at run-time, as well as connections and disconnections. We have discussed in (Douin and Gilliot, 2003) main characteristics of embedded devices, we will here focus on the extensions that are specific to smartcards and include them in our framework.

Our approach is to give interoperability to smartcards and to provide at the same time a lightweight infrastructure. This infrastructure is based on Java as card manufacturers chose to migrate to it in smart cards to achieve cross-platform compatibility. Our first choice was to adopt as collaboration schemes, the basic patterns proposed in litterature (Gamma et al., 1994). Those patterns are sufficient to implement various schemes, such as polling or notification. The second choice was to use a minimal transport layer for communications. HTTP is now widely adopted, and it is current that even a small device implements a web server. Last choice was to make an adaptation to web at the hosting device level and not at Java Card level as proposed in (Rees and Honeyman, 1999).

In a first section, we define the key design concepts and the possible directions to develop such middleware. Main points concern: full interoperability and integration by making Java Card Applets accessible in an unified way, lightweight approach to be compatible with every device that accepts smartcards. Then, in the next section, we describe mechanims to turn Java Card Applets into webservices and show how to implement them in a lightweight infrastructure. In section 4, we propose a mechanism, based on a collaboration pattern to make JavaCards active and to permit spontaneous discovery of services. Last section is dedicated to the structuration of numerous Java Cards and Java Card Applets, to give opportunity to access and manage them dynamically.

2.   Design considerations

The development of a distributed embedded solution imposes to make some tradeoffs. We hypothetize that resources are minimal and that portability has to be ensured. Scarce resources impose a lightweight approach on the Java Card side but also on the hosting device side.

2.1   Basic architecture

A Java Card, to be accessible, must be hosted on an acceptance or hosting device. The functionality can be provided by other appliances such as rings (DS1957B, 2004) or GSM SIM wich look different but provide the same features. We used them as classical smart cards. Each card can contain some Java Card Applets or applets, that are pieces of executable code on the JavaCard, provided it is called with a specific protocol named APDU.

On an acceptance device, different Java Cards can be connected. We will assume in the sequel of this paper that the hosting device is connected to the Internet. We don't suppose any specific kind of connection. The hosting device can be of course a PC, but we propose that this device can be any kind of smart object, or embedded device. This is resumed in figure 1.

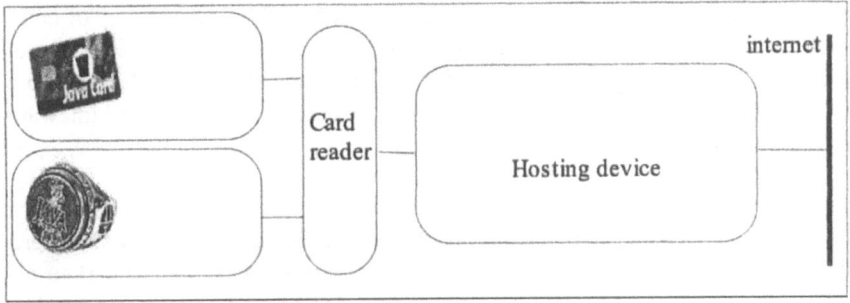

Figure 1.   Physical architecture

2.2   Possible approaches

TCP/IP in Smart Card.   A first choice is to provide directly in the smart card the basic TCP/IP communication standard to enable any network development (Muller and Deschamps, 2002). The purpose is to give full interoperability, adaptability and autonomy to the Smart Card. However, this approach is limited to basic connected communications, and do not propose extensible services for distributed integration.

WebServer in Smart Card.   A second choice is the support of the HTTP protocol. Some authors have proposed specific webserver on the SmartCard (Rees and Honeyman, 1999),(Guthery et al., 2000). This simplifies the support of basic communications and provides real possibility of integration as the Card becomes a full address on the Internet and thus can be accessed remotely. But still, the approach is limited to the low computing power of the smartcard, which do not allow additional features. We can also notice that some works propose an opposite approach attempting to externalize even the information dedicated to a smartcard (Biget, 1999).

RMI approach.   JavaCard RMI was first proposed in (Vandewalle and Vétillard, 1998), and is accepted as a standard (Inc., 2002). This give remote access to JavaCard, but only as synchronous call. As we thought it doesn't ensure open interoperability to non Java applications.

Proxy Server in Card Acceptance Device.   As an acceptance device is mandatory for use of the Smart Card and the event of the connection to the device has to be acknowledged. So it is obvious to propose some software on the acceptance device side.

A first step is to propose a translation from the APDU standard to a more standard protocol. This ensures the encapsulation of this specific way of communication. Providing this server as a web server allows to take advantage of the interoperability at HTTP level which is clearly as open as TCP/IP, and more readable. We will see how to extend such server to provide full middleware requirements.

MOM .   Message-oriented middleware provides the right abstractions, such as discovery, asynchronous messaging and so on. However, the use of standard middlewares like JMS (Donsez et al., 2001) or Jini (Kehr et al., 2000a) suffer from extensive needs of resources on the Card Acceptance Device. As emphasized in (Vogt et al., 2004), the major requirements for smart card middleware are encapsulation of specific communications, interoperability, and the integration of the systems. In the sequel of this article, we propose a such middleware approach but based on a lightweight implementation enabling its execution in most embedded connected devices.

2.3   Mobility of Cards

Cards are mobile devices : the card may connect and disconnect anytime. As they are highly portable, they are carried by their owners anywhere they go. Those user may wish to connect and disconnect, anytime, anywhere, to allow the use of features offered by SmartCard. This implies some asumptions that should be taken into account :

- On arbitrary connections means that should function spontaneously and be integrated with the surrounding distributed environment as soon as it is inserted in a reader.

- Anytime disconnections means that one cannot rely on connected communications. As stated in (Mascolo et al., 2002), this implies that connections are asynchronous.

- Localization is another important feature. But, as the Java Card must be connected to a hosting device, it can be delegated to it.

2.4 Hosting device is a web server

Providing a web server as an URL programming interface (Giorgio, 1999) has already been proposed in the context of smartcard for payment applications (Barber, 1999). As we already noticed in (Douin and Gilliot, 2003), this approach can be embedded with relatively small web servers like (Brazil, 2000). Extending URL programming enables, as we will see, many interesting features for smart cards. We chose generic and embeddable web servers available (Giorgio, 2000).

The Java language proposes standard APIs for patterns (Observer, ...), thus we adapt them to our implementation choices. Let's notice that with the same pattern abstraction, we can propose different implementations with different behaviors as described in (Douin and Gilliot, 2003).

2.5 Abstractions for the system

In order to allow an uniform manipulation of the different elements of a JavaCards based system, we define them as objects. We use patterns to handle dynamics in the system and to be able to manage the scalability. We use Web connections in an uniform way. As a consequence, applets are presented as servlets which hides specificities like APDU protocol. Hence generality can be achieved in accordance with modern protocols and service descriptions (Donsez et al., 2001). We have chosen to focus on functionalities and how to achieve them with scarce resources, both at the Java Card and hosting device levels. Notice that abstraction also permits to vary the implementation transparently in order to take into account non functional requirements or to improve performance.

3. Accessing and Managing remotely a JavaCard

As JavaCard must be connected to a device to be able to compute, we argue that this device may play the role of a web proxy for the JavaCard. Then this device translates a HTTP request into APDUs and retransmit the response. Hence, we decouple the transmission part from the application part included in the JavaCard.

3.1 Abstract the JavaCard

APDUS are the standard access for smartcards. Their use is very specific, hence we propose the use of web servers as wrappers. Connections of the smartcard to a host initiate the web server as proxy for datas and Java Card Applets are proposed as servlets. Cards are then presented as a collection of services. Java Card Applet being those services available individually.

(Donsez et al., 2001) propose to provide access to those services with the standard models of communications, namely MOM for asynchronous messaging, SOAP as protocol, UDDI for discovery and WSDL for service description. This would allow a full integration in the Information System.

Another direction is to integrate Java Card in a Jini environment thanks to specific surrogate servers on the hosting device, as proposed in (Kehr et al., 2000b).

We have chosen to investigate more minimal protocols to make sure that a minimal web server is sufficient to ensure an interface to connected Java Cards. Hence we handle simple URL and provide dedicated definitions. An access to a Java Card Applet (naming it as cardlet) with GET method will then simply be something like : `http://server/JavaCardId/cardletId/?cardletParam=value`

With this simple feature, the embedded Java Card Applet become a *servlet* of the hosting device. To realize this, a simple http web server, is needed on the hosting device. A specific translation in APDU calls is provided through handlers. In the rest of the paper we show that it is sufficient to provide a dynamic management of numerous connected Javadards in a system.

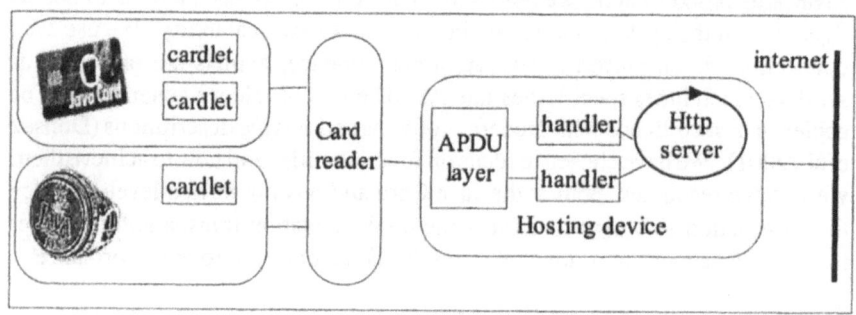

Figure 2.   Basic software architecture to abstract Java Card

3.2   Access to Java Card Applets

The first step to access carldets as servlets is to provide a basic description of the different Java Card Applets and their parameters. This description should be integrated in the Java Card Applet itself, as a signature, giving syntaxic call parameters and possibly the version. We have chosen to provide it with a definition compliant to the HTML forms definition in a compact format. A signature request will look like `http://server/JavaCardId/cardletId/?sig=0` and will return an answer such as `5TaText20S`, interpreted as a parameter named `aText` of type `Text` with a 20 character size. This basic signature offers a compression factor of 80 % essential for scarce resource.

It also gives the opportunity to provide more advanced services :

- Discovery of Java Card Applet description and corresponding construction of dynamic web pages or dynamic use on the client side. As our hypothesis is that the server can be embedded on arbitrary small device, it is obviously a coherent feature.

- Construction of a dynamic web page on the server side enabling the use of the Java Card Applet. Brazil (Brazil, 2000) provides a lightweight mechanism called BSL, offering comparable basic features like the JSP (Java server Pages) technology. The template of the page can even be embedded in the JavaCard. In this way the use and the construction of a web page access is provided by the card itself and adapted to the hosting device.

- Construction of description in a more widely used language just as WSDL or Jini discovery mechanism is possible and can be delegated to a device playing the role of gateway.

3.3   Mutual exclusion

Multiple clients can address to the same URL i.e. to the same servlet, this fact requires for some applications a mutual exclusion mechanism (Kehr et al., 2000b). This is achieved by addressing the URL with the parameter beginMutex to enter in critical section and to exit from this section with the parameter end-Mutex. The host computer provides this mutual exclusion between distributed clients of smartcard's servlet. These two URL are an example of use:

- `http://server/JavaCardId/cardletId/?beginMutex`

- `http://server/JavaCardId/cardletId/?endMutex`

In our current implementation, on the host computer, a client is identified by its cookie and a lock mechanism is associated to a servlet. If a client uses or decides to block a URL for its use, a second client can wait with or without a timeout. A possible problem is that a client can block indefinitely an URL. In our implementation presented in this paper, we chose to ignore this problem. There are solutions, with other problems, one of these consists in using a fixed maximum amount of time for a client or the client installs in advance a latency time.

3.4   Dynamic reconfiguration of the JavaCard

Java Card will be owned by users who don't want to worry about code evolution, or technical ways to adapt services or options on their javacard. A basic distant mechanism can be provided to update Java Card Applets remotely. It is just an adaptation of APDU commands at the HTTP level.

3.5   Security considerations

Quoting Baber (Barber, 1999) : " Responsibility for security now lies in the card server, not the browser. The card server can implement the appropriate model for its intended use, with the Java class library providing enough functionality to implement mechanisms such as code signing, access control lists and so on." In other words security should be possible in our infrastructure, but we didn't explore specifically this issue specifically.

3.6   Summary

Connecting the Java Card Applets on the web has reifed them as distributed objects which can be called like servlets, identified by checking their parameters and updated. As these services are interfaced in accordance to HTTP protocol they provide the widest interoperability with other systems.

4.   Distributed applications with nomadic Cards

In this section, we will discuss the adaptation of the Observer pattern in order to solve the plug and play requirement for an easy use of Java Cards. We will show that implementing this pattern at the HTTP level gives the opportunity to acknowledge events such as connections and disconnections of a Java Card in a distributed infrastructure.

4.1   Handling connections and disconnections

As smartcards users just come and go, connections and disconnections are normal events of the system. Those events may be used by applications for specific functions such as acknowledging the presence of users in a room.....
Additionnaly, on connections, the content of the smartcard has to be synchronized with the content of the different groups it belongs to.

Disconnections may represent different kinds of events. It may be intentional, meaning that the owner of the card is leaving the place. Or it may be unintentional in case of resource problems (connection, battery exhausted ...) as it is common in mobile context (Chen and Kotz, 2000)

4.2   Making JavaCard active on the Web

As Cards may come and go, one must provide a simple asynchronous service to give the opportunity to other services to be aware of the connection or the disconnection of JavaCards. This may be achieved by giving the opportunity to the device hosting the card to inform the potentially interested services of those events. Those services may vary in time. Thus, they need to be able to suscribe and to be acknowledged of date modifications. Moreover, we want to provide a distant access to the card, to make datas and Java Card Applets available. This is exactly the definition of the Observer pattern. This pattern or its extension, the Model-View-Controller pattern (Gamma et al., 1994) is based on logically asynchronous acknowledgement of model modification to Observers or Views. Those Observers can be added or removed at anytime. At the Internet level, Observers are translated as URLs to which the updates can be sent directly (see (Douin and Gilliot, 2003) for details). Controllers and Models can also be located on different hosts as shown in figure 3.

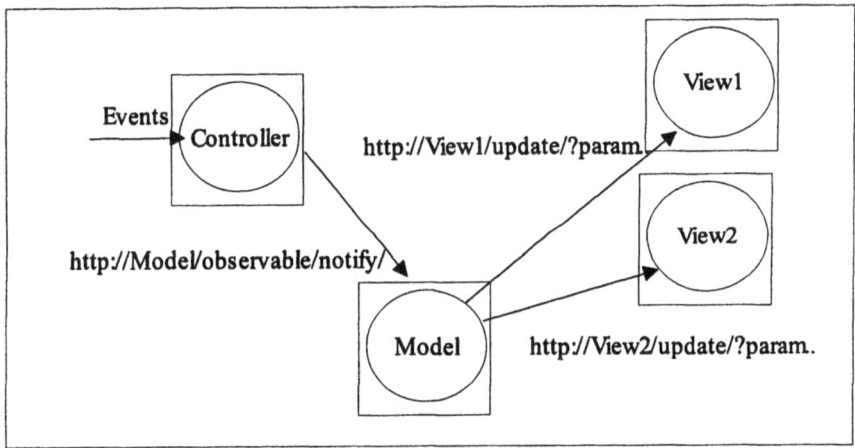

Figure 3.   Basic collaboration for MVC

We have defined the following basic Views :

- At a first level, we propose to maintain a local list of servlets available on a hosting device, generally located on the device itself.

- A discovery service can be defined as a View of a set of machines, to propose servlets available on fly.

- By adding a card reader on every workstation in a classroom, defined as controllers, a model of all those machines can be aware of students whose JavaCard are currently connected. As an illustrative example, we have included a web server in an applet which can visualize all events from a Model as soon it is declared as View of this model. Because in our system, every device is a web server this feature is possible as long as the standard java applet is downloaded from the server of the Model to a traditional browser. A screenshot of this applet is shown in figure 4.

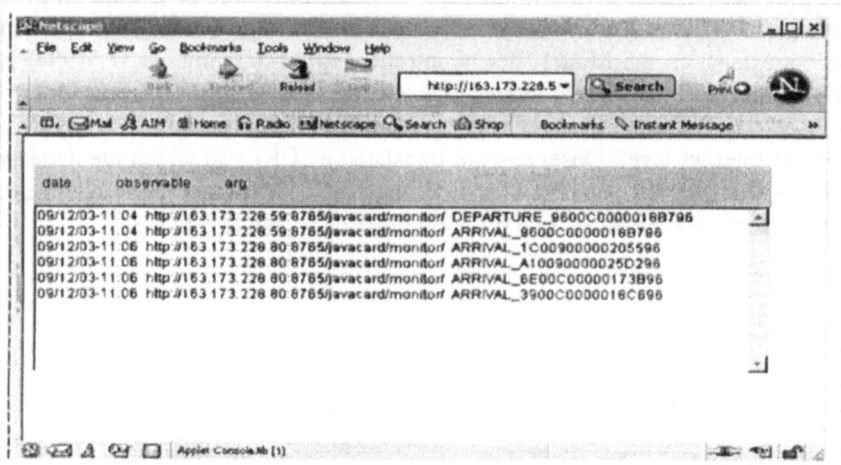

Figure 4.   An applet as a View

Information of different servlets available on a device is published to every application listening to a Model. This way any reactive behavior (the Plug and Play effect) can be implemented in a distributed way. A collaboration diagram explaining the diffusion of events is given in figure 5.

This scheme is simple and lightweight to develop in a web environment as we explained in (Douin and Gilliot, 2003). This could be adapted to a full Publish and Subscribe communication scheme (Eugster et al., 2003) if necessary.

5.   Applying Design Patterns to manage numerous devices

Usual applications are based on a few servers delivering datas or computation to many clients. When managing sensor or embedded device networks, we face many data providers to a possibly limited set of services/applications. Moreover, different schemes of communications must be handled, Client/Server is one of them, but it is useful to provide event-based communications.

In the previous section, we have proposed as a first step to install a MVC scheme in order to handle the event notification of asynchronous events. Yet, JavaCards and Java Card Applets still need to be organized to enable the management. For this, one need to adress logically synchronously sets of JavaCard or Java Card Applets. This is the purpose of the second pattern implementation presented here.

In this section, we will detail the use of the coupling of Composite and Visitor patterns to demonstrate how to define a structure for the system and a flexible way to cross the structure and enhance the collaboration between elements.

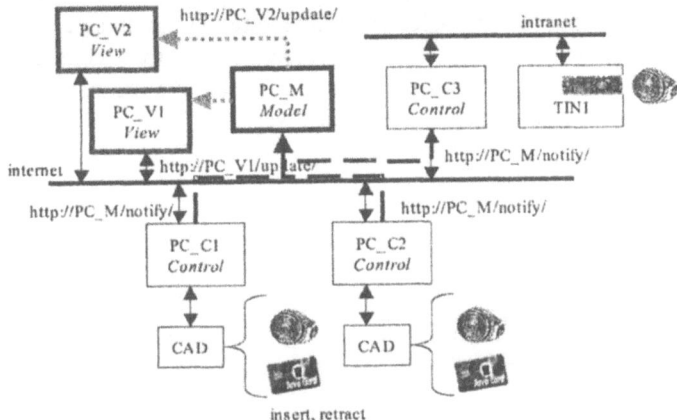

Figure 5. functional handling of connections

5.1 Structure your JavaCards and Java Card Applets: the Composite pattern

In order to structure our objects, we have to take into account the following points :

- exhibit the physical structure, such as the fact that a Java Card embed a set of Java Card Applets or that a device can connect several JavaCards at the same time

- enable logical grouping such as all Java Card Applets corresponding to a specific service, like each student's schedule of lecture

- give the possibility to add new basic or structural elements

- define the structure hierarchically as it is obvious in the first point

The Composite pattern is the classical hierarchical answer for the design of such structure. Our proposition in (Douin and Gilliot, 2003) is a distributed composite implementation wih URLs as references for the objects constituting the structure. This enables full and lightweight interoperability with any connected device. In figure 6, we propose a composite structure adapted to Java Card, the left side of composite diagram exhibits the different basic elements constituting the system. Notice that a new element may be added by specializing exisiting elements to take into account new products. The specialization can be carried out on three main directions, either by taking into account the nature of the service (agenda modification, quotation, . . .) or the specificity of the device or by taking into account the selected architecture. The right side of the diagram, shows the possible combination of elements to construct coherent structures. Structures can be mixed to meet application needs. Notice that in this structure, a SmartCard is a collection of Java Card Applets.

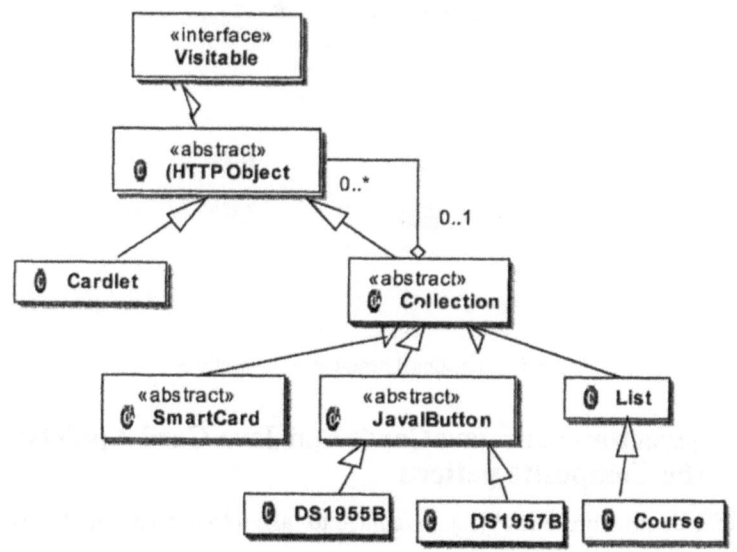

Figure 6.   Composite Structure for Smart cards

This means a more systematic way to aggregate different devices in an application.

5.2   Accessing elements of the structure: the Visitor pattern

To be useful, a structure like a Composite structure should be crossable in a systematic and uniform way. This is the goal of Visitor pattern. The purpose is to develop a visitor whose behavior is adapted to each object visited and ensures to visit every element of a structure. Hence it is possible to visit composite objects such as previous collections or other structures and final element in an uniform way. The Composite and Visitor patterns couple can be seen as an extension of Collection and Iterator patterns, just like a tree is an extension of an array.

In Fig. 7, we show a possible hierarchy of visitors, one being able to print the list of the elements in a console, the other gathering status datas in a html format. Services can then easily be extended by adding new Visitor classes. For example, we have implemented several visitors, one of them checks the firmware version of javacards, another allows information from the configuration of the moment , yet another installs a new version of a servlet, ...

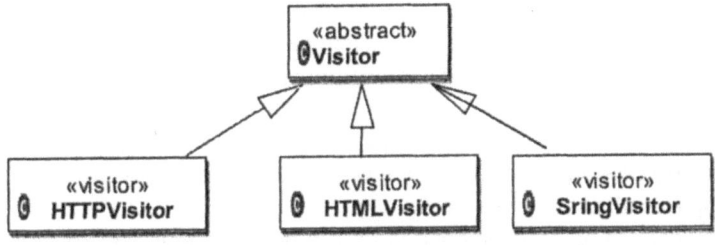

Figure 7.   Visitors of Smart cards Composite

5.3   Managing the structure

As JavaCards may come and go, as well as environment may vary in time, different mechanisms must be installed to maintain the correctness of the structure. In a first approach, we have developped two basic complementary mechanisms :

- Controllers of the MVC scheme collect informations of upcoming Java Cards. Structures can be defined either as models of the MVC (for example the local collections of Java Cards and corresponding Java Card Applets present on the device), and being directly updated, or as views (for example the room or the building containing the devices). Hence, in this configuration it is sufficient to install a view on a classroom structure to be aware of students currently connected and then to access to each of them. Active disconnections of JavaCards can be propagated the same way.

- As passive disconnections, possibly transient, may happen silently, we have defined a visitor who polls the structure in order to check if declared Java Card Applets/servlets are still currently connected. Different policies can be easily defined to remove definitely or not, information from the structure (checking them specifically for example to treat recovery).

6.   Conclusion

In this paper, we have motivated the need for a lightweight integration of smartcards into distributed systems. We have shown that a loose coupling system with HTTP connections and small webservers at hosting level can be extended using basic design patterns as key for design. Thanks to this solution, we have provided a plug and play mechanism as well as asynchronous and asynchronous collaborations. This is achieved in a structured vision of host and JavaCards managed.

Basic examples of potentiality have been exposed and could be adapted to other applications. For example, deployement of new services or other down-loadings or updates are possible in a systematic way.

An on-line prototype is available at the following adress : `http://vivaldi.cnam.fr:8765/javacard/monitor/C100900000126496/Cardis04/`, where `http://vivaldi.cnam.fr:8765/javacard/monitor` is the url for the server, `C100900000126496` the identity of the JavaCard and `Cardis04` is a servlet/Java Card Applet embedded in the card.

Our model can be easily supported by any device able to embed a minimal Web server, thus being deployable in most infrastructures. As communications are based on HTTP protocol, it can interoperate with any other platform. For example, interaction with JiniCard and use of kSOAP are under investigation.

References

Barber, J. (1999). The smart card url programming interface. In *1st Gemplus Developer Conference, Paris, France*.

Biget, P. (1999). How smarcards can benefit from internet technologies to break memory constraints. In *1st Gemplus Developer Conference, Paris, France*.

Brazil (2000). Brazil documentation and api, http://www.experimentalstuff.com.

Chen, G. and Kotz, D. (2000). A survey of context-aware mobile computing research. Technical Report TR2000-381, Dept. of Computer Science, Dartmouth College.

Donsez, D., Jean, S., Lecomte, S., and Thomas, O. (2001). (a)synchronous use of smart cards services using soap and jms. In *Gemplus Developer Conference*.

Douin, J. and Gilliot, J. (2003). Collaboration patterns for networked embedded servers. Lisbon, Portugal. ETFA2003 9th IEEE International Conference on Emerging Technologies and Factory Automation.

DS1957B (2004). Java powered ibutton.

Eugster, P. T., Felber, P. A., Guerraoui, R., and Kermarrec, A. (2003). The many faces of publish/subscribe. *ACM Computing Surveys (CSUR)*, 35(2):114–131.

Gamma, E., Helm, R., Johnson, R., and Vlissides, J. (1994). *Design Patterns : Elements of Reusable Object-Oriented Software*. Addison Wesley.

Giorgio, R. D. (1999). An introduction to the url programming interface. *JavaWorld*.

Giorgio, R. D. (2000). Serve clients' specific protocol requirements with brazil, part 1-4. *JavaWorld*.

Guthery, S., Kehr, R., and J., P. (2000). How to turn a gsm sim into a web server. In *CARDIS 2000, Bristol*. IFIP.

Inc., S. M., editor (2002). *Java Card 2.2 Runtime Environment (JCRE) Specification, chapter 8: Remote Method Invocation Service*, chapter 8, pages 53–68.

Kehr, R., Rohs, M., and Vogt, H. (2000a). *Issues in Smartcard Middleware*, chapter Java on Smart Cards: Programming and Security., pages 90–97. Springer-Verlag LNCS 2041.

Kehr, R., Rohs, M., and Vogt, H. (2000b). Mobile code as an enabling technology for service-oriented smartcard middleware. In *Proc. 2nd International Symposium on Distributed Objects and Applications DOA'2000, Antwerp, Belgium, IEEE Computer Society*, pages 119–130.

Mascolo, C., Capra, L., and Emmerich, W. (2002). Mobile computing middleware. *Lecture Notes in Computer Science*, 2497:20–??

Muller, C. and Deschamps, E. (2002). Smart cards as fist-class network citizens. In *4th Gemplus Developer Conference, Singapore*.

Rees, J. and Honeyman, P. (1999). Webcard: a java card web server. In *CARDIS 2000, Bristol*. IFIP.

Vandewalle, J.-J. and Vétillard, E. (1998). Smart card-based applications using java card. In *Proceedings of the 3rd Smart Card Research and Advanced Application Conference (CARDIS 98)*.

Vogt, H., Rohs, M., and Kilian-Kehr, R. (2004). *Middleware for Communications*, chapter 16, Middleware for Smart Cards. John Wiley and Sons.

CARD-CENTRIC FRAMEWORK - PROVIDING I/O RESOURCES FOR SMART CARDS

Pak-Kee Chan, Chiu-Sing Choy, Cheong-Fat Chan, and Kong-Pang Pun
Department of Electronic Engineering, The Chinese University of Hong Kong
{pkchan, cschoy, cfchan, kppun}@ee.cuhk.edu.hk

Abstract: The Intelligent Adjunct (IA) model, proposed by Balacheff et al in [1], is a novel paradigm for smart card applications, in which off-card resources are provided for smart cards to run application logics. This paper presents the Card-Centric Framework as an evolution of the conceptual IA model to provide a more rigorous solution for smart cards to access off-card I/O resources. It consists of a system model and a communication protocol. With the Card-Centric Framework, smart cards can run any applications that involve user interactions. A system prototype constructed with the current smart card technologies showed only reasonable performance. To cater for performance issues, another demo system that made use of enhanced smart card technologies was implemented. It not only shows a significant improvement in performance, but also proves the feasibility of the framework in the future.

Key words: Card-Centric Framework, Console, I/O, middleware, protocol, smart card, system, user interaction

1.   INTRODUCTION

Smart card has a tamper-proof property that makes it an ideal authentication device of the connected terminal. However, it is often equipped with limited processing power that is just enough for security usage. This has been, and will be, limiting smart card's potential of use in the conventional client-server security model, in which smart card is just a slave of the connected terminal. With the continual advancement in microelectronics technologies, the processing power of smart card is being enhanced, and a sufficient condition is available for a change in smart card application model.

In [1], Balacheff et al proposed the Intelligent Adjunct (IA) model to replace the traditional client-server topology by a peer-to-peer one. The model involves two entities: smart card and the connected terminal. The smart card acts as an independent processor, whereas the terminal provides essential resources for use by smart card. They cooperate with each other to run applications, which may involve the use of off-card I/O resources for user interaction.

However, the details of implementations were not clearly defined in [1].

Therefore, we move from concept to implementation, and propose the Card-Centric Framework as an evolution of the IA model. To allow for smart cards to handle off-card resources, we introduce an entity called Console to provide smart card with connected I/O devices. With the framework, any applications that require basic user interactions could be run on smart cards, thus the limitation of smart card as a security device could be relieved.

The rest of this paper is organized as follows. The next section provides a brief image of the Card-Centric Framework. The framework consists of a system model and a communication protocol, which will be described in sections 3 and 4 respectively. Section 5 demonstrates a system prototype based on the framework. Due to limitations of existing smart card infrastructures, it showed only reasonable results. To cater for performance issues, we employ the technologies predicted by RESET roadmap [2] and implement an improved system, which will be described in section 6. Section 7 discusses security issues in employing the framework. The last section sums up with a brief conclusion.

2.   CARD-CENTRIC FRAMEWORK

The Card-Centric Framework adopts some of the principles from the Intelligent Adjunct model [1], including:

a) Migration of application logic from the card-connected terminal to the smart card;
b) Adoption of active role by smart card, and;
c) Provision of terminal and network resources for smart card.

It is designed for smart cards to handle off card I/O resources. It describes a more rigid guideline for implementation, including a system model and a communication protocol.

The system model, which conforms to the criteria of middleware for system integration in RESET roadmap [3], consists of two entities: Smart Card and Console. Smart Card, which is interpreted as IA in [1], runs all the application logic. Console is the essential portion of the card-connected terminal that provides Smart Card with off-card I/O and network resource entities.

The resource entities are objects to be handled by the smart card. The smart card not only sends out manipulation commands for them, but also awaits interrupts from them when events arise. To achieve this, an object-oriented protocol, titled Card-Centric Protocol (CCP), is proposed for the communication between Smart Card and Console.

Details of the system model will be described in section 3, whereas the protocol in section 4.

Under the framework, software developers can employ user interactions through I/O resources, and thus enhance user-friendliness. Provided that both the smart card and the card-connected terminal support the Card-Centric Protocol, they can develop any applications to run on the smart card. Consider the application of IA as an example. It is originally proposed by [1] for users to configure user-automated tasks by means of programming. However, smart card owners are often not capable of programming works. With the Card-Centric Framework, the users can get rid of coding and input the tasks through user-friendly interfaces such as display, sound, mouse, etc.

Unlike the IA model, the Card-Centric Framework considers only the case where the smart card makes use of resources on the connected terminal. Therefore, instead of peer-to-peer model, a master-slave one is adopted, but in the inverse direction. In other words, smart card swapped the role with the terminal and becomes the master. Whenever they need to make use of the resources of each others, we suggest adopting the Proactive SIM method as proposed in [1], and running CCP on top of it, such that both Smart Card and Console could become peers.

3. SYSTEM MODEL

As shown in figure 1, the system consists of two major components: the Smart Card and the Console. The Smart Card, similar to a brain, is the core processing unit of the whole system. The Console, like a mere body, provides various off-card resources for use by Smart Card. These resources, including I/O peripherals and Services, are coordinated by a Bridge embedded in the Console. In this section, each of the system components will be introduced.

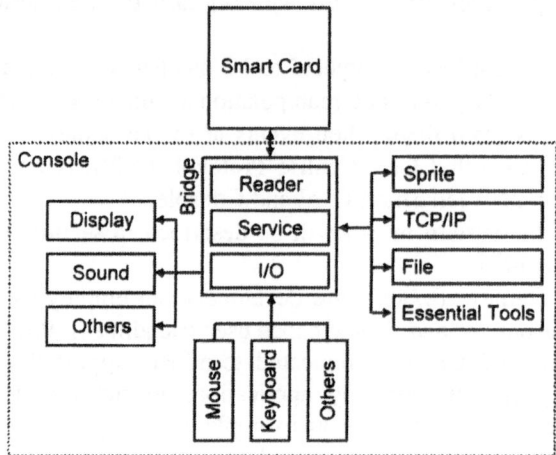

Figure 1. System model of Card-Centric Framework

3.1   Smart Card

The Smart Card not only handles the core applications of the whole system, but also the off-card resources on the Console. It manipulates them by sending commands to the Console. On the other hand, it handles I/O interrupt signals sent from the Console, and then takes the corresponding actions.

In order to achieve a mapping between the on-card resources and the off-card counterparts, interface objects and methods describing the features of Console are available on the card side. When the on-card application calls those methods, the interface objects will send requests to the Console, and then wait for returned data. This mechanism is similar to conventional smart card Remote Method Invocation (RMI) [4][5], but in a reverse manner, such that the card can invoke any remote methods supported by the Console rather than being invoked.

3.2   Reader

The Reader is the serial communication channel between Smart Card and Console. It should be ISO7816 [6] and PC/SC [7] compatible. ISO7816 is a communication protocol designed for smart card, whereas PC/SC is a standard for the integration of smart cards to personal computers. Conventional card readers fix the link speed to only 10kbps. Therefore, it may become a burden of the whole system, especially when the on-card application makes use of resources from fast networks.

3.3   I/O Peripherals

I/O is the major user interface with the on-card application. Each I/O resource is an object. Examples of input are mouse and keyboard, whereas examples of output are display and sound. The output objects could be accessed using the on-card interface objects, whereas the input ones involve the use of interrupts.

On arrival of user event, which may be a mouse move or a key press, the Console will send an interrupt packet to the Smart Card. After receiving the interrupt, the Smart Card will send requests to the Console for more information on the I/O generating the interrupt. The information, such as the coordinates of the mouse cursor or the key being pressed, are abstracted by the Console and then sent back to the Smart Card for further processing.

3.4   Services

In order to save the precious on-card processing power, heavy-weight off-card resources should not be manipulated in detail. For instance if a 24-bit display of 640x480 resolution is manipulated directly by the card dot by dot on a 25fps basis, it will require at least a 180 Mbps serial link for smooth display disregard traffic and processing overheads, and the link that is in Kbps order will then become a bottleneck. Since each Send/Receive operation involves the movement of data into or out of the input/output buffer that are often flash memory devices, larger bandwidth means higher processing power requirements. To cater for this, we make use of abstract instructions, in which the vast amount of operations required to finish a certain task are organized to form one general instruction. Related instructions are further grouped as a Service. Similar to I/O Peripherals, each Service entity is an object. Through Service, the Smart Card can indirectly manipulate I/O resources with lower hardware requirements.

There are four featured groups of services in the Console: Sprite, TCP/IP, Files and Essential Tools. Sprite, an essential technique of 2D game programming, is responsible for basic graphics and text manipulations. With Sprite, the Smart Card can off-load the tedious job of graphics manipulations to Console, and lowers the link usage as well as on-card processing power requirements. TCP/IP adapts Smart Card to the internet, such that the Smart Card can off-load the vast amount of handshaking, packetization and similar jobs necessary for TCP/IP connections to the Console, and handles them by simple pointers to sockets and read/write/listen commands. Files, which are used as temporal storage of data, release the pressure of on-card storage requirements. Essential tools help in the management of resources (e.g. allocation and deallocation) and may sometimes be useful in debugging.

3.5   Bridge

The core of Console is the Bridge that coordinates communications between the reader, I/O and Services. It functions as a message router that runs all traffic over the ISO7816 communication protocol [6]. It parses commands sent from the Smart Card, routes them to the appropriate Service or I/O objects, and then passes back the returned data to the Smart Card. In other words, it is similar to the resource manager of the Intelligent Adjunct model [1]. Moreover, it passes the interrupts generated by resource objects to the Smart Card, and then awaits the subsequent I/O requests. With the Bridge, the Smart Card gains accessibility to any connected off-card resources.

4.   CARD-CENTRIC PROTOCOL

4.1   Protocol Layering

Application
Card-Centric Protocol
ISO7816 Transport
ISO7816 Physical

Figure 2. Protocol layer model in Card-Centric Framework

Figure 2 shows the protocol layer model involved in Card-Centric Framework. The CCP layer is located in between the Application and ISO7816 Transport layers, where the former is developer dependent and the latter handles data transfer between Smart Card and Console. In other words, we embed the CCP packets as payloads of ISO7816 messages, leaving the implementations of the underlying layers unaltered. Therefore, it is simple to integrate the framework into existing smart card infrastructures.

4.2   Packet Format

There are two types of CCP packets: Request and Response.

Request packets flow from Smart Card to Console, by which the Smart Card initiates requests and acquires an active role. As mentioned before, each I/O or Service entity is an object. In order to access those off-card objects, the Smart Card must provide the *Pointer* to identify the object, the instruction byte *INS* and the necessary *Parameters* to manipulate the object.

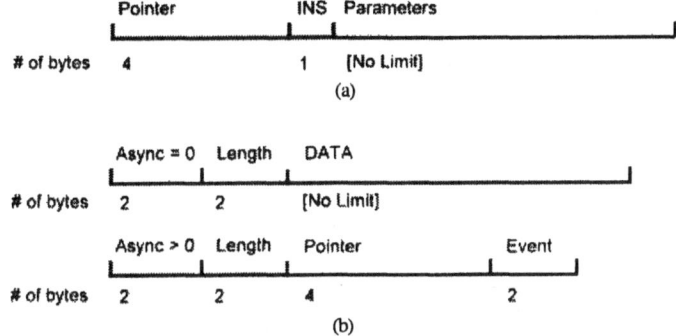

Figure 3. Packet format (a) from Smart Card to Console and (b) from Console to Smart Card

They are arranged in the Request packet sent from Smart Card to Console, as shown in figure 3.

In reply to each Request, there is a Response packet flowing in the reverse way. There are two types of response packets: Synchronous and Interrupt. Synchronous packets ($Async = 0$) are responses to request packets. It consists of the corresponding results (*DATA*). Interrupt packets ($Async > 0$) are issued when there are events from the resource objects. It tells which object is involved (*Pointer*) and how it is involved (*Event*). An alternative to the Interrupt packet is the ENVELOPE command of SIM APDUs, which allows Smart Card to obtain interrupt events from Console [8][9]. In such case, the Pointer and Events are arranged in the ENVELOPE APDU. In order to support non-SIM environments, the Interrupt packet will be more preferable. The *Async* word not only indicates if a packet is an interrupt, but also the type of the resource generating the interrupt. In case where several objects share the same type, such as various TCP/IP connections sharing the type Socket, the *Pointer* can indicate the exact object involved.

4.3   Handshaking Sequence

The general handshake sequence between Smart Card and Console is illustrated in figure 4. Each handshake cycle consists of two sequences: request and response. In reply to the Nth request from Smart Card, the Console sends synchronous response #N. Suppose a user event arrives right after the issue of the N+1th request, the Console suspends the process of the current request, and generates an interrupt packet. Receiving the preemptive interrupt, the Smart Card handles the I/O object involved by request #N+2, and then sends a "No Operation" request at the end. The Console detects the request, and then restores the suspended request #N+1 and continues the communication sequence.

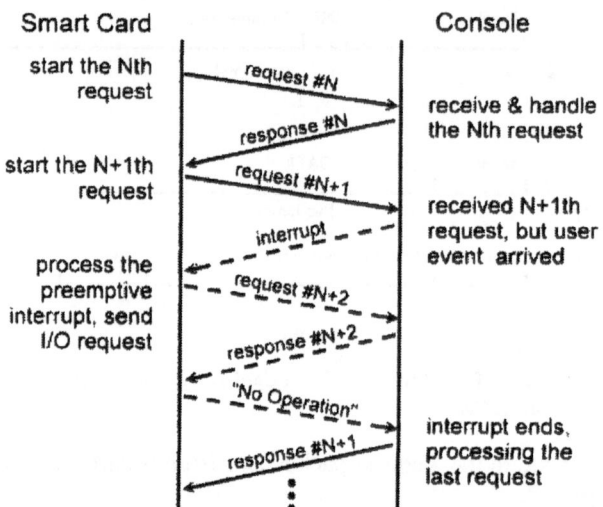

Figure 4. General communication sequence between Smart Card and Console, dashed lines indicates the sequence of interrupt requests.

ISO7816 is originally designed for passive smart card access, in which the card-connected terminal initiates a request whereas the smart card replies with the corresponding results [6]. To initiate smart card-active communication, we employ a slight modification on the smart card's conventional response to the ISO7816 SELECT command at the very beginning of communication, as shown in figure 5. Instead of sending only the conventional "No Error" message (90 00h) and then waiting for incoming requests as in conventional ISO7816 applications, the Smart Card immediately initiates the request-response sequence and takes the active role. In the Intelligent Adjunct model that makes use of SIM APDUs, a similar mechanism was proposed [1], as shown in figure 5(b). However, before starting the request-response sequence, the smart card (or IA) has to wait for the FETCH request sent from the Terminal after the "No Error" message (91 XXh). Therefore, an extra handshake cycle is required for the smart card to take an active role.

In our proposed protocol as shown in figure 5(c), we use neither FETCH nor "No Error" but let the Smart Card issue request immediately after SELECT. Therefore, one request-response cycle is saved.

4.4   Object Allocation

The Smart Card must allocate objects on the Console before manipulating them; otherwise it will receive an error warning that the referred object is invalid. Objects could be either pre-allocated or

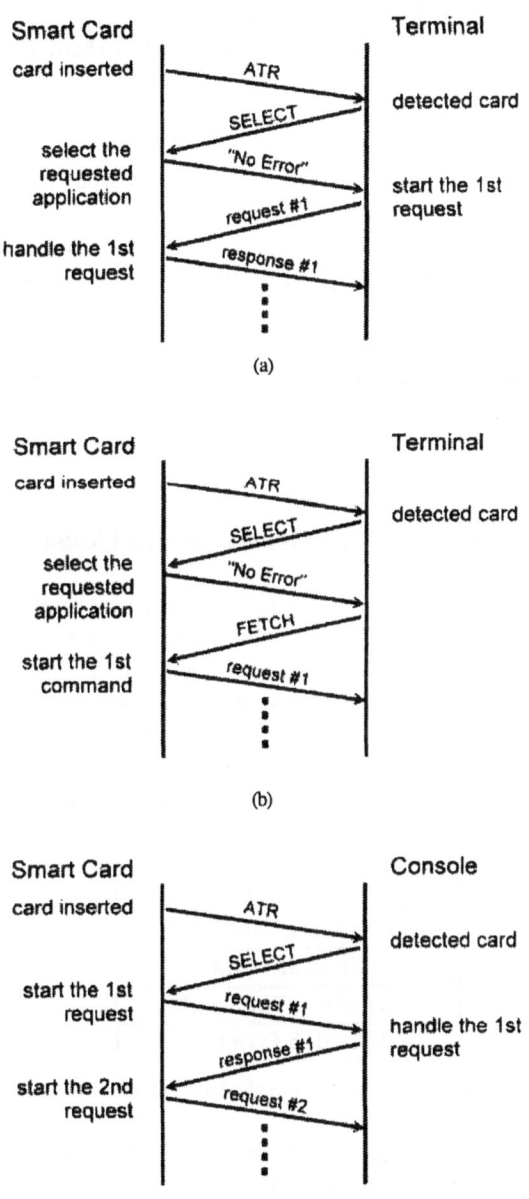

Figure 5. Initialization phase of (a) conventional ISO7816 application, (b) Intelligent Adjunct model and (c) modified communication sequence. ATR is the Answer-To-Reset string sent from smart card right after power-up.

dynamically allocated. Pre-allocation, which is performed automatically since the power-up of Console, results in an intrinsic NULL Pointer. The only object that requires pre-allocation is the Essential Tools Service. Dynamic allocation results in a non-null *Pointer* to be referred by the Smart Card. It is handled by the Object-allocation Instruction (INS = 00h) of the Essential Tools Service object.

The framework supports a maximum of 65536 types of resource objects. The first 1024 types (0-1023) are reserved for standardization, and their manipulation instructions should be the same across different Console implementations. Application specific objects should be implemented with type 1024 or above.

Depending on the need from applications, more than one object could be allocated for each type of resource. For example, several sockets may be required in some applications, whereas only one mouse is needed in most cases. After allocation, the properties of those objects could then be altered by the instructions supported.

4.5 Accessing On-card Resources by Console

The Card-Centric Protocol is designed for on-card applications to make use of resources on the Console. However, there are circumstances where the Console needs to make use of resources on the Smart Card. A similar case is Proactive SIM: it not only initiates requests to the connected handset (or Console), but also accepts requests from the network operator for administrative features [9], where the requests are routed trough the handset. In such case, the on-card application has to issue requests to as well as accept requests from the Console.

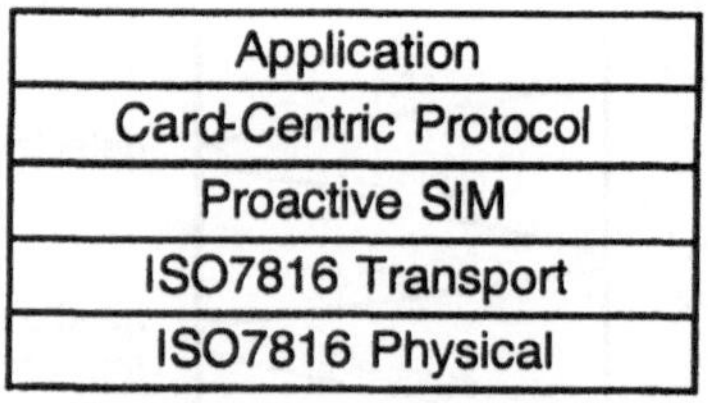

Application
Card-Centric Protocol
Proactive SIM
ISO7816 Transport
ISO7816 Physical

Figure 6. Protocol layer model with Proactive SIM

To cater for this, we suggest running the Card-Centric Protocol over the Proactive SIM protocol [8][9], as illustrated in figure 6.The CCP Request and Response packets should be encapsulated in the Proactive commands and responses, whereas Interrupt packets in ENVELOPE commands, as shown in figure 7.

Note that adapting CCP to the Proactive SIM Protocol introduces three extra messages for each request-response pairs: status word 91 XX, FETCH

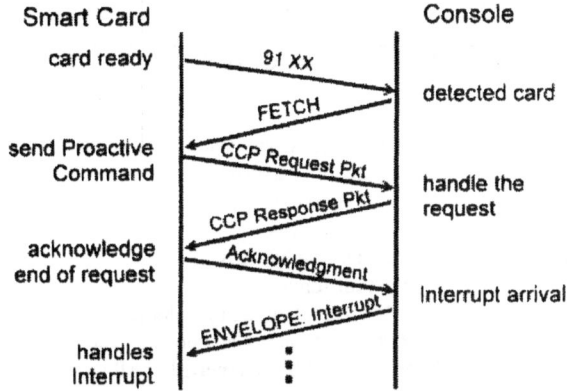

Figure 7. Communication sequence of CCP over Proactive commands

and Acknowledgment. Despite the Acknowledgment message that may consist of the status word 91 XX, this solution results in a 50% overhead. Each message transmitted not only occupies bandwidth, but also precious processing power, especially for smart cards that are equipped with slow flash memories. Therefore, it should be applied only when necessary. In case where only Smart Card issues requests to Console, we suggest eliminating the Proactive SIM layer, and adopting the simplified model of Figure 2.

5. SYSTEM PROTOTYPE

A system prototype is implemented to demonstrate the feasibility of the Card-Centric Framework. It consists of a typical smart card and a software-emulated Console. The smart card specifications are available in table 1. It cannot store high resolution graphics in just 64KByte of memory. In order to reduce on-card memory usage, the graphics are pre-stored in a web server on the same LAN as the Console, and then fetched and stored as off-card files whenever needed. The Console emulator is written in Visual C++ for x86 computers. It provides mouse, keyboard and display for the smart card to handle user interactions. It also adapts the smart card to the internet by

Table 1. Specification of the Smart Card of system prototype

Model	Gemplus GXP PRO-R3
Processor speed	4 MHz
Memory type	Flash Memory
Memory size	64 KByte
On-card OS	JAVACard 2.1.1
Communication port	10 Kbps serial
Communication protocol	CCP /ISO7816 transport /ISO7816 physical

Table 2. Specification of the Console of system prototype

General Information	
Development Environment:	Microsoft Visual C++ 6.0

Off-card Resources

Type #	Name	Features Supported
0	Essential Tools	Object allocation and deallocation; Unlimited number of objects; Debugging by message box.
1	Mouse	Interrupt; 3 buttons.
2	Keyboard	Interrupt; 101 keys.
3	Bitmap	Context for raw bitmaps.
4	Font	Context for fonts.
5	Sprite	512x384 resolution; 24-bit color depth.
6	Off-card Files	Max. 4 GB file size; Clear on end of session.
7	Socket	Interrupt; TCP/IP/Ethernet@100 Mbps; Support create, accept, read, write.

Table 3. Performance of system prototype

Max. application time	120 s
Min. application time	100 ms
Transport layer throughput	1.5 Kbps

TCP/IP socket objects. More details of Console could be found in table 2.

The on-card application is a Tic-Tac-Toe game that handles both game arithmetic and user interactions. The game waits for the first move from human player, and then counterattacks using 6-ply minimax procedure and alpha-beta algorithm [10]. User interactions are accessible through the on-card interfaces as described in table 2.

Table 3 shows the performance of the system prototype. The performance parameter involved is the application time measured by the Console between two successive requests disregard the time required for fetching off-card graphics. The results could be affected by two factors: communication overhead of the serial link and the processing power of the smart card. Communication overhead of the 10Kbps serial link, which is in the order of 10^{-4} s, is insignificant when compared with the measured result. Therefore, the result is due to the poor processing power of the smart card, especially the slow write speed of the on-chip flash memory. The minimax procedure involves a vast amount of assignment instructions to the slow flash memory, and thus the 120s required to accomplish one request-response cycle is reasonable. Even the fastest program segment that involves only the copy and transmission of memory contents requires 100ms of application time. These measured values, when compared with the maximum 230 μ s for a 6000MIPS x86 processor (Pentium 4 running at 2.4GHz) to run the same application, are unacceptable for users.

Therefore, it is obvious that the current smart card technologies are not suitable for memory intensive applications.

6. DEMO SYSTEM BASED ON TECHNOLOGIES FROM THE FUTURE

Fortunately, European smart card industry and academic stakeholders noticed that smart card has yet to improve its performance and features for being acknowledged as a new generation element in the consumer domain, and therefore introduced the RESET roadmap to identify and address the major challenges of the smart card industry [11]. According to the roadmap, the smart card should, in the near future, be equipped with faster memory such as floating gate memory, FeRAM or MRAM [2]. The fastest of them could achieve an access time as short as 10ns.

In order to demonstrate the feasibility of Card-Centric Framework over the future technology, the smart card in our demo system is replaced by a system-on-programmable-chip that makes use of comparable memory (SRAM). It is an FPGA-based development board from Altera [12] configured to simulate the behavior of a smart card, and thus called *Smart Card* hereafter. Its specifications are listed in table 4. Similar to smart card, it has two types of on-board memory: SRAM and Flash. Flash memory is used for application storage, whereas SRAM for program execution. For full utilization of the 50MHz SRAM, a processor running at the same speed is implemented on the FPGA. With this configuration, the application time required for those memory intensive procedures in the game could be significantly reduced.

With a great enhancement in processing power, the application time is predicted to be in the order of 10^{-4} s, thus comparable to the communication overhead of the serial link that is in Kbps order. In order to reduce communication overheads for more reasonable measurements, the original serial link is now replaced by 100Mbps Ethernet.

The Console Emulator and on-card application used in the system prototype of section 5 were adopted. To cater for the change in communication link, the communication drivers are modified to run over TCP/IP. To achieve this, both Smart Card and Console are equipped with a TCP/IP stack, and offered a unique IP address. With the assumption that the

Table 4. Specification of the Smart Card in the 2nd system

Model	Altera Nios development board, Stratix Ed.
Processor speed	50 MHz
Memory type	SRAM, Flash
Memory speed	50 MHz
Memory size	1 Mbyte (SRAM)
	8 Mbyte (Flash)
On-card OS	N /A
Communication port	100 Mbps Ethernet
Communication protocol	CCP /ISO7816 transport /TCP /IP /Ethernet

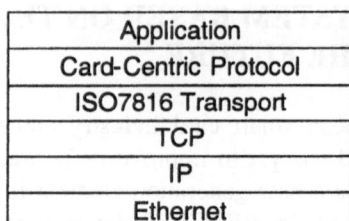

Figure 8. Protocol layer model in 2nd system

Smart Card's address is known, the Console can connect to the Smart Card as usual, and then encapsulate ISO7816 messages in TCP/IP packets. This is illustrated in the protocol layer diagram in figure 8. On top of ISO7816, the Card-Centric Protocol is run without modifications.

Table 5 shows the performance of the 2nd demo system. With an enhancement in smart card hardware technologies, the application speed increases drastically. The link utilization increases 200 times, therefore the choice of high speed Ethernet is appropriate.

The application speed is not as fast as Pentium 4, yet the application delay is unnoticeable by end users. Namely, the user interface is smooth enough. In fact, applications other than games usually require lower processing power. Consider the example application of [1], where the end user configures the smart card to run user-defined tasks through Console,. The on-card application may only need to receive a command string from the keyboard and then store it in the on-card memory. These simple procedures require far lower processing power to fulfill than games. Therefore, the near-future technologies predicted by [2] are enough for the Card-Centric Framework.

Table 5. Performance of 2nd system

Max. application time	1 ms
Min. application time	520 μ s
Transport layer throughput	304 Kbps

7.   SECURITY ISSUES

Smart card, originally designed for security, has a tamper proof property that favors the protection of data stored inside. The Card-Centric Framework inherits this benefit, but does not consider the security of data transferred outside the smart card. Therefore, software developers are required to handle the security of such case.

In the framework, the information transferred between smart card and Console are mainly I/O signals, which may sometimes be critical

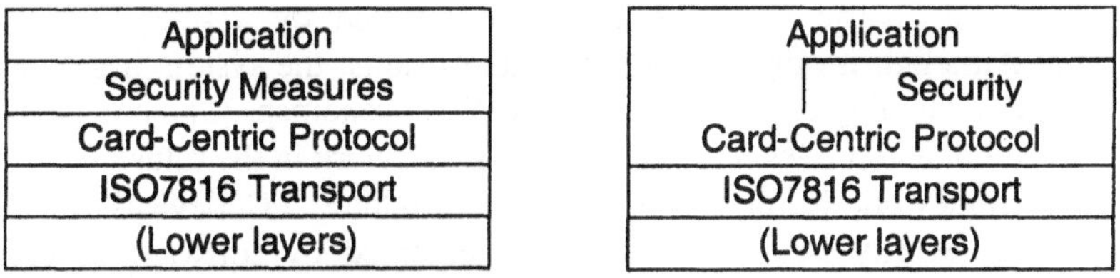

Application
Security Measures
Card-Centric Protocol
ISO7816 Transport
(Lower layers)

(a)

Application
Security
Card-Centric Protocol
ISO7816 Transport
(Lower layers)

(b)

Figure 9. Protocol layer model for security: (a) A separate layer for security, (b) Security measures integrated into the application layer

information displayed on Console or keyed in by the user. It may become insecure when the compromise of Console or the interception of communication link becomes possible.

In order to enhance the security of such case, we suggest two non-intrusive measures to the Card-Centric Framework:

a) A custom layer could be introduced between the application and CCP layer, as shown in figure 9(a). In this layer, any security related measures could be implemented.

b) The security features are fully integrated into the application layer, over which the smart card has full control. This is illustrated in figure 9(b). In other words, the security measures could be accessed by the application when necessary.

For the security features, unilateral or mutual authentication could be implemented for Smart Card and Console to verify if they have the right to access each other [13]. Provided trust is established, they could communicate by CCP in plain text within the session. In case a packet could not be sent in plain text, such as when critical information is transmitted, the on-card application could protect the data by cryptography [14]. Note that cryptographic measures may introduce serious processing overhead to the system and degrade user perception. It should be employed only when necessary.

8. CONCLUSIONS

The Intelligent Adjunct model initiates a new usage model for smart cards, allowing them to run application logics. We move one step further, and proposed the Card-Centric Framework for smart cards to handle off-card I/O resources. Under this framework, any on-card applications that require user interactions could be developed, thus becomes more user-friendly. To enable this, we propose a system model and a set of communication protocol as guidelines for implementation.

The Card-Centric Framework is demonstrated by a smart card-based system. Although it only shows reasonable results, the system based on the enhanced technologies predicted by RESET roadmap [2] shows drastic improvement and proves feasibility of the framework in the future.

The framework is for the purpose of providing I/O devices for smart cards, yet security issues are to be implemented by software developers depending on the application involved. In the future, some more rigorous solutions might evolve, such as more appropriate measures for the security layer, and a compromised mechanism for I/O resource allocation.

REFERENCES

[1] B. Balacheff, B. Van Wilder, and D. Chan, "Smartcards – from security tokens to intelligent adjuncts", in Proceedings of Cardis98, Louvain-la-Neuve, Belgium, pp. 71-84, September 14-16, 1998.

[2] ERCIM and Eurosmart, "Micro-electronics", RESET Final Roadmap, Mar. 2003, pp.52-59.
[http://www.ercim.org/reset/]

[3] ERCIM and Eurosmart, "Systems and Software", RESET Final Roadmap, Mar. 2003, pp.19-20.
[http://www.ercim.org/reset/]

[4] J. –J. Vandewalle and E. Vétillard, "Developing smart card-based applications using Java Card", in Proceedings of Cardis98, Louvain-la-Neuve, Belgium, pp. 105-124, September 14-16, 1998.

[5] Sun Microsystems, Inc., Java Card 2.1.1 API, Rev. 1.0, May. 2000.
[http://java.sun.com/products/javacard/specs.html]

[6] International Organization for Standardization, International Standard ISO/IEC 7816: Integrated circuit(s) cards with contacts, parts 1-4, 1987-1998.

[7] PC/SC Workgroup, PC/SC Workgroup Specifications 1.0, parts 1-7, Dec. 1997.

[8] T. M. Jurgensen, and S. B. Guthery, "SIM APDUs", Smart Cards: The Developer's Toolkit, Upper Saddle River, NJ: Prentice Hall PTR, 2002, pp. 267-287.

[9] Proactive SIM, GSM 11.14 specification, July 1997.

[10] P. W. Frey, "An introduction to computer chess", Chess Skill in Man and Machine, 2nd ed., Springer-Verlag: New York, 1983, pp.61-68.

[11] ERCIM and Eurosmart, "Introduction", RESET Final Roadmap, Mar. 2003, pp.8.
[http://www.ercim.org/reset/]

[12] Altera, Nios Development Board – Reference Manual, Stratix Ed., Rev. 1.1, Jul. 2003.
[http://www.altera.com/literature/manual/mnl_nios_board_stratix_1s10.pdf]

[13] W. Rankl, and W. Effing, "Protection: authentication", Smart Card Handbook, 3rd ed., Hoboken, NJ : Wiley, 2003, pp.559.

[14] W. Rankl, and W. Effing, "Protection: secure data transmission", Smart Card Handbook, 3rd ed., Hoboken, NJ : Wiley, 2003, pp.558-559.

ON THE SECURITY OF THE *DEKART* PRIMITIVE

Gilles Piret
UCL Crypto Group, Laboratoire de Microelectronique, Universite Catholique de Louvain,
Place du Levant, 3, B-1348 Louvain-la-Neuve, Belgium.
piret@dice.ucl.ac.be

Francois-Xavier Standaert
UCL Crypto Group, Laboratoire de Microelectronique, Universite Catholique de Louvain,
Place du Levant, 3, B-1348 Louvain-la-Neuve, Belgium.
fstandae@dice.ucl.ac.be

Gael Rouvroy
UCL Crypto Group, Laboratoire de Microelectronique, Universite Catholique de Louvain,
Place du Levant, 3, B-1348 Louvain-la-Neuve, Belgium.
rouvroy@dice.ucl.ac.be

Jean-Jacques Quisquater
UCL Crypto Group, Laboratoire de Microelectronique, Universite Catholique de Louvain,
Place du Levant, 3, B-1348 Louvain-la-Neuve, Belgium.
jjq@dice.ucl.ac.be

Abstract   *DeKaRT* primitives are key-dependent reversible circuits presented at CHES 2003. According to the author, the circuits described are suitable for data scrambling but also as building blocks for block ciphers. Data scrambling of internal links and memories on smart card chips is intended for protecting data against probing attacks. In this paper, we analyze the *DeKaRT* primitive using linear cryptanalysis. We show that despite its key-dependent behavior, *DeKaRT* still has strongly linear structures, that can be exploited even under the particular hypothesis that only one bit of the ciphertexts is available to the attacker (as it is the case in the context of probing attacks), and using very few plaintext-ciphertext pairs.

The attack methodology we describe could be applied to other data scrambling primitives exhibiting highly biased linear relations.

Keywords: Smart Card, Probing Attacks, Data Scrambling, Linear Cryptanalysis.

1.   Introduction

Probing attacks on smart cards are invasive techniques consisting in introducing a conductor in some point of a tamper-resistant chip to monitor the electric signal, in order to recover secret information passing through this point [4, 5]. For example, the bus connecting the RAM and the microprocessor is particularly vulnerable. Using classical block ciphers like DES or AES seems to provide a natural solution to this problem; however it is simply not realistic, because of the very high throughput and small size requirements.

It is why primitives have been developed offering hastier and lighter solutions, but at the cost of a lower security level. In [2] implementations of keyed bit permutations are proposed, trying to simultaneously achieve small logical depth and large key space. Nevertheless this type of primitive has the drawback of being perfectly linear. At CHES2003, the $DeKaRT$ construction was presented [1], which was aimed at providing non-linearity but at the cost of a heavier structure than in [2].

In this paper, we analyze the security of the $DeKaRT$ primitive in accordance with the usual properties of block ciphers. Despite the complex structure of the primitive and its key-dependent behavior, it is underlined that the use of such a scrambling function does not efficiently prevent probing attacks.

In practice, we illustrate the strongly linear structure of the $DeKaRT$ block. The main observation is that, although the suggested block size of $DeKaRT$ primitives is very small compared to block ciphers (which reduces the total number of plaintext-ciphertext pairs a priori available), it is still possible to do linear predictions of its inputs. Moreover, these predictions may only involve a very limited number of output bits, what is actually relevant in the probing attack context. As a block cipher building block, $DeKaRT$ exhibit even stronger weaknesses as more linear relationships are available to the attacker.

2.   Specification of a Concrete Instance of $DeKaRT$ and Notations

In [1] general building principles are given for construction of a $DeKaRT$ data scrambling function, but there is no precise instance defined. We believe it is a mistake, as security analysis requires a completely specified cipher; and security of a cipher which has not been subject to a substantial effort regarding security analysis can be questioned (except if a security proof is given, which is never the case for block ciphers). This is why we first specify a cipher, based on the design principles given in [1].

A *generic building block* acts on a small number of input data bits which are divided into two groups of m and n bits. The m input bits are used for control and are passed to the output intact, like in the Feistel structure. They are used to select k out of $2^m k$ key bits by the multiplexer (MUX) circuit with m control

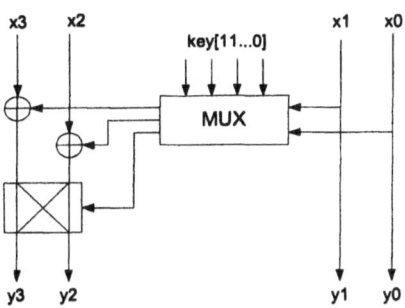

Figure 1.   Elementary *DeKaRT* building block.

bits, $2^m k$ input bits and k output bits. In the original paper, the author suggests an *elementary DeKaRT building block* with parameters $(m, n, k) = (2, 2, 3)$ as shown in Figure 1 where after a XOR with 2 key bits, x_3 and x_2 pass through a conditional switch. We will use this building block in our specification. Each box requires 12 key bits. We will denote these bits by

$$(k^{(11)}, k^{(10)}, k^{(01)}, k^{(00)}) =$$
$$(k_X^{(11)}, k_{\oplus 1}^{(11)}, k_{\oplus 2}^{(11)}; k_X^{(10)}, k_{\oplus 1}^{(10)}, k_{\oplus 2}^{(10)}; k_X^{(01)}, k_{\oplus 1}^{(01)}, k_{\oplus 2}^{(01)}; k_X^{(00)}, k_{\oplus 1}^{(00)}, k_{\oplus 2}^{(00)})$$

where $k^{(i)}$ denotes the three key bits selected by control bits $(x_1, x_0) = i$; $k_X^{(i)}$ is conditioning the switch, while $k_{\oplus 1}^{(i)}$ is XORed with x_3 and $k_{\oplus 2}^{(i)}$ is XORed with x_2. The set of such 12 key bits is called a *subkey*.

The instance of *DeKaRT* we will consider acts on blocks of 16 bits. Thus the key dependent layer, denoted KT for *K*eyed *T*ransform, consists in the parallel application of 4 such elementary blocks. The number of rounds considered is 5 (this number is given as an example in [1], p.104). We will denote the subkeys parameterizing the four elementary blocks of the rth round by $RK_3^r, RK_2^r, RK_1^r, RK_0^r$, from left to right. The set of 4 such subkeys is called a *round key* (thus it has 48 bits).

The KT layer alternates with a *B*it *P*ermutation layer (noted BP). Two design rules regarding BP were given in [1]:

- The control bits ((x_1, x_0) in Fig. 1) in each layer should be used as the transformed bits ((x_3, x_2) in Fig. 1) in the next layer.

- For each elementary block, input bits should come from the maximum possible number of blocks in the previous KT layer. In the instance considered, it means that each of the 4 input bits to an elementary block comes from a different block in the previous KT layer (which also implies that the 4 output bits of an elementary block are sent to 4 different

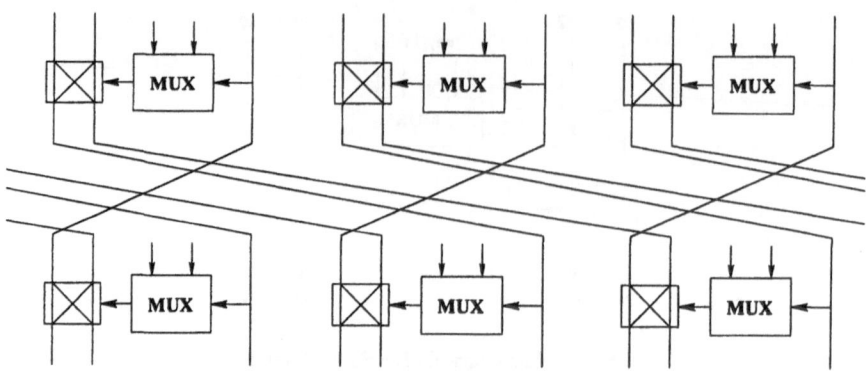

Figure 2.   a part of the Key Expansion

blocks in the next KT layer).

We arbitrarily chose a bit permutation following these rules. It is given in Annex A.1 (see also Fig. 3).

Finally, as suggested in [1], the key expansion algorithm also alternates "keyed" layers (with the key being an arbitrarily chosen constant), with bit permutations layer. The 48-bit data obtained every two rounds is used as a round key. The keyed layer is made out of the parallel application of 16 *reduced DeKaRT building block* (as proposed in [1]) with 3-bit input and output, alternating with a bit permutation layer complying with the design rules given above. Fig. 2 pictures part of this algorithm. Note that the bit permutation layer could have been chosen less "regular", but in fact its influence on the results of our attack is negligible.

3.   Analysis of an elementary $DeKaRT$ block

A $DeKaRT$ building block generates key-dependent boolean functions. If a (2,2,3) block is used, 4096 substitution tables (4 × 4 bits) can be generated, as this type of block is parameterized by 12 key bits.

In this section, we investigate the possible linear approximations of an elementary $DeKaRT$ block. For each output bit, there exist $2^4 - 1$ possible non-trivial input masks (i.e. $2^4 - 1$ possible linear combinations of input bits) and if we combine output bits together, we have $(2^4 - 1) \times (2^4 - 1)$ possible non-trivial linear approximations of the $DeKaRT$ block. For such a small block size, the problem of finding good linear approximations is therefore easily solved by exhaustive search.

Table 1. Linear approximations of a single $DeKaRT$ block.

ϵ	1/2	3/8	1/4
Nbr. Approximations	2304	1024	768

In this paper, we define the *bias* of a linear approximation that holds with probability p as $\epsilon = |p - 1/2|$. We also denote a linear approximation with probability $p = 0$ or $p = 1$ as a *perfect* linear approximation.

As the $DeKaRT$ block defines 4096 substitution tables, we first investigated the best linear approximations of each individual table. It is summarized in Table 1, where we rejected approximations involving bits y_0, y_1 only as they are obviously perfectly linear. We observe that more than one half of the generated tables presents perfect linear approximations and may therefore be considered as very weak from a cryptographic point of view.

On the basis of this first experiment, we may therefore assess that a large number of keys will generate weak building blocks for the scrambling function or block cipher as they are perfectly approximated by a linear approximation. This motivated a more general analysis.

4. The Attack

The scenario of the attack is the following: we consider the case where the data scrambling function $DeKaRT$ is used to protect communication between the smart card microprocessor and the RAM; this is the most common use for data scrambling functions. We assume the attacker is allowed to play with the microprocessor, which implies that he can send known data to the memory. Moreover he has access to a small number of bits of the encrypted data via probing attack. Formally, this means that the attacker can obtain a certain number m of pairs (P, c), where P is a known plaintext, and c is one fixed bit of the corresponding ciphertext C. His goal is to obtain information about a secret data (such as an RSA private key) present in the card and read from the RAM at some time. Thus it it **not** a key recovery attack, in the sense that we do not intend to retrieve the key used for $DeKaRT$ data scrambling (it can be changed at each use of the card anyway). This security model is the one described in [2].

Due to the building blocks of $DeKaRT$ being implemented using key-dependent MUXs, the probability of a linear relation between a given bit of the ciphertext and some bits of the plaintext is highly key-dependent. In our attack, pairs (P, c) are used to identify a linear relation between one only bit

of the ciphertext and a linear combination of bits of the plaintext, that holds
with a high bias for the key used. Then when the secret data pass through the
channel between the RAM and the processor, the relation we just identified
permits probabilistic information about it to be retrieved.

Let λ be the linear relation we are considering. Knowing m pairs (P, c), we
count how many of these satisfy the linear relation λ; let n_λ^m be the random variable corresponding to this number. We would like to compute the probability
that λ holds for a random plaintext, provided n_λ^m takes value B:

$$P_{K,P}[\lambda \text{ holds}|n_\lambda^m = B] \tag{1}$$

$|P_{K,P}[\lambda \text{ holds}|n_\lambda^m = B] - 1/2|$ gives the reliability of the prediction we will
make. Let us define the random variable N_λ as being the number of plaintexts
for which λ holds, out of all 2^{16} plaintexts this time. Then we have:

$$P_{K,P} \ [\lambda \text{ holds}|n_\lambda^m = B]$$

$$= \sum_{A=0}^{2^{16}} P[\lambda \text{ holds}|N_\lambda = A] \cdot P[N_\lambda = A|n_\lambda^m = B]$$

$$= \sum_{A=0}^{2^{16}} A/2^{16} \cdot P[N_\lambda = A|n_\lambda^m = B]$$

$$= \sum_{A=0}^{2^{16}} A/2^{16} \cdot P[n_\lambda^m = B|N_\lambda = A] \cdot \tag{2}$$

$$\frac{P[N_\lambda = A]}{\sum_{A'=0}^{2^{16}} P[N_\lambda = A'] \cdot P[n_\lambda^m = B|N_\lambda = A']}$$

$$\cong \frac{\sum_{A=0}^{2^{16}} A/2^{16} \cdot P[Bi(m, A/2^{16}) = B] \cdot P[N_\lambda = A]}{\sum_{A'=0}^{2^{16}} P[Bi(m, A'/2^{16}) = B] \cdot P[N_\lambda = A']}$$

$$\tag{3}$$

where $Bi(.,.)$ denotes the binomial distribution law, and the last approximation assumes $m \ll 2^{16}$. Thus computation of (1) requires knowledge of the
probability distribution of N_λ, i.e. $\{P_K[N_\lambda = A]\}_{A=0}^{2^{16}}$. Next section will show
how such a distribution can be computed.

5.   Computing Probability Distribution for a Linear Relation Through a 5-Round Cipher

Consider 5 rounds of $DeKaRT$, beginning and ending with a keyed layer
(Fig. 3). We denote the plaintext: $(p_{15}, p_{14}, ., p_1, p_0)$ or $(a_{15}^{(1)}, a_{14}^{(1)}, ., a_1^{(1)}, a_0^{(1)})$.

Also $KT((a_{15}^{(i)}, a_{14}^{(i)}, a_{13}^{(i)}, ..., a_1^{(i)}, a_0^{(i)})) =: (b_{15}^{(i)}, b_{14}^{(i)}, b_{13}^{(i)}, ..., b_1^{(i)}, b_0^{(i)})$

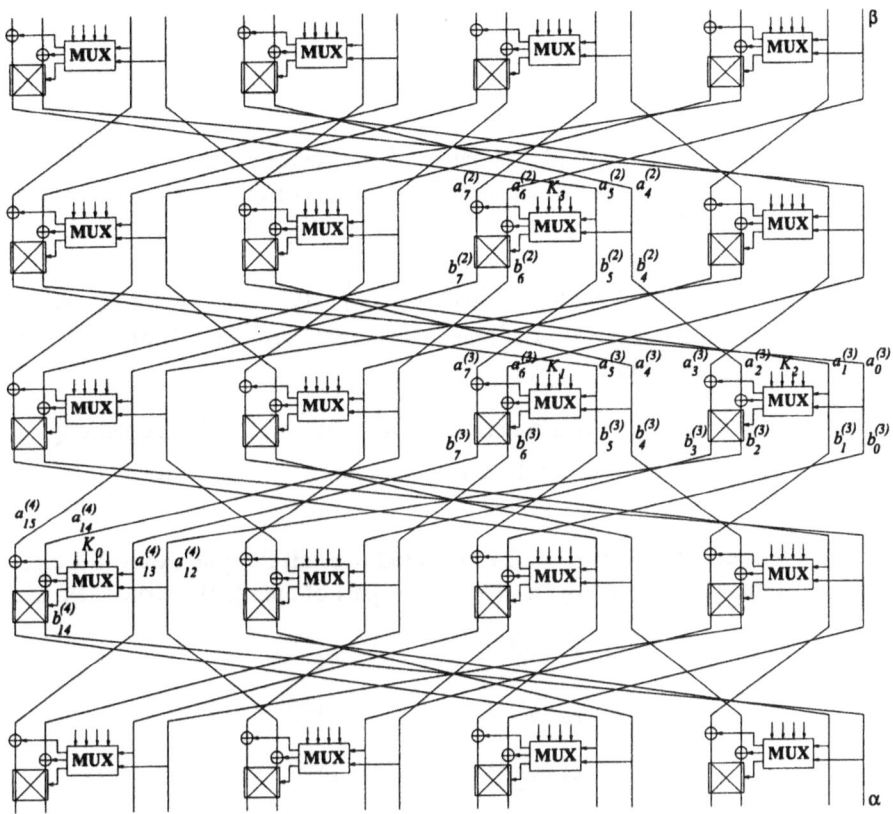

Figure 3. Linear Approximation through a 5-Round $DeKaRT$

And:

$$BP((b_{15}^{(i)}, b_{14}^{(i)}, b_{13}^{(i)}, ..., b_1^{(i)}, b_0^{(i)})) =: (a_{15}^{(i+1)}, a_{14}^{(i+1)}, a_{13}^{(i+1)}, ..., a_1^{(i+1)}, a_0^{(i+1)})$$

The exponent denotes the round number and the ciphertext will be denoted by

$$(b_{15}^{(5)}, b_{14}^{(5)}, ..., b_1^{(5)}, b_0^{(5)}) \text{ or } (c_{15}, c_{14}, ..., c_1, c_0)$$

Consider one bit $\alpha := b_0^{(5)}$ of the ciphertext after a 5-round cipher. As an example, we will analyze the linear relation between α and $\beta := a_0^{(1)}$ (see Fig. 3). Other relations can be analyzed similarly. We write successively:

- $\alpha = b_{14}^{(4)}$

- $b_{14}^{(4)}$ is a function of $(a_{15}^{(4)}, a_{14}^{(4)}, a_{13}^{(4)}, a_{12}^{(4)})$ depending on $K_0 := RK_3^4$ (see Fig. 3).

- $(a_{15}^{(4)}, a_{14}^{(4)}) = (b_7^{(2)}, b_6^{(2)})$

- $a_{13}^{(4)} = b_7^{(3)}$ is a function of $(a_7^{(3)}, a_6^{(3)}, a_5^{(3)}, a_4^{(3)})$ depending on $K_1 := RK_1^3$. As for β fixed $a_6^{(3)}$ is balanced (i.e. takes values 0 and 1 equally often) and as this bit affects α only by means of the block keyed by K_1, key bits $k_{\oplus 2}^{(i)}$ of K_1 do not affect the probability of equation $\alpha = \beta$.

- $a_{12}^{(4)} = b_2^{(3)}$ is a function of $(a_3^{(3)}, a_2^{(3)}, a_1^{(3)}, a_0^{(3)})$ depending on $K_2 := RK_0^3$. For the same kind of reason as before, key bits $k_{\oplus 1}^{(i)}$ of K_2 do not affect the probability.

- Finally, $(b_7^{(2)}, b_6^{(2)})$ is a function of $(a_7^{(2)}, a_6^{(2)}, a_5^{(2)}, a_4^{(2)})$ depending on $K_3 := RK_1^2$. Still using the same arguments, we note that key bits $k_{\oplus 1}^{(i)}$ of K_3 do not affect the probability.

Thus the probability of $\alpha = \beta$ (computed over all 2^{16} plaintexts) depends on the 4 subkeys K_0, K_1, K_2, K_3 (or at least part of them). We write them as:

$$K_0 = (k_0^{(11)}, k_0^{(10)}, k_0^{(01)}, k_0^{(00)})$$

$$K_1 = (k_1^{(11)}, k_1^{(10)}, k_1^{(01)}, k_1^{(00)})$$

$$K_2 = (k_2^{(11)}, k_2^{(10)}, k_2^{(01)}, k_2^{(00)})$$

$$K_3 = (k_3^{(11)}, k_3^{(10)}, k_3^{(01)}, k_3^{(00)})$$

There are $12 + 3 \times 8 = 36$ subkey bits implied. A priori this does not allow easy exhaustive computation of the probability for every key (complexity $2^{36} \times 2^{16}$). However there are groups of values for which the associated probability is the same. Let us say that 2 subkeys K_1 and K_{1*} are *equivalent* if for any K_0, K_2, K_3, the probability associated to (K_0, K_1, K_2, K_3) and (K_0, K_{1*}, K_2, K_3) is identical; Equivalence between 2 subkeys K_2 and K_{2*} is defined similarly. We can make the following observations:

- For β fixed $(a_5^{(3)}, a_4^{(3)})$ is balanced. Therefore two subkeys K_1 and K_{1*} such that there exists a permutation $\pi : \{0,1\}^2 \to \{0,1\}^2$ satisfying $k_1^{(i)} = k_{1*}^{(\pi_i)} (\forall i \in \{00, 01, 10, 11\})$ are equivalent.

- The same argument can be used for K_2.

- As from the output of the block keyed by K_1 only bit $b_7^{(3)}$ matters, if $k_{1,X}^{(i)} = 1$ (for some $i \in \{00, 01, 10, 11\}$), then $k_{1,\oplus 1}^{(i)}$ does not affect the probability. Otherwise stated, if K_1 and K_{1*} are such that

$k_1^{(i)} = (1,0,0)$ while $k_{1*}^{(i)} = (1,1,0)$ (other bits being the same), they are equivalent.

- With the same kind of argument, $k_2^{(i)} = (1,0,0)$ and $k_{2*}^{(i)} = (1,0,1)$ are equivalent.

- Similarly, as output $b_{15}^{(4)}$ of the block keyed by K_0 does not matter, we have:

$$k_0^{(i)} = (0,0,0) \sim k_0^{(i)} = (0,1,0) \qquad k_0^{(i)} = (0,0,1) \sim k_0^{(i)} = (0,1,1)$$
$$k_0^{(i)} = (1,0,0) \sim k_0^{(i)} = (1,0,1) \qquad k_0^{(i)} = (1,1,0) \sim k_0^{(i)} = (1,1,1)$$

- Suppose $\exists i, j : k_1^{(i)} = (0,0,0)$ and $k_1^{(j)} = (0,1,0)$. Then, the key bit added to $a_7^{(3)}$ is 0 for the 2^{14} plaintexts for which $(a_5^{(3)}, a_4^{(3)}) = i$; it is 1 for the 2^{14} plaintexts for which $(a_5^{(3)}, a_4^{(3)}) = j$. Thus (taking into account that for $(a_5^{(3)}, a_4^{(3)})$ fixed $a_7^{(3)}$ is balanced) when $(a_5^{(3)}, a_4^{(3)}) \in \{i, j\}$, $b_7^{(3)} = 1$ one half of the times.
  Now if we replace $k_1^{(i)}$ and $k_1^{(j)}$ by $(1,0,0)$, when $(a_5^{(3)}, a_4^{(3)}) \in \{i, j\}$ we have $b_7^{(3)} = a_6^{(3)}$. As $a_6^{(3)}$ is balanced (for fixed $(a_5^{(3)}, a_4^{(3)})$), we still have that $b_7^{(3)} = 1$ one half of the times.
  The conclusion is that if a given key is such that $\exists i, j : k_1^{(i)} = (0,0,0)$ and $k_1^{(j)} = (0,1,0)$, then if $k_1^{(i)}$ and $k_1^{(j)}$ are replaced by $(1,0,0)$ the key obtained is equivalent to the former one.

- Similarly, if $\exists i, j : k_2^{(i)} = (0,0,0)$ and $k_1^{(j)} = (0,0,1)$, replacing these bits by $k_2^{(i)} = k_2^{(j)} = (1,0,0)$ does not change the probability.

Putting all these observations together, there are 9 equivalence classes for K_1 as well as for K_2. They are given in Table 2, with the number of elements in each class (out of 2^{12}).

Finally, the number of different quadruples (K_0, K_1, K_2, K_3) to explore in order to compute the probability distribution $\{P_K[N_{\alpha=\beta} = A]\}_{A=0}^{2^{16}}$ is $9^2 \cdot (2^8)^2 \cong 2^{22}$. This number could be further reduced, by exploiting more complex equivalences such as: "if K_0 has such value, then value of K_1 does not matter". However it is not necessary as with complexity $2^{22} \cdot 2^{16}$, the probability distribution of $N_{\alpha=\beta}$ is computable. It is roughly given in Annex A.2.

It is worth mentioning that in the previous discussion we made the (classical) hypothesis that the round keys are independent and uniformly distributed, while in practise they are derived from the master key using the key expansion

Table 2.   Equivalence classes for K_1 and K_2 with their cardinalities

K_1	#	K_2	#
$(000; 000; 000; 000)$	16	$(000; 000; 000; 000)$	16
$(000; 000; 000; 010)$	448	$(000; 000; 000; 001)$	448
$(000; 000; 000; 100)$	128	$(000; 000; 000; 100)$	128
$(000; 000; 010; 010)$	1120	$(000; 000; 001; 001)$	1120
$(000; 000; 010; 100)$	896	$(000; 000; 001; 100)$	896
$(000; 010; 010; 010)$	448	$(000; 001; 001; 001)$	448
$(000; 010; 010; 100)$	896	$(000; 001; 001; 100)$	896
$(010; 010; 010; 010)$	16	$(001; 001; 001; 001)$	16
$(010; 010; 010; 100)$	128	$(001; 001; 001; 100)$	128

Table 3.   Families of linear relations with the best mean bias

Mean bias	Number of lin. rel.	Output bit
$7 \cdot 10^{-2}$	16	$\in S_1$
$4 \cdot 10^{-2}$	64	$\in S_2$
$2,5 \cdot 10^{-2}$	192	$\in S_1$
$1,5 \cdot 10^{-2}$	64	$\in S_2$

described in section 2; in fact some quadruples (K_0, K_1, K_2, K_3) simply cannot be derived from a master key. Computation of the value taken by $N_{\alpha=\beta}$ for a small number of random keys from which round keys are derived perfectly validated the hypothesis.

6.   Searching for other Linear Relations Through a 5-Round Cipher

The procedure described in the previous section to compute the distribution of N_λ for a given linear relation λ is complicated and has non-negligible time complexity. It is however possible to identify linear relations having a big mean bias by evaluating this bias using only a part of the 2^{16} plaintexts, and this for a relatively small number of keys. Doing this, we observed that there are "families" of linear relations having about the same mean bias. Moreover linear relations from some families have their output bit belonging to $S_1 \equiv \{c_0, c_1, c_4, c_5, c_8, c_9, c_{12}, c_{13}\}$, while those from the other families have their output bit belonging to $S_2 \equiv \{c_2, c_3, c_6, c_7, c_{10}, c_{11}, c_{14}, c_{15}\}$. This is due to the fact that the last round of $DeKaRT$ need not be approximated if the output bit $\in S_1$.

Details about the families with the best mean bias are given in Table 3.

The linear relations of the first family (with bias $\sim 7 \cdot 10^{-2}$) are given in Table 4.

Table 4. Linear relations through 5-round $DeKaRT$ with the highest mean bias

$p_0 \oplus c_0$	$p_4 \oplus c_1$	$p_8 \oplus c_8$	$p_{12} \oplus c_9$
$p_0 \oplus c_5$	$p_4 \oplus c_4$	$p_8 \oplus c_{13}$	$p_{12} \oplus c_{12}$
$p_1 \oplus c_1$	$p_5 \oplus c_0$	$p_9 \oplus c_9$	$p_{13} \oplus c_8$
$p_1 \oplus c_4$	$p_5 \oplus c_5$	$p_9 \oplus c_{12}$	$p_{13} \oplus c_{13}$

7.   Implementation of the Attack

As explained in Section 4, we assume that the attacker knows for example 128 pairs $\{(P^j, c_i^j)\}_{j=0\ldots127}$, where P^j is a plaintext and c_i^j is the ith bit of the corresponding ciphertext. One attack strategy could be:

1 Consider all $2^{16} - 1$ possible input masks μ (i.e. all possible linear combinations of input bits).

2 For each of them, compute the bias of $\mu \bullet P = c_i$ over the 128 pairs ($\bullet$ denotes the scalar product over $\mathbb{Z}_2^{16}$).

3 The attacker intercepts a ciphertext bit c_i^* of which he does not know the corresponding plaintext P^*. The input mask μ^* with the highest bias (computed at step 2) is used to predict the unknown bit $\mu^* \bullet P^*$.

The efficiency of this algorithm is measured by the bias associated to μ^*, **computed over all 2^{16} plaintexts** this time. Indeed, it gives the reliability of the guess made at step 3. Practical experiments show that we have a mean bias of 0,059 when the ciphertext bit considered $\in S_1$ and of 0,022 when it is in S_2.

However it is possible to do better if in step 1 of the attack, we restrain ourself to the 336 relations mentioned in Section 6 (or more precisely, to those of them concerning bit c_i). Then the bias obtained is 0,107 when the ciphertext bit considered $\in S_1$ and 0,074 when it is in S_2. Moreover the probability computed over 128 plaintexts almost always "goes in the same direction" than the one computed on all 2^{16} plaintexts (i.e. suggests the same value for $\mu \bullet P \oplus c_i$). This significant improvement is due to the fact that if we consider all possible input masks, it is often the case that estimation on 128 plaintexts happens to emphasize a linear relation which in fact has a small (or null) bias when computed over all 2^{16} plaintexts; a pre-selection of "a priori good" input masks greatly reduces this phenomenon. It is this improvement that motivated the research of *a priori* good linear relations described in Section 6.

In Table 5 we give mean biases for different numbers of pairs (P^j, c_i^j) known by the attacker. We insist on the fact that these figures are *mean* biases. This means that sometimes the bias associated to μ^* will be 0, which

Table 5. Mean bias as a function of the number of pairs known by the attacker

# pairs	if $c_i \in S_1$	if $c_i \in S_2$
64	0,095	0,067
128	0,107	0,074
256	0,118	0,083
512	0,123	0,087
1024	0,127	0,092

means that the attack has completely failed. Other times the bias will be 1/4 (or even 1/2) and the information gained by the attacker is real. The attacker must be able to compute a priori the bias he can expect. It is given by equation (2) in Section 4.

As an improvement to the attack, it could also be possible to consider several approximations (implying the same ciphertext bit c_i) simultaneously in order to retrieve more information. However this is far from trivial, as the possible correlation between these approximations must be taken into account. The paper from Biryukov&al. [6] could help in this context.

Also, we assumed only one bit of the ciphertext was available. If several are, this allows more bits of information about the plaintext to be retrieved (moreover linear approximations implying linear combinations of these ciphertext bits can be considered, which can improve the efficiency of the attack).

8.   Conclusion

In this paper we have seen that $DeKaRT$, despite its structure being significatively more complex than previous primitives, is vulnerable to linear cryptanalysis. Even using one only bit of the ciphertext (as it is often the case in the context of probing attacks), it is possible to obtain information about an unknown plaintext using very few known (Plaintext, Ciphertext bit) pairs. We do not claim $DeKaRT$ is useless for data scrambling: indeed, the requirements for such a type of primitive can be relaxed in comparison with usual requirements for block ciphers. More than the overall structure, some proposals for the number of rounds provided in the original paper seem to be too optimistic for a really strong security. The purpose of this paper is rather showing that such type of key-dependent transform still has strongly linear structures. It is why we believe it is possible to construct a better primitive with the same throughput and size constraints; probably a structure nearer the classical paradigms of block cipher design (constant and highly non-linear S-boxes) could achieve it. There is place for research effort in this direction.

Table A.1. Probability Distribution of $N_{\alpha=\beta}$

$N_{\alpha=\beta}$	P_K
$\in [0, 16383]$	0,006
$= 16384$	0,010
$\in [16385, 22527]$	0,029
$= 22528$	0,013
$\in [22529, 24575]$	0,013
$= 24576$	0,053
$\in [24577, 26623]$	0,024
$= 26624$	0,028
$\in [26625, 27647]$	0,009
$= 27648$	0,022
$\in [27649, 28671]$	0,014
$= 28672$	0,045
$\in [28673, 29695]$	0,012
$= 29696$	0,026
$\in [29697, 30719]$	0,013
$= 30720$	0,050
$\in [30721, 31743]$	0,015
$= 31744$	0,016
$\in [31745, 32767]$	0,013
$= 32768$	0,178

Appendix

1.   The Bit Permutation layer BP

The bit permutation we chose is:

Inp. Bit Pos.	15	14	13	12	11	10	9	8	7	6	5	4	3	2	1	0
Outp. Bit Pos.	5	0	15	10	1	4	11	14	13	8	7	2	9	12	3	6

2.   Probability Distribution of $N_{\alpha=\beta}$

Out of the $2^{16} + 1$ a priori possible values for $N_{\alpha=\beta}$, only 199 occur with a non-zero probability. In Table A.1 we only mention the ones having probability $\geq 0,01$. Moreover we give probabilities associated with intervals. As it is easy to show that $P[N_{\alpha=\beta} = A] = P[N_{\alpha=\beta} = 2^{16} - A]$, the table only goes from 0 to 2^{15}.

References

[1] J.D.Golic, *DeKaRT: A New Paradigm for Key-Dependent Reversible Circuits*, Proceedings of CHES 2003, Lecture Notes in Computer Science, vol. 2779, pp. 98 − 112, 2003.

[2] E. Brier, H. Handschuh, C. Tymen, *Fast Primitives for Internal Data Scrambling in Tamper Resistant Hardware*, Proceedings of CHES 2001, Lecture Notes in Computer Science, vol.

2162, pp. 16 – 27, 2001.

[3] M. Matsui, *Linear Cryptanalysis Method for DES Cipher*, Advances in Cryptology - EU-ROCRYPT 93, Lecture Notes in Computer Science, vol. 765, pp. 386 – 397, 1994.

[4] R. Anderson and M. Kuhn, *Tamper resistance - a Cautionary Note*, second USENIX Workshop on Electronic Commerce Proceedings, pp. 1 – 11, Oakland, California, November 1996.

[5] O. Kømmerling and M. Kuhn, *Design principles for Tamper-Resistant Smartcard Processors*, USENIX Workshop on Smartcard Technology, Chicago, Illinois, USA, May 1999.

[6] A. Biryukov, C. De Canniere, M. Quisquater, *On Multiple Linear Approximations*, Available at http://eprint.iacr.org/, 2004/057.

AN OPTIMISTIC FAIR EXCHANGE PROTOCOL FOR TRADING ELECTRONIC RIGHTS

Masayuki Terada
Network Management Development Dept., NTT DoCoMo
te@rex.yrp.nttdocomo.co.jp

Makoto Iguchi
Information Sharing Platform Labs., NTT
iguchi@isl.ntt.co.jp

Masayuki Hanadate
Information Sharing Platform Labs., NTT
hanadate@isl.ntt.co.jp

Ko Fujimura
Information Sharing Platform Labs., NTT
fujimura@isl.ntt.co.jp

Abstract   Reliable electronic commerce systems must offer fairness. In this paper, we propose a fair exchange protocol for trading electronic vouchers, which are the representation of rights to claim goods or services. This protocol enables two players to exchange vouchers stored in their smartcards fairly and efficiently. The players can exchange vouchers through a 4-round mutual communication protocol between their smartcards as long as the protocol is performed properly. If the protocol becomes unable to proceed due to misbehavior of the partner or for any other reason, the player that has fallen into the unfair condition (i.e. the player who sent its voucher but didn't receive the desired voucher) can recover fairness by performing a recovery protocol with a trusted third-party. Since the recovery protocol doesn't need the cooperation of the partner, fairness can be recovered without identifying or tracking the partner; one can trade vouchers securely even if the trading partner cannot be identified.

1. Introduction

Fairness is essential for realizing secure electronic commerce in which both consumers and merchants can participate with a sense of safety. From the consumers' viewpoint, payment should not be committed without receiving the merchandise purchased, while the merchants may not want to send the goods or services prior to being paid.

Both of the above requirements can be achieved easily at face-to-face transactions in real shops. However, they are difficult in electronic commerce because payments and the delivery of merchandise are rarely conducted simultaneously.

Once people are disadvantaged, they need to identify the trading partner and acquire compensation. Unfortunately, identifying the partner is not easy in the Internet[4]. Since this difficulty is likely to become even worse in the consumer-to-consumer (C2C) market that is forming, security that depends on an identification process is impractical.

In order to realize secure electronic commerce systems that dispense with identification, we focus on "electronic rights", irreproducible and transferable digital data representing the rights to claim services or goods[14]. Their real world equivalents are pieces of paper, i.e. tickets, coupons and vouchers. Following RFC3506[5], we refer to such data as (electronic) vouchers hereafter. Since electronic vouchers must be transferred among users and redeemed in an off-line environment like real tickets and coupons, we assume that they are stored and managed in tamper-resistant devices like smartcards to prevent illegal acts like reproduction and forgery.

A voucher can represent diverse types of rights. For instance, it can represent the right to claim goods like "one hamburger", or to claim services like "one night accommodation". Moreover, it can be used as it were money by representing the right to claim conversion into currency or a certain amount of some valuable item (e.g. gold); electronic money can be treated as a sort of voucher. The fair exchange of vouchers, therefore, enables us to realize not only secure barter trading but also secure purchase transactions[7].

Identification of the trading partner is not required in these transactions provided that the fairness of the exchange is guaranteed. The only thing they need to certify is the genuineness of the vouchers being exchanged. Transaction security, therefore, doesn't depend on the trustworthiness of the trading partner, but rather on that of the voucher's issuer. This simplifies the certification process because there would be many fewer voucher issuers than participants. Fair exchange of vouchers is thus the key component to realize fair and secure electronic commerce while dispensing with the need for identification[6].

Up to now, however, there has been no efficient method to exchange vouchers with fairness. A mediating (or active) trusted third-party (TTP) can exchange vouchers fairly, but this approach has drawbacks in terms of availability and scalability because the TTP has to be synchronously involved in every exchange, which creates potential bottlenecks[11]. On the other hand, a number of optimistic fair exchange protocols have been proposed in which the TTP participates only if errors occur during the exchange[1, 11, 8, 17]. These protocols are much more efficient and practical for exchanging digital signatures (aka contract signing) or payments and receipts, but they cannot be used to exchange vouchers because they fail to prevent the reproduction of vouchers.

In this paper, we propose a new protocol that enables the fair and effective exchange of electronic vouchers that represent rights. This protocol exchanges vouchers in an optimistic manner; the trading participants first try to exchange vouchers through mutual communication, and they activate the TTP to recover fairness if the exchange is suspended or interrupted and can not be continued. Since this protocol guarantees "strong fairness" in any exchange, fairness can be recovered without identifying or tracking the trading partner.

The rest of the paper is organized as follows: Section 2 states the definitions and requirements for representing vouchers and fairness in voucher trading. This section also describes issues of the legacy method that uses a mediating TTP and previous optimistic fair exchange protocols. Section 3 details the protocol proposed in this paper. Section 4 discusses the proposed protocol by reference to the requirements stated in Section 2.

2. Preliminaries

This section states the definition and requirements of electronic vouchers and fairness in voucher exchanges. In addition, this section discusses drawbacks of the previous approaches.

2.1 Electronic voucher

An electronic voucher is a digital representation of the right to claim services or goods. In this context, the right is created by the issuer, such as a supplier of merchandise or provider of services, as a promise to the right holder. According to RFC3506[5], a voucher is defined as follows:

Electronic voucher Let I be a voucher issuer, H be a voucher holder, and P be the issuer's promise to the right holder. An electronic voucher is defined as the 3-tuple of (I, P, H).

Similar to paper tickets and current money, vouchers should be transferable. The voucher holder H can transfer the voucher (I, P, H) to another participant H'. This transfer is represented as the rewriting of the tuple $(I, P, H) \rightarrow (I, P, H')$. The right lapses from holder H as a result of the transfer.

Vouchers must be protected from illegal acts like forgery, alteration, and reproduction. These security requirements are defined as follows:

Preventing forgery A voucher (I, P, H) must be generated (issued) only by issuer I and must not be generated by any other participant [1]. In addition, it must not be possible for anyone to alter issuer I once the voucher is issued.

Preventing alteration Once a voucher (I, P, H) is issued, it must not be possible for anyone to alter promise P.

Preventing reproduction It must not be possible for voucher (I, P, H) to be reproduced. In particular, in a transfer from H to H', both (I, P, H) and (I, P, H') must not exist simultaneously throughout the transfer.

2.2   Exchange of vouchers

As mentioned in Section 1, diverse types of commerce transactions can be mapped into exchanges of vouchers. An exchange of vouchers consists of a pair of mutual transfers of vouchers.

To exchange vouchers fairly, it must be assured that no trading participant loses its voucher without receiving the desired voucher; each participant must be able to recover its voucher or the desired voucher in a given period as long as it behaves properly. An exchange of vouchers that satisfies this property is defined as a fair voucher exchange. Details are given below:

Fair voucher exchange Assume that there are two electronic vouchers (I_1, P_1, H_A) and (I_2, P_2, H_B). An exchange constructed of the pair of transfers $(I_1, P_1, H_A) \to (I_1, P_1, H_B)$ and $(I_2, P_2, H_B) \to (I_2, P_2, H_A)$ is defined as a fair voucher exchange, if all of the following conditions are satisfied:

1 Provided that H_X (i.e. H_A or H_B) executes the exchange in proper manner, it is assured that H_X will own either voucher as a result of the exchange. That is, either (I_1, P_1, H_X) or (I_2, P_2, H_X) is assured to exist at the termination of the exchange.

2 Provided that H_X executes the exchange in proper manner, it is assured that H_X can terminate the exchange in finite time.

The following conditions must also be satisfied to guarantee the prevention of voucher reproduction:

3 Throughout the exchange, (I_i, P_i, H_A) and (I_i, P_i, H_B) $(i \in \{1, 2\})$ must not exist simultaneously.

---

[1] Another participant I' may generate voucher (I', P, H) since the issuer is I'.

2.3   Implementation model of vouchers

Although several models can be used to implement electronic vouchers[6], we will focus on the "stored value" model like FlexToken[14]because this type of model has advantages in terms of usability and scalability.

This implementation model assumes that users (owners of vouchers) have tamper-resistant devices like smartcards to store and manage vouchers; it enables vouchers to be transferred among users and redeemed in off-line environments securely without involving banks or other third-parties. In this paper, we refer to these devices as PTDs (personal trusted devices). A PTD protects vouchers from illegal acts like forgery and reproduction from anyone, including its owner.

In this model, a voucher (I, P, H) is considered to exist when user H has PTD U storing digital data v that corresponds to (I, P). In FlexToken, for example, PTDs store 2-tuple entries $(h(PkI), h(R))$, called a token for v, where $h()$ is a secure hash function like SHA-1[10], PkI is a public key of issuer I, and R is a document called rights definition, which defines the contents of promise P.

The voucher transfer $(I, P, H_A) \rightarrow (I, P, H_B)$ is realized by transferring v from sender's PTD U_A to receiver's PTD U_B, which concludes by deleting v from U_A and storing v in U_B.

2.4   Previous works

While the previous methods of implementing vouchers realize secure circulation of vouchers including issuance, transfer, and redemption with preventing illegal acts like forgery and reproduction, they don't realize fairness in the exchange of vouchers. Although FlexToken discusses fairness in circulating rights, it only aims to ensure the non-repudiation of the fact of circulation; ensuring fairness in voucher exchanges was not considered.

To achieve the fair exchange of vouchers, a voucher trading system that uses a mediating TTP has been introduced[7]. This system satisfies the fairness and security requirements described above, provided that the TTP acts honestly. However, it is weak in terms of availability because every exchange requires synchronous interactions between the TTP and both trading partners; its scalability is also suspect because traffic would be heavily focused on the TTP.

A number of fair exchange protocols for digital signatures and digital data have been proposed[1, 11, 8, 17]. In particular, protocols called optimistic protocols[2, 16, 15]or off-line TTP protocols[18]use a TTP only if errors occur in the exchange process. Under the assumption that errors rarely occur, these protocols relax the problems with using a TTP and enable exchanges to be performed efficiently. These protocols realize, in a practical manner, fair contract signings, certified mails, and fair exchanges of a payment and its receipt as demonstrated in SEMPER[9, 12], however, no optimistic protocol has been proposed or can be applied for exchanging vouchers.

In order to exchange vouchers fairly without identifying and tracing the trading partner, it is required to ensure fairness without an external dispute resolution process. This level of fairness is called strong fairness[11, 1].

According to [11], strong fairness can be achieved in an optimistic protocol if one of the items to be exchanged ensures "strong generatability" which means the item (or an equivalent item) is generatable by the TTP or "strong revocability" which means the item is revocable by the TTP[2].

We first discuss protocols that use strong generatability and are based upon the protocol introduced in [2]. A voucher is strongly generatable if the TTP could generate a message equivalent to the message that ensures that a voucher is stored while preventing illegal acts on the voucher. However, a voucher is not strongly generatable in this sense because voucher reproduction is possible by replaying the first message of the protocol[3]. This message enables its recipient to perform the resolve protocol with the TTP, which may allow replay of the exchange[4]. This might be harmless for applications like contract signing because it only brings another evidence of confirmation of the same contract concluded in the previous exchange (assuming that contents of the contract are assured to be unique). However, it causes the recipient of the message to reproduce a voucher already received in the previous voucher exchange.

Applying a protocol that uses strong revocability[15]for voucher exchanges is possible but rather impractical. This protocol requires a means by which the TTP can ensure revocation that would prevent the receipt of improperly exchanged vouchers. This is easy for closed-loop electronic money systems (referred to "ECash-like system" in [15]) which involve a bank to confirm payments, but it is difficult for vouchers which have off-line capabilities similar to real tickets or current money; it is impractical to inform revocation to all participants who may receive the revoked voucher before the voucher is transferred or redeemed to them.

We therefore propose a new optimistic protocol for exchanging vouchers that is not based upon either type of protocol mentioned above. Our protocol prevents the replay attacks that make voucher reproduction possible. In addition, the protocol is simple and efficient; its main protocol consists of four messages, two of which are signed, and is as simple as a naive voucher exchange involving the mutual transfer of two vouchers using challenge and response.

---

[2] The other item is required to be "weakly generatable", which is rather easy to achieve.

[3] Another replay attack against this protocol is pointed out in [13], but it can be fixed easily as described in that paper.

[4] Additional message exchanges could prevent this replay attack, but it would make the whole protocol much more complicated.

3.   Fair Exchange Protocol for Vouchers

This section describes a fair exchange protocol that satisfies the requirements stated in Section 2.

In this section, we assume that the vouchers to be exchanged are (I_1, P_1, H_A) and (I_2, P_2, H_B), the implementation model is stored value model, (I_i, P_i) is represented by token v_i, and voucher (I_i, P_i, H_X) exists when v_i is stored in PTD U_X held by user H_X. The set of tokens stored in U_X is referred to as V_X. PTD U_X is a tamper-resistant device like a smartcard which is capable of preventing illegal alteration of V_X or the program performing the exchange, as well as keeping its signing key and n_2 (described in 3.2) secret. Each process in the PTD is assumed to be performed atomically.

User H_X would also have a user terminal device such as a mobile phone or a PDA to interact with its PTD U_X if the PTD doesn't have a user interface (like a smartcard). The user terminal would facilitate the generation of communication channels among PTDs and TTP T (see below) as well, but it doesn't have to be trusted by anyone but its owner; the owner might try to cheat its PTD or another PTD by forging or replaying messages using the terminal. The terminals aren't shown explicitly hereafter since they can be merely treated as a part of (insecure) communication channels in the proposed protocol.

This protocol consists of three sub-protocols: a main protocol, an abort, and a resolve protocol. An exchange starts by performing the main protocol. The exchange completes only within the main protocol whenever both participants act honestly and there is no trouble in the communication channel between them. If there are any troubles in the main protocol, either participant can recover fairness and terminate the exchange with the abort protocol or the resolve protocol using TTP T, which is a third-party trusted by both participants (and also the issuers of the vouchers). This protocol exchanges vouchers fairly and optimistically.

3.1   Definitions

The other definitions and assumptions needed to describe the proposed protocol are given below (details are given in Section 4).

U_A has public key certificate $Cert_A$ including public key PkA, a signing key which generates a signed message $(m)_{PkA}$ verifiable with $Cert_A$, and a verify function $Verify((m)_{PkX}, Cert_X)$ which verifies the signed message $(m)_{PkX}$ using the corresponding certificate $Cert_X$; and U_B and T likewise.

$Cert_A$ and $Cert_B$ represent that their keyholders U_A and U_B are certified as proper PTDs, while $Cert_T$ represents that its keyholder T is certified as a proper TTP. Note that these certificates don't have to be identity certificates to identify individuals, but have to be issued by a party who is trusted by the issuer of the vouchers being exchanged. It is easy to ensure this requirement if the

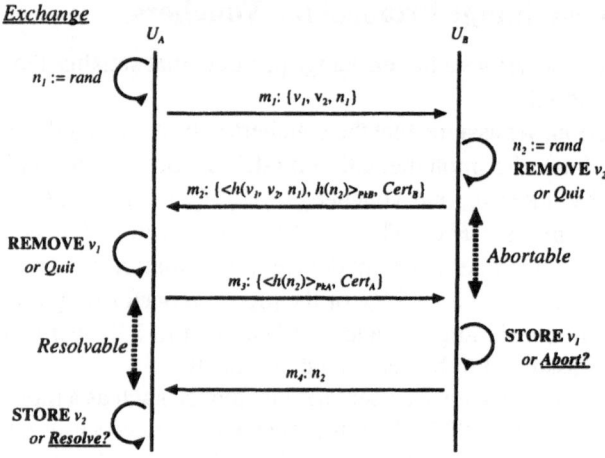

Figure 1.   Main protocol.

existence of a certificate authority trusted by all participants can be assumed. If not, additional certificates by the issuer like accredit information in FlexToken can be used (see details in [14]).

In addition, U_A and U_B manage sets of sessions S_A and S_B respectively, and T manages sets S_{abort} and $S_{resolve}$ which include sessions that have been aborted or resolved.

3.2   Main protocol

Figure 1 shows the main protocol. At the initial point, tokens v_1 and v_2 are stored in U_A and U_B respectively ($v_1 \in V_A, v_2 \in V_B$). If this protocol terminates without failure, v_1 and v_2 are swapped ($v_1 \in V_B, v_2 \in V_A$). The exchange is performed without intervention of the TTP in this case. In the explanation hereafter, it is assumed that H_A knows the content of v_2 at the initial point for simplicity.

H_B may order U_B to abandon this protocol at any time in the abortable section in Figure 1 (after sending m_2 and before receiving m_3) and terminate the exchange by performing the abort protocol with T which recovers fairness for H_B. Likewise, H_A may order U_A to terminate the exchange by performing the resolve protocol with T throughout the resolvable section (after sending m_3 and before receiving m_4). Before these sections, H_A and H_B can merely quit the exchange without losing fairness.

H_A starts the main protocol by ordering U_A to exchange v_1 in U_A and v_2 in U_B. The main protocol is performed in the following way:

1 After receiving the order from H_A, U_A performs the following:

    (a) Generates a random number n_1 and adds it to set S_A.

    (b) Sends $m_1 : \{v_1, v_2, n_1\}$ to U_B, which is the offer of the exchange.

2 U_B receives m_1 and performs the following:

    (a) Confirms m_1 if it is an acceptable offer for H_B. If not, U_B quits the exchange (and may inform U_A of this event).

    (b) Generates n_2 and adds it to set S_B; n_2 has to be kept in secret until it is sent as m_4.

    (c) Removes v_2 from V_B, which causes the corresponding rights to lapse from H_B; H_B temporarily falls into unfair condition.

    (d) Calculates $s_1 := h(v_1|v_2|n_1)$ and $s_2 := h(n_2)$.

    (e) Sends $m_2 : \{(s_1|s_2)_{PkB}, Cert_B\}$ to U_A.

3 U_A receives m_2 and performs the following:

    (a) Confirms that all of the following equations are satisfied.

        i $s_1 = h(v_1|v_2|n_1)$
        ii $Verify(m_2) = true$

    If not, U_A waits m_2 again or quits the exchange by removing n_1 from S_A.

    (b) Removes n_1 from S_A and adds s_2 to S_A.

    (c) Removes v_1 from V_A, which causes the corresponding rights to lapses from H_A; H_A temporarily falls into unfair condition (That is, H_A and H_B fall into unfair condition at this time).

    (d) Sends $m_3 : \{(s_2)_{PkA}, Cert_A\}$ to U_B.

4 U_B receives m_3 and performs the following:

    (a) Confirms that all of the following equations are satisfied.

        i $s_2 = h(n_2)$
        ii $Verify(m_3) = true$

    If not, U_B waits m_3 again or abandons the main protocol and performs the abort protocol to recover fairness.

    (b) Removes n_2 from S_B and adds v_1 included in m_1 to V_B. H_B enters fair condition again.

    (c) Sends $m_4 : n_2$ to U_A, and U_B terminates the exchange.

5 U_A receives m_4 and performs the following:

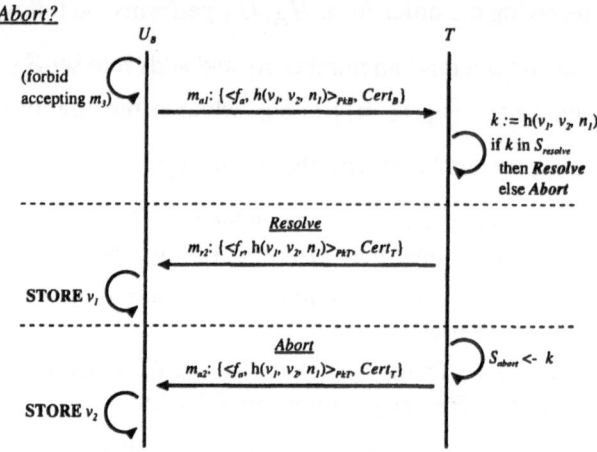

Figure 2.   Abort protocol.

(a) Confirms $h(m_4) = s_2$. If not, U_A waits m_4 again or performs the resolve protocol to recover fairness.

(b) Removes s_2 from S_A and adds v_2 to V_A. H_A enters fair condition again, and U_A terminates the exchange.

3.3   Abort protocol

Figure 2 shows the flow of the abort protocol. As mentioned above, U_B may abandon the main protocol at any time in the abortable section by performing this protocol.

The abort protocol enables U_B to recover fairness by sending an abort request m_{a1} to T and receiving the abort admission m_{a2} which allows U_B to restore v_2, or the resolve admission m_{r2} to store v_1 if T has already received the resolve request from U_A. If U_B cannot receive either m_{a2} or m_{r2} from T in a given period, U_B may resend m_{a1}.

The abort protocol is performed as follows:

1   U_B abandons the main protocol, and is prohibited from receiving m_3 in the main protocol.

2   U_B sends $m_{a1} : \{(f_a|s_1)_{PkB}, Cert_B\}$ to T. Herein, f_a is a flag which represents the process of aborting.

3   T receives m_{a1} and performs the following:

     (a) Confirms $Verify(m_{a1})$. If not, waits m_{a1} or m_{r1} (described in the resolve protocol) again.

(b) Let $k := s_1$,

     i If $k \in S_{resolve}$, then send the resolve admission.

     ii If $k \notin S_{resolve}$, then add k to S_{abort} and send the abort admission.

The procedures U_B performs when U_B receives the abort admission or the resolve admission are as follows:

Receiving abort admission

1 T sends $m_{a2} : \{(f_a|k)_{PkT}, Cert_T\}$ to U_B.

2 U_B receives m_{a2} and performs the following:

     (a) Confirms $k = s_1$ and $Verify(m_{a2})$. If not, waits m_{a2} or m_{r2} again.

     (b) Removes n_2 from S_B.

     (c) Adds v_2 to V_B and terminates the exchange.

Receiving the resolve admission

1 T sends $m_{r2} : \{(f_r|k)_{PkT}, Cert_T\}$ to U_B.

2 U_B receives m_{r2} and performs the following:

     (a) Confirms $k = s_1$ and $Verify(m_{r2})$. If not, waits m_{a2} or m_{r2} again.

     (b) Removes n_2 from S_B.

     (c) Adds v_1 to V_B and terminates the exchange.

3.4   Resolve protocol

Figure 3 shows the flow of the resolve protocol. U_A may perform this protocol at any time in the resolvable section of the main protocol.

Similar to the abort protocol, the resolve protocol enables U_A to recover fairness by sending a resolve request m_{r1} to T and receiving the abort admission m_{a2} which allows U_A to restore v_1, or the resolve admission m_{r2} to store v_2 if T has already received the abort request from U_B. If U_A cannot receive either m_{a2} or m_{r2} from T in a given period, U_A may resend m_{r1}.

The resolve protocol is performed as follows:

1 U_A sends $m_{r1} : \{(f_r|s_1)_{PkA}, Cert_A\}$ to T. Herein, f_r is a flag that represents the process of resolving.

2 T receives m_{r1} and performs the following:

Resolve?

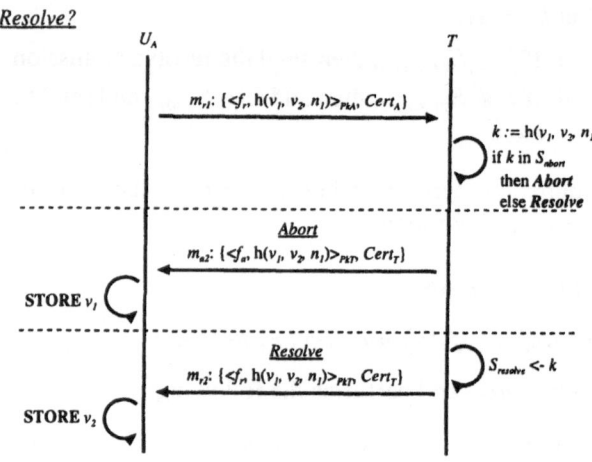

Figure 3.   Resolve protocol.

(a) Confirms $Verify(m_{r1})$. If not, waits m_{a1} or m_{r1} again.

(b) Let $k := s_1$,

    i If $k \in S_{abort}$, then send the abort admission.

    ii If $k \notin S_{abort}$, then add k to $S_{resolve}$ and send the resolve admission.

U_A processes the abort admission or the resolve admission in a similar way to U_B in the abort protocol, except for the following differences:

- U_B is replaced by U_A.

- Deleting n_2 from S_B is replaced by deleting s_2 from S_A.

- v_1 and v_2 stored by receiving m_{r2} or m_{a2} are swapped.

- U_A doesn't have to abandon the main protocol. U_A may conclude the exchange by receiving m_4 in the main protocol before receiving m_{a2} or m_{r2}, because it can be assumed that U_B has successfully concluded the exchange when U_A receives m_4 and therefore there is no chance for U_A to receive the abort admission m_{a2}.

4.   Discussions about Security and Fairness

This section discusses how the proposed protocol ensures the security requirements for vouchers and the fairness requirement.

In the discussion hereafter, the following assumptions are made:

1 Users H_A and H_B may try to cheat their PTDs; they might replay or forge any messages.

2 Tamper-resistant capability of both PTDs, U_A and U_B, is not compromised; they process the input data m_i properly and their signing keys are kept secret. In addition, n_2 is kept secret until it is sent as m_4.

3 TTP T properly processes abort requests and resolve requests in a given period.

4 The communication channels linking U_A, U_B and T may be insecure, i.e. attacks including eavesdropping, replaying and alteration are possible.

5 The communication channel between U_A and U_B may be lost permanently (i.e. either partner might escape in the middle of the exchange and the other partner is not assured of catching him).

6 The communication channel between T and U_A or U_B may be lost, but it recovers in a given period.

7 The hash function and signatures used in the protocol are sufficiently secure; e.g. the risk of collision of the hash function or forgery of the signatures can be ignored.

8 The certificate practice is secure enough; e.g. no certificate involved in the protocol can be forged or improperly issued.

4.1 Preventing forgery and alteration

In the proposed protocol, forgery or alteration is possible if it is possible to store a token v' different from either v_1 or v_2 included in offer message m_1.

There is no chance to store a different $v'(\notin \{v_1, v_2\})$ to U_A because the offer m_1 originates from U_A itself. H_A, H_B or another faulty party may try to store $v_i(\notin \{v_1, v_2\})$ to U_B by altering v_1 or v_2 included in m_1, but this would be detected by its inconsistent hash value at step 3 in the main protocol.

Assuming v_1 in m_1 is altered to v', s_1 included in the m_2 (protected by signature of U_B) becomes $s_1 = h(v', v_2, n_1)$, which is inconsistent with $h(v_1, v_2, n_1)$ as indicated by the comparison in step 3 of the main protocol.

The risk of forgery and alteration therefore depends on the collision resistance of hash function $h()$ and the strength of the signature, which are assumed to be sufficient.

Forgery and alteration are, therefore, prevented in the proposed protocol.

4.2 Preventing reproduction

In order to prevent reproduction, there must not be any token storing operation without the preceding corresponding removal operation of the same token.

In the main protocol, any storing operation of token v_i is performed after removing v_i. In addition, both m_2 and m_3, the evidence of the removal of v_2 and v_1, respectively, are protected against forgery or alteration by the attached signatures, and they cannot be replayed because storing v_1 and v_2 always involves the removal of n_2 and s_2 respectively, which prevents the overlapped storing of tokens. The main protocol thus prevents the reproduction of vouchers.

A discussion about abusing the recovery protocols is more complex. We discuss the possibility of reproducing v_1 and v_2 separately.

Preventing reproduction of v_1. v_1 is reproduced when both of the following are performed:

1 U_A stores v_1 in the resolve protocol.

2 U_B stores v_1 in the main protocol or the abort protocol.

In order for U_A to store v_1 in the resolve protocol, U_A needs to receive abort admission m_{a2}, which is signed and sent from T only if T has received abort request m_{a1} from U_B before receiving resolve request m_{r1} from U_A.

Assuming U_B sent m_{a1}, U_B cannot store v_1 in the main protocol because the main protocol is abandoned when U_B starts the abort protocol. Neither can U_B store v_1 in the abort protocol because U_B needs to receive resolve admission m_{a2}, which is sent from T only if T received m_{r1} before receiving m_{a1}. Since this contradicts the above condition for U_A to store v_1, v_1 cannot be reproduced by abusing the recovery protocols.

Preventing reproduction of v_2. v_2 is reproduced when both of the following are performed:

1 U_A stores v_2 in the main protocol or the resolve protocol.

2 U_B stores v_2 in the abort protocol.

For U_A to store v_2 in the main protocol, U_A needs to receive m_4, which is only known by U_B and cannot be guessed by any other participant. If U_B sent m_4, U_B cannot store v_2 because U_B ought to have stored v_1 and already terminated the exchange before sending m_4.

If U_A is to store v_2 in the resolve protocol, U_A needs to receive resolve admission m_{r2}, which is signed and sent from T only if T has *not* received abort request m_{a1} before receiving resolve request m_{r1}. Meanwhile, in order to let U_B store v_2 in the abort protocol, U_B needs to receive abort admission m_{a2}, which is signed and sent from T only if T has *not* received resolve request m_{r1} before receiving abort request m_{a1}. Since these two conditions contradict each other, v_2 cannot be reproduced by abusing the recovery protocols.

4.3   Ensuring fairness

In order to ensure fairness in voucher exchange, the following must be satisfied according to the definition of fairness described in Section 2:

1 Both U_A and U_B can obtain either token v_1 or v_2 (in the set V_A and V_B) when the exchange is terminated.

2 Both U_A and U_B can terminate the exchange in finite time.

3 v_1 must not be obtained by both U_A and U_B simultaneously throughout the exchange; v_2 likewise.

The last condition is derived from the requirement to prevent reproduction as already discussed in Section 4.2. We therefore discuss the remaining two conditions hereafter.

In the main protocol, U_A falls into unfair condition in which it has neither v_1 nor v_2 in V_A while in the resolvable section, and U_B falls into unfair condition while in the abortable section. Except in these sections, they have either v_1 or v_2 and can instantly quit the exchange at will.

While in the resolvable section, U_A is assured of recovering fairness and can terminate the exchange in a given period, because U_A may perform the resolve protocol at any time in the resolvable section, and the resolve protocol concludes in a given period provided that assumptions 3 and 6 described at the beginning of this section are satisfied. U_B in the abortable section is also assured of recovering fairness in a similar manner.

The proposed protocol therefore guarantees fairness for both U_A and U_B.

5.   Conclusion

In this paper, we stated the definitions and requirements for vouchers and fairness in voucher exchanges as a basis of realizing reliable electronic commerce, and proposed a new practical protocol that enables fair and efficient (optimistic) exchanges of vouchers stored in trusted devices like smartcards, while preventing illegal acts on vouchers such as reproduction and forgery.

This protocol enables the trading participants to exchange vouchers without recourse to a TTP provided there is no accident in the exchange; it guarantees the recovery of fairness and safe termination of the exchange in a given period by accessing a TTP even if there are problems with the communication channel or dishonest acts by the partner. These properties eliminate the problems inherent in the previous method that uses a mediating TTP, including the scalability problem (traffic concentrated on the TTP) and the availability problem (TTP involved in each trade).

This protocol is thus practical and enables people participating in electronic commerce to trade with unidentified partners in complete confidence.

References

[1] N. Asokan. *Fairness in Electronic Commerce.* PhD thesis, University of Waterloo, 1998.

[2] N. Asokan, V. Shoup, and M. Waidner. Asynchronous protocols for optimistic fair exchange. In *Proceedings of the IEEE Symposium on Research in Security and Privacy,* pages 86–99. IEEE Computer Society Press, May 1998.

[3] N. Asokan, V. Shoup, and M. Waidner. Optimistic fair exchange of digital signatures. *IEEE Journal on Selected Areas in Communications,* 18(4):593–610, Apr. 2000.

[4] C. Ellison and B. Schneier. Ten risks of PKI: What you're not being told about public key infrastructure. *Computer Security Journal,* 16(1):1–7, 2000.

[5] K. Fujimura and D. Eastlake. *RFC 3506: Requirements and Design for Voucher Trading System (VTS),* Mar. 2003.

[6] K. Fujimura and M. Terada. Trading among untrusted partners via voucher trading system. In *Proceedings of the 1st Conf. on e-Commerce, e-Business, e-Government.* IFIP, Oct. 2001.

[7] M. Iguchi, M. Terada, Y. Nakamura, and K. Fujimura. Voucher-integrated trading model for C2B and C2C e-commerce system development. In *Proceedings of the 2nd Conf. on e-Commerce, e-Business, e-Government.* IFIP, Oct. 2002.

[8] S. Kremer, O. Markowitch, and J. Zhou. An intensive survey of fair non-repudiation protocols. *Computer Communications,* 25(17):1606–1621, Nov. 2002.

[9] G. Lacoste, B. Pfitzmann, M. Steiner, and M. Waidner, editors. *SEMPER — Secure Electronic Marketplace for Europe,* volume 1854 of *LNCS.* Springer-Verlag, New York, NY, USA, 2000.

[10] NIST. *FIPS 180-1: Secure Hash Standard,* Apr. 1995.

[11] H. Pagnia, H. Vogt, and F. C.Gärtner. Fair exchange. *The Computer Journal,* 46(1):55–75, Jan. 2003.

[12] M. Schunter. *Optimistic Fair Exchange.* PhD thesis, Universität des Saarlandes, 2000.

[13] V. Shmatikov and J. C. Mitchell. Analysis of a fair exchange protocol. In *Proceedings of the 1999 FLoC Workshop on Formal Methods and Security Protocols,* July 1999.

[14] M. Terada, H. Kuno, M. Hanadate, and K. Fujimura. Copy prevention scheme for rights trading infrastructure. In *Proceedings of the 4th Working Conference on Smart Card Research and Advanced Applications (CARDIS),* pages 51–70. IFIP, Sept. 2000.

[15] H. Vogt. Asynchronous optimistic fair exchange based on revocable items. In *Proceedings of the 7th International Financial Cryptography Conference,* pages 208–222. IFCA, Jan. 2003.

[16] H. Vogt, H. Pagnia, and F. C. Gärtner. Modular fair exchange protocols for electronic commerce. In *Proceedings of the 15th Annual Computer Security Applications Conference,* pages 3–11, Dec. 1999.

[17] J. Zhou. Achieving fair nonrepudiation in electronic transactions. *Journal of Organizational Computing and Electronic Commerce,* 11(4):253–267, Dec. 2001.

[18] J. Zhou, R. Deng, and F. Bao. Evolution of fair non-repudiation with TTP. In *Proceedings of the 4th Australasian Conference on Information Security and Privacy (ACISP'99),* pages 258–269, Apr. 1999.

ACCOUNTABLE RING SIGNATURES: A SMART CARD APPROACH

Shouhuai Xu
Department of Computer Science, University of Texas at San Antonio

shxu@cs.utsa.edu

Moti Yung
Department of Computer Science, Columbia University

moti@cs.columbia.edu

Abstract

Ring signatures are an important primitive for protecting signers' privacy while ensuring that a signature in question is indeed issued by some qualified user. This notion can be seen as a generalization of the well-known notion of group signatures. A group signature is a signature such that a verifier can establish its validity but not the identity of the actual signer, who can nevertheless be identified by a designated entity called group manager. A ring signature is also a signature such that a verifier can establish its validity but not the identity of the actual signer, who indeed can *never* be identified by any party. An important advantage of ring signatures over group signatures is that there is no need to pre-specify rings or groups of users.

In this paper, we argue that the lack of an *accountability* mechanism in ring signature schemes would result in severe consequences, and thus accountable ring signatures might be very useful. An accountable ring signature ensures the following: anyone can verify that the signature is generated by a user belonging to a set of possible signers that may be chosen on-the-fly, whereas the actual signer can nevertheless be identified by a designated trusted entity – a system-wide participant independent of any possible ring of users. Further, we present a system framework for accountable ring signatures. The framework is based on a compiler that transforms a traditional ring signature scheme into an accountable one. We also conduct a case study by elaborating on how a traditional ring signature scheme is transformed into an accountable one while assuming a weak trust model.

1.   Introduction

The notion of ring signatures was formally introduced to resolve the following problem [26]: Suppose that Bob (also known as "Deep Throat") is a member of the cabinet of Lower Kryptonia, and that Bob wishes to leak a juicy fact to a journalist about the escapades of the Prime Minister, in such a way that Bob remains anonymous, yet such that the journalist is convinced that the leak was indeed from a cabinet member. On one hand, it should be clear that some straightforward solutions would not really solve the problem. It seems that even group signatures [14] may not suffice because they assume that the cabinet has deployed a group signature scheme, which means that there has been a designated entity (called group manager) specific to the cabinet. Even if the cabinet has deployed a group signature scheme, Bob still cannot send the journalist a group signature on the message, because it is likely that the group manager is under the control of the Prime Minister. On the other hand, ring signatures sufficiently solve the above problem because they allow Bob to send a message to the journalist without assuming any manager (meaning that the Prime Minister has absolutely no way to figure out that the signature is produced by Bob), where the message is accompanied with a ring signature indicating that it is really from a cabinet member.

In this paper, we argue that the lack of an *accountability* mechanism in a ring signature scheme would result in severe consequences, because Bob could rumor anything without being held accountable. Therefore, it is desirable to have accountable ring signatures while assuming no manager (an advantage of ring signatures over group signatures) and being able to hold misbehaving parties accountable (an advantage of group signatures over ring signatures) simultaneously. Moreover, it would also be desirable that such a scheme can be seamlessly integrated with a standard public key infrastructure (PKI).

1.1   Our Contributions

We introduce the concept of accountable ring signatures. An accountable ring signature ensures the following: anyone can verify that the signature is generated by a user belonging to a set of possible signers that may be chosen on-the-fly, whereas the actual signer can nevertheless be identified by a designated trusted entity – a system-wide participant independent of any possible set of users. Accountable ring signatures would be more appropriate than ring signatures for real-world utilization because: (1) a message accompanied with an accountable ring signature would be more reliable than a message accompanied with a traditional unaccountable ring signature, and (2) the information-theoretic anonymity provided by traditional ring signature schemes would hurdle their deployment in the real world – a similar scenario has happened in the context of anonymous e-cash schemes [13, 29].

Further, we present a system framework for accountable ring signatures. The framework is based on a compiler that transforms a traditional ring signature scheme into an accountable one while utilizing tamper-resistant smart cards, and can be seamlessly integrated with a standard PKI (including the traditional application of smart cards for protecting people's private keys, and the certificate revocation methods). We also conduct a case study to show how a traditional ring signature scheme is transformed into an accountable one, while assuming a *weak* trust model. This is particularly important in settings where smart cards may not be completely trusted (e.g., the manufacturers may be interested in getting information about the users' cryptographic keys via various subliminal channels, although they will not compromise the tamper-resistance of the smart cards – after all, that's how they make revenues).

1.2   Related Work

The notion of group signatures was introduced in [14]. A group signature can be seen as a normal digital signature with the following extra properties: anyone can verify that a signature is generated by a legitimate group member, while the actual signer can nevertheless be identified by a designated trusted entity called group manager. The basic idea underlying most group signature schemes is the following: In order for a group member (Alice) to sign a message, she needs to construct an *authorization-proof* to show that she has a legitimate membership certificate, and an *ownership-proof* to demonstrate knowledge of the secret corresponding to the membership certificate. Group signatures have been intensively investigated in the literature early on (cf. [15, 11, 10, 3, 9, 28] and the references therein). However, a group signature scheme requires an initialization procedure for specifying a group, which may not be possible or desirable under certain circumstances.

The notion of ring signatures was explicitly introduced in [26], although the basic idea had been mentioned several times [14, 8, 17, 18]. For example, [17] showed how one can produce witness-indistinguishable interactive proofs, which can be naturally transformed into a ring signature scheme via the Fiat-Shamir heuristics [20]. The scheme presented in [26] is based on the RSA public key cryptosystem [25], and its security is analyzed in the ideal cipher model [4]. This scheme was improved in

[7] where the authors presented a variant ring signature scheme whose security is analyzed in the random oracle model [5]. [1] proposed a general ring signature scheme that accommodates both RSA and discrete logarithm based cryptosystems. However, none of the above-mentioned ring signature schemes provides a revocation mechanism to hold a misbehaving party accountable.

In regard to utilizing smart cards, [12] showed how to implement a group signature scheme based on tamper-resistant smart cards. The basic idea underlying [12] is as follows: A group manager chooses two pairs of public and private keys, namely $\langle pk_1, sk_1 \rangle$ with respect to a public key cryptosystem and $\langle pk_2, sk_2 \rangle$ with respect to a digital signature scheme, where both pk_1 and pk_2 are publicly known. Each group member has a unique identity U and holds a tamper-resistant smart card that is equipped with pk_1 and sk_2. A group signature is indeed a digital signature obtained by applying sk_2 to the concatenation of a message m and a ciphertext c that is the encryption of U under public key pk_1 (i.e., c allows the group manager to revoke anonymity of the actual signer of a given group signature). Besides that this scheme is an implementation of traditional group signatures (i.e., the groups are pre-specified), it has the drawback that compromise of a single smart card (after all, tamper-resistance is still heuristic [2]) or revoking a user's membership (e.g., due to group dynamics) would force all the rest of smart cards to participate in a key-update process. As we will see, we achieve a higher assurance that a compromise of a smart card does not result in such a complex procedure, while making the revocation of a smart card or certificate completely transparent to the rest smart cards.

1.3   Organization

In Section 2, we briefly review some cryptographic preliminaries. In Section 3, we define accountable ring signatures. In Section 4 we present a system framework for accountable ring signatures, which is based on a generic compiler that transforms a traditional ring signature scheme into an accountable one in a *strong* trust model. In Section 5 we conduct a case study on transforming a traditional ring signature scheme into an accountable one while assuming a *weak* trust model. We conclude in Section 6.

2.   Cryptographic Preliminaries

Besides using standard public key cryptosystems, $\mathcal{E} = (GEN, E, D)$, that are semantically secure [22], and digital signature schemes, $\mathcal{S} = (GEN, S, V)$, that are existentially unforgeable under adaptive chose-message attacks [23], we use a threshold symmetric key cryptosystem specified below.

Threshold symmetric key cryptosystems. In order to revoke anonymity of accountable ring signatures while ensuring optimal resilience, we need to facilitate the following task: an entity (called dealer) encrypts a message so that the ciphertext can be decrypted by a set of a constant number n of servers $\{S_1, \cdots, S_n\}$, of which at most $t = \lfloor \frac{n-1}{2} \rfloor$ servers can be corrupted.[1] Below we give a concrete threshold symmetric key cryptosystem DSKC=(DSKC.Setup, DSKC.Enc, DSKC.Dec), which is adapted from a secure distributed pseudorandom function [24, 21] that can be based on a block cipher for performance reason.

DSKC.Setup. Define $d = \binom{n}{t}$, and define d subsets $\{S_j\}_{j=1}^d$ as all the subsets of $n - t$ of the n servers. The dealer chooses a key $\Theta = \{\theta_1, \cdots, \theta_d\}$, and defines a function F_Θ as $F_\Theta(x) = \oplus_{j=1}^d f_{\theta_j}(x)$, where $f_\alpha(\cdot) : \{0,1\}^\kappa \times \{0,1\}^l \rightarrow \{0,1\}^l$ is a pseudorandom function keyed by α of length κ. Finally, let all the servers in subset S_j hold θ_j, which means that server S_i ($1 \le i \le n$) holds $\Theta_i = \{\theta_j | S_i \in S_j\}$ such that $\bigcup_{i=1}^n \Theta_i = \{\theta_1, \cdots, \theta_d\} = \Theta$. Note that the dealer always holds Θ.

DSKC.Enc. The dealer encrypts a message m of length l as follows: choose $R \in_R \{0,1\}^l$ and set the ciphertext $\phi = \langle R, c = m \oplus F_\Theta(R) \rangle$.

DSKC.Dec. Given a ciphertext $\phi = \langle R, c \rangle$, the servers jointly decrypt it as follows: let S_i ($1 \le i \le n$) contribute $f_{\theta_j}(R)$ for every $\theta_j \in \Theta_i$, then a straightforward algorithm can decide the correct evaluations and recover the plaintext m.

3. Definition of Accountable Ring Signatures

In this section we review the definition of traditional (i.e., unaccountable) ring signatures and then give a definition of accountable ring signatures.

3.1 Definition of Ring Signatures

A ring signature scheme (RS) specifies a set of possible signers and a proof that is intended to convince a verifier that the actual signer of the

---

[1]Technically, a threshold message authentication scheme also suffices, but it would incur a linear communication complexity because of the size of the signatures. Note that a threshold message authentication scheme can be easily derived from a threshold symmetric key cryptosystem.

signature belongs to the set, while preserving her anonymity. A crucial property of ring signatures is that they are setup-free (i.e., there is no need for any special procedure beyond a standard PKI). Note that the size of a ring signature grows linearly with the size of the specified ring, given that the dynamic ring membership is not known in advance and has to be provided as part of a signature.

DEFINITION 1 *Suppose each user U_s is associated (via a PKI or certificate) with a pair of public and private keys $\langle pk_{U_s}, sk_{U_s} \rangle$. A ring signature scheme* RS *has the following two procedures.*

RS.Sign. *This is a probabilistic algorithm which, on input a message m, the public keys $pk_{U_1}, \cdots, pk_{U_z}$ of z ring members (i.e., possible signers) $U_1, \cdots, U_z$, and the secret key sk_{U_s} of U_s ($1 \leq s \leq z$), produces a ring signature σ (including $pk_{U_1}, \cdots, pk_{U_z}$) for the message m.*

RS.Ver. *This is a deterministic algorithm which takes as input (m, σ) and outputs either* TRUE *or* FALSE.

DEFINITION 2 *A ring signature scheme* RS *is said to be secure if it possesses the following properties:*

Correctness. *A ring signature produced by an honest user is always accepted as valid with respect to the specified ring.*

Unforgeability. *It must be infeasible for any user, except for a negligible probability, to generate a valid ring signature with respect to a ring he does not belong to.*

Anonymity. *No verifier is able to guess the actual signer's identity with probability greater than $1/z + \epsilon$, where z is the size of the ring and ϵ is a negligible function.*

3.2   Definition of Accountable Ring Signatures

DEFINITION 3 *Suppose each user U_s is associated with a pair of public and private keys $\langle pk_{U_s}, sk_{U_s} \rangle$. An accountable ring signature scheme (ARS) consists of the following algorithms.*

ARS.Sign. *This is a probabilistic algorithm which, on input a message m, the public keys $pk_{U_1}, \cdots, pk_{U_z}$ of z respective ring members $U_1, \cdots, U_z$, the secret key sk_{U_s} of U_s ($1 \leq s \leq z$), and some other information that will be used to produce a tag ϕ that allows some designated authority to recover the identity U_s, produces a ring signature σ (including $pk_{U_1}, \cdots, pk_{U_z}$ and ϕ) for the message m.*

ARS.Ver. *This is a deterministic algorithm which takes as input (m, σ) and outputs either* **TRUE** *or* **FALSE**.

ARS.Open. *This is a deterministic algorithm which takes as input (m, σ) as well as the authority's secret, and outputs the identity of the actual signer U_s.*

DEFINITION 4 *An accountable ring signature scheme* ARS *is said to be secure if it possesses the following properties:*

Correctness. *A ring signature produced by an honest user is always accepted as valid with respect to the specified ring.*

Unforgeability. *It must be infeasible for any user, except for a negligible probability, to generate a valid ring signature with respect to a ring he does not belong to.*

Anonymity. *No verifier should be able to guess the actual signer's identity with probability greater than $1/z + \epsilon$, where z is the size of the ring, and ϵ is a negligible function.*

Unlinkability. *No verifier is able to associate two ring signatures produced by the same user, even if the user's identity remains to be unknown to the adversary.*[2]

Revocability. *There is an anonymity revocation authority that is able to identify the actual signer of a given ring signature.*

No-misattribution. *It is infeasible for the anonymity revocation authority to convince an honest verifier that a given ring signature is produced by an honest user, who is actually not the signer though.*

Coalition-resistance. *A colluding subset of users (even all users) cannot generate a ring signature that the anonymity revocation authority cannot trace back to at least one of the colluding users.*

4. A Compiler

The basic idea underlying the compiler is to introduce into the system model an Anonymity Revocation Authority (ARA) that recovers the identity of the actual signer of a given accountable ring signature when the need arises, and tamper-resistant smart cards that are issued to the

---

[2]Previous ring signature schemes [26, 7] indeed achieve unconditional anonymity, which immediately imply unlinkability because $\epsilon = 0$. In accountable ring signature schemes, unconditional anonymity is still possible but would render them impractical.

users who want to produce accountable ring signatures. In order to make the compiler concise and generic, here we assume a *strong* trust model (We will show through a case study in Section 5 how one can weaken the trust model, particularly the trust on smart cards, to a more practical level.):

- The Certificate Authority (CA) is trusted in issuing public key certificates, which means that it will not issue a certificate to a user without conducting an appropriate authentication. This assumption is inherited from traditional ring signatures based on a standard PKI.

- The smart cards are tamper-resistant, while we must minimize the damage due to the compromise of a smart card. Such a cautious measure seems necessary given that tamper-resistance is currently only heuristic.

- The smart cards are trusted not to leak any information about its secrets (e.g., it will not utilize any channel to leak the holder's private key to the CA, the ARA, or the holder).

- The ARA is trusted in preserving the users' privacy.

Suppose each user U_s holds a smart card, while the unique identity U_s may be publicly known. The compiler takes as input a secure ring signature scheme RS=(RS.Sign, RS.Ver), and outputs a secure accountable ring signature scheme ARS=(ARS.Sign, ARS.Ver, ARS.Open) as specified below:

1 There is a setup procedure that is extended from the setup procedure in a standard PKI. (If users would utilize smart cards in the PKI, then this setup procedure does not incur any significant extra complexity.)

   (a) It establish an Anonymity Revocation Authority (ARA), which initializes a crypto-context whereby the identities of the card holders can be "embedded" into signatures in a certain way so that the ARA can recover them.

   (b) Suppose U_s intends to produce accountable ring signatures, the CA gives her a smart card that will generate a pair of public and private keys $\langle pk_{U_s}, sk_{U_s} \rangle$ with respect to an appropriate digital signature scheme, where pk_{U_s} is certified by the CA but sk_{U_s} is known only to the smart card.

2 ARS.Sign works as follows: Suppose U_s $(1 \leq s \leq z)$ wants to generate an accountable ring signature on message m with respect to the ring members' public keys $pk_{U_1}, \cdots, pk_{U_z}$.

  (a) U_s's smart card generates a token ϕ that embeds U_s. We require that the token ϕ does not leak any information (in either a computational or information-theoretic sense) about U_s, but does somehow allow ARA to recover U_s.

  (b) U_s's smart card executes RS.Sign on the message $m||\phi$, where "$||$" means string concatenation. Denote the output of RS.Sign by σ, which is the resulting accountable ring signature.

3 ARS.Ver is the same as RS.Ver, except that the message is $m||\phi$.

4 ARS.Open works as follows: Given a valid accountable ring signature σ on message $m||\phi$, ARA extracts the identity U_s of the actual signer from ϕ (perhaps via an inversion to the procedure that ϕ is generated).

5. Case Study

In this section we conduct a case study to show how the compiler transforms a secure ring signature scheme RS into a secure accountable ring signature scheme ARS while assuming a *weak* trust model. (Recall that, in order to make the compiler concise and generic, we assumed a strong trust model in Section 4.)

This section is organized as follows. In Section 5.1 we elaborate on the system model. In Section 5.2 we present the input, namely a secure ring signature scheme, and in Section 5.3 we present the output, namely a secure accountable ring signature scheme, whose security is analyzed in the full version of this paper due to space limitation.

5.1 System Model

Participants. There are three explicit categories of participants that are modeled as probabilistic polynomial-time Turing machines.

- As in a standard PKI, we assume that there is a certificate authority (CA) that certifies the users' public keys. (The extension to accommodating multiple CAs is straightforward.) The CA gives smart cards to individual users after an appropriate procedure (e.g., initializing some cryptographic parameters and installing some software programs).

- A user U_s may utilize her public key certificate as in a standard PKI for generating normal signatures. However, if she needs to generate any accountable ring signature, she must apply for a tamper-resistant smart card from the CA. The same pair of public and private keys $\langle pk_{U_s}, sk_{U_s} \rangle$ may be used to produce normal signatures and/or ring signatures. Besides holding a smart card, the user also possesses a computer that typically interacts with her smart card.

- There is an Anonymity Revocation Authority (ARA) that is responsible for identifying the actual signers of accountable ring signatures.

Trust. In order to simplify the system, we claim the following trust relationship (which is weaker than the trust model adopted by the generic compiler in Section 4).

- The smart card hardware is tamper-resistant in the sense that it will erase the secrets stored in it if there is an attempt at breaking into it. However, since the notion of tamper-resistance is only heuristic, we must ensure that the damage due to the compromise of a smart card is minimized. In particular, it would be highly desirable that compromise of a smart card does not allow the adversary to frame any other honest user for generating any accountable ring signature.

- The smart card software (e.g., operating system) is secure in the following sense: (1) it will not tamper with any application software program installed in the smart card; (2) there is no back-door for leaking the secrets stored in the smart card; (3) it provides a secure pseudorandom generator, if necessary.

- The CA will not collude with any user to issue him or her a certificate on a public key that will be accepted as eligible for generating accountable ring signatures. This also implies that the software program installed in the smart card by the CA (e.g., for generating temporary certificates) will not leak any information about the private keys to the card holders.

- The users' computers (including the software programs) are secure. In particular, the randomness output by the software program is guaranteed to be chosen uniformly at random, and there are no Trojan Horses in their computers.

- The ARA is trusted to preserve the anonymity of the users. To mitigate the trust, it is natural to implement the functionality via a distributed cryptosystem. Again, the ARA cannot be simply based on a tamper-resistant hardware because we must maintain certain access structure for restricting the access to it.

Communication channels. Once a user U_s obtained her smart card from the CA, the smart card cannot communicate with the outside world except U_s's computer. We assume that the communication channel between U_s's computer and smart card is physically secure, and that the channel through which U_s publishes an accountable ring signature is anonymous (the same as what is assumed in any ring signature scheme). All the other communication channels (including the one between U_s and the CA) are public and thus subject to eavesdropping by any party.

Adversary. Given the above trust model, we consider the following potential attacks.

- A dishonest user U_s manages to obtain a certificate for his public key pk_{U_s} corresponding to the private key sk_{U_s} that is known to U_s (e.g., she may try to forge a certificate on pk_{U_s}).

- A dishonest user U_s manages to obtain the private key sk_{U_s} corresponding to the public key pk_{U_s} that is generated by the software program installed by the CA in her smart card and thus certified by the CA.

- Although the CA is trusted to behave honestly in issuing certificates, it may not be trusted in any other sense. For example, the software program installed by the CA for generating "proofs" may not be trustworthy and may intend to leak information via a subliminal channel [27] or a kleptographic channel [30]. As a result, the private key may be completely compromised after further cryptanalysis. A similar channel may be established between a smart card and the ARA to compromise the secrets on the smart card. For simplicity, we assume that the software program on a smart card will not adopt the *halting* strategy [19] to construct a subliminal channel for leaking secrets. By *halting* strategy we mean that the software program installed by the CA decides whether or not to generate a valid message that is requested by the smart card holder. For example, the software program does generate the requested message only when it has embedded certain information into the subliminal channel known to the CA or ARA. We also assume that the software program will not adopt the *delaying* strategy to construct a subliminal channel. By *delaying* strategy we mean that the software program outputs a valid response to

a request from the smart card holder only at the time that co-
incides with a predefined subliminal time channel. For example,
a message generated in the morning meaning the bit of 0, and a
message generated in the afternoon meaning the bit of 1. In short,
we assume no covert or subliminal channels.

5.2   Input: A Secure Ring Signature Scheme

This scheme is adapted from [7], where it is shown to be secure with respect to Definition 2. Let l, l_b be security parameters, and $\mathcal{H} : \{0,1\}^* \to$
$\{0,1\}^l$ be a hash function that is modeled as a random oracle. Suppose
that each user U_i has a regular signature scheme built on a trapdoor
one-way permutation g_i (e.g., RSA) and that the modulus has length
$l_b < l$. Since the involved RSA moduli N_i are different, an adaptation
has to be made in order to combine them efficiently. This can be done by
extending the trap-door permutation g_i over $\mathbb{Z}_{N_i}$ to an permutation g'_i
over $\{0,1\}^l$ as follows. For any l-bit input x define non-negative integers
q_i and r_i such that $x = q_i N_i + r_i$ and $0 \le r_i < N_i$. Then, define

$$g'_i(x) = \begin{cases} q_i N_i + g_i(r_i) & \text{if } (q_i + 1)N_i \le 2^l \\ x & \text{otherwise} \end{cases}$$

Intuitively, g'_i is defined by using g_i to operate on the low-order digit of
the N_i-ary representation of x, leaving the higher order bits unchanged.
The exception is when this might cause a result larger than $2^l - 1$, in
which case x is unchanged. If we choose $l - l_b \ge 160$, the chance that a
randomly chosen x is unchanged by the extended g'_i becomes negligible.
The function g'_i is clearly a one-way trapdoor permutation over $\{0,1\}^l$.
The scheme is described below:

RS.Sign. Suppose user U_s ($1 \le s \le z$) is to generate a ring signature
with respect to the ring $U_1, \cdots, U_z$.

  1 U_s chooses $\delta \in_R \{0,1\}^l$, and computes along the ring as fol-
    lows (where $z + 1$ is treated as 1):

$$\begin{aligned} v_{s+1} &= \mathcal{H}(m, \delta), \\ v_{s+2} &= \mathcal{H}(m, v_{s+1} \oplus g'_{s+1}(r_{s+1})), \\ &\cdots \\ v_s &= \mathcal{H}(m, v_{s-1} \oplus g'_{s-1}(r_{s-1})). \end{aligned}$$

Just before closing the ring, the signer uses her secret key to
compute r_s such that $\delta = v_s \oplus g'_s(r_s)$.

2 In order to make the signature anonymous, U_s chooses at random an index i_0 ($1 \leq i_0 \leq z$), and outputs the signature $(i_0, v_{i_0}, r_1, \cdots, r_n)$.

RS.Ver. The verifier accepts if $v_{i_0} = \mathcal{H}(m, v_{i_0-1} \oplus g'_{i_0-1}(r_{i_0-1}))$, where $v_j = \mathcal{H}(m, v_{j-1} \oplus g'_{j-1}(r_{j-1}))$ for $j = i_0 + 1, \cdots, i_0 - 1$.

5.3 Output: A Secure Accountable Ring Signature Scheme

The basic idea underlying the setup procedure is to let a user U_s hold a pair of public and private keys $\langle pk_{U_s}, sk_{U_s} \rangle$ such that sk_{U_s} is only known to U_s's smart card, and let U_s's computer initialize an instance of a distributed symmetric key cryptosystem DSKC so that the card holds Θ and server ARA_i ($1 \leq i \leq n$) holds Θ_i such that $\bigcup_{i=1}^{n} \Theta_i = \Theta$. However, the effort to prevent various possible subliminal channels makes it a bit complicated. Below is the scheme.

1 A user U_s applies for a certificate on her public key pk_{U_s} that is eligible for generating accountable ring signatures, where the corresponding private key sk_{U_s} is known to U_s's smart card but not U_s. Note that a public key is always assumed to be publicly known (e.g., certified by the CA).

(a) The ARA is implemented by n servers $ARA_1, \cdots, ARA_n$, among them at most $t = \lfloor \frac{n-1}{2} \rfloor$ servers can be corrupted. Suppose ARA_i ($1 \leq i \leq n$) has two pairs of public and private keys: $\langle pk_{ARA_i}, sk_{ARA_i} \rangle$ for a secure public key cryptosystem and $\langle pk'_{ARA_i}, sk'_{ARA_i} \rangle$ for a secure digital signature scheme. Suppose ARA_i maintains a database for storing the secrets that the users will provide.

(b) U_s applies for a smart card and a certificate for her public key pk_{U_s}.

    i U_s applies for a smart card from the CA, just as she applies for a certificate in a standard PKI (meaning that there is an appropriate authentication procedure).

    ii The CA returns U_s a smart card equipped with a pair of public and private keys $\langle pk_{CARD}, sk_{CARD} \rangle$ with respect to a deterministic signature scheme (for the purpose of avoiding subliminal channels), and $(pk_{ARA_1}, \cdots, pk_{ARA_n})$.

    iii U_s's computer executes DSKC.Setup to obtain $\Theta = (\theta_1, \cdots, \theta_d)$ for U_s's smart card and computer, and Θ_i for ARA_i where $1 \leq i \leq n$. In order to securely send Θ_i to

ARA_i, U_s's computer encrypts Θ_i using pk_{ARA_i}; denote the resulting ciphertext by η_i. Then, U_s's computer sends $(\Theta; \Theta_1, \eta_1, \cdots, \Theta_n, \eta_n)$ to U_s's smart card via the physically secure communication channel, and keeps Θ for itself.

iv U_s's card checks if the Θ_i's and η_i's are correctly generated. If so, U_s's card generates a pair of public and private keys $\langle pk_{U_s}, sk_{U_s} \rangle$ and a signature ρ on $(U_s, pk_{U_s}, \eta_1, \cdots, \eta_n)$ using its private key sk_{CARD}, keeps Θ, and sends $\Upsilon = (pk_{CARD}, U_s, pk_{U_s}, \eta_1, \cdots, \eta_n, \rho)$ to U_s's computer.

v U_s's computer checks the validity of Υ, and forwards it to the CA and the ARA_i's via the public channels.

vi CA checks the validity of ρ and returns a certificate $cert_{U_s}$ on pk_{U_s}. On the other hand, ARA_i $(1 \le i \le n)$ verifies the validity of ρ, decrypts η_i to obtain Θ_i, keeps (U_s, Θ_i) in its secret database, and generates (using $sk_{ARA_i'}$) and returns a signature ζ_i on (pk_{CARD}, U_s) back to U_s's computer.

vii U_s's computer sends valid $cert_{U_s}$ and $\{\zeta_i\}_{i=1}^n$ to her smart card, which is now ready for generating accountable ring signatures (and may erase sk_{CARD}).

2 ARS.Sign works as follows. Suppose U_s $(1 \le s \le z)$ is to generate an accountable ring signature with respect to a ring consisting of $U_1, \cdots, U_z$, whose certificates have not be revoked.

   (a) U_s's computer executes DSKC.Enc to encrypt U_s (for the purpose of avoiding a subliminal channel), and sends the resulting ciphertext $\phi = \langle R, F_\Theta(U_s) \rangle$ to her smart card.

   (b) U_s's card checks the validity of ϕ, and then executes RS.Sign to generate a ring signature σ on $m' = m||\phi$. The signature σ is sent back to her computer, which then publishes it via an anonymous channel.

3 ARS.Ver is the same as RS.Ver, except that $m' = m||\phi$.

4 Suppose ϕ is embedded in a valid ring signature σ, ARS.Open works as follows: For $j = 1$ to z, the ARA's jointly execute DSKC.Dec to check whether ϕ is an encryption of U_j. If so, U_j is the actual signer.

THEOREM 5 *Suppose that the adopted signature schemes are existentially unforgeable under adaptive chosen-message attacks, that adopted public key cryptosystems are semantically secure, and that the security parameters are chosen such that the birthday attack is avoided. The above* ARS *scheme is secure with respect to Definition 4.*

The proof is left to the full version of this paper.

Remark. It is clear that our accountable ring signature scheme is essentially as efficient as the underlying ring signature scheme, except that it additionally involves some symmetric key operations (i.e., evaluating d pseudorandom functions). Although the computational overhead on a card may not be an issue, the communication overhead may be a serious problem (e.g., $z = 100$). We observe that both computational and communication overhead on a smart card can be substantially released, because we can let U_s's computer execute ARS.Sign until the point that it is supposed to close the ring, and then send $(m||\phi, \delta, v_{s-1}, r_{s-1})$ to U_s's card. U_s's card checks if ϕ is correctly computed. If so, it returns r_s such that $\delta = v_s \oplus g'_s(r_s)$, where $v_s = H(m||\phi, v_{s-1} \oplus g'_{s-1}(r_{s-1}))$. We stress that this performance optimization does not jeopardize security, because $\delta \oplus v_s$ is uniformly distributed over $\{0,1\}^l$ and U_s is entitled to query the oracle for inverting g'_s at such points.

6. Conclusion

We argued that accountable ring signatures would be very useful. We presented a system framework for accountable ring signatures, and conducted a case study showing how a traditional ring signature scheme is transformed into an accountable ring signature scheme while assuming a weak trust model.

References

[1] M. Abe, M. Ohkubo, and K. Suzuki. 1-out-of-n Signatures from a Variety of Keys. Asiacrypt'02.

[2] R. Anderson and M. Kuhn. Low Cost Attacks on Tamper Resistant Devices. Security Protocol'97.

[3] G. Ateniese, J. Camenisch, M. Joye, and G. Tsudik. A Practical and Provably Secure Coalition-Resistant Group Signature Scheme. Crypto'00.

[4] M. Bellare, D. Pointcheval, and P. Rogaway. Authenticated Key Exchange Secure against Dictionary Attacks. Eurocrypt'00.

[5] M. Bellare and P. Rogaway. Random Oracles Are Practical: A Paradigm for Designing Efficient Protocols. ACM CCS'93.

[6] D. Boneh and M. Franklin. Efficient Generation of Shared RSA Keys (Extended Abstract). Crypto'97.

[7] E. Bresson, J. Stern, and M. Szydlo. Threshold Ring Signatures and Applications to Ad-Hoc Groups. Crypto'02.

[8] J. Camenisch. Efficient and Generalized Group Signatures. Eurorypt'97.

[9] J. Camenisch and A. Lysyanskaya. Dynamic Accumulators and Application to Efficient Revocation of Anonymous Credentials. Crypto'02.

[10] J. Camenisch and M. Michels. A Group Signature Scheme based on an RSA-variant. Tech. Report RS-98-27, BRICS. Preliminary version appeared at Asiacrypt'98.

[11] J. Camenisch and M. Stadler. Efficient Group Signature Schemes for Large Groups (Extended Abstract). Crypto'97.

[12] S. Canard and M. Girault. Implementing Group Signature Schemes with Smart Cards. Cardis'02.

[13] D. Chaum. Blind Signatures for Untraceable Payments. Crypto'82.

[14] S. Chaum and E. van Heyst. Group Signatures. Eurocrypt'91.

[15] L. Chen and T. Pedersen. New Group Signature Schemes. Eurocrypt'94.

[16] J. Coron, M. Joye, D. Naccache, and P. Paillier. Universal Padding Schemes for RSA. Crypto'02.

[17] R. Cramer, I. Damgard, and B. Schoenmakers. Proofs of Partial Knowledge and Simplified Design of Witness Hiding Protocols. Crypto'94.

[18] A. De Santis, G. Di Crescenzo, G. Persiano, and M. Yung. On Monotone Formula Closure of SZK. FOCS'94. pp 454–465.

[19] Y. Desmedt. Simmons' Protocol Is Not Free of Subliminal Channels. Computer Security Foundation Workshop'96.

[20] A. Fiat and A. Shamir. How to Prove Yourself: Practical Solutions to Identification and Signature Problems. Crypto'86.

[21] O. Goldreich, S. Goldwasser, and S. Micali. How to Construct Random Functions. J. ACM, Vol. 33, No. 4, 1986, pp 210-217.

[22] S. Goldwasser and S. Micali. Probabilistic Encryption. JCSS, 1984.

[23] S. Goldwasser, S. Micali, R. Rivest. A Digital Signature Scheme Secure against Adaptive Chosen-message Attacks. SIAM J. Computing, 17(2), 1988.

[24] M. Naor, B. Pinkas, and O. Reingold. Distributed Pseudo-Random Functions and KDCs. Eurocrypt'99.

[25] R. L. Rivest, A. Shamir, and L. Adleman. A Method for Obtaining Digital Signatures and Public-Key Cryptosystem. Communication of the ACM, Vol. 21, No. 2, 1978.

[26] R. Rivest, A. Shamir, and Y. Tauman. How to Leak a Secret. Asiacrypt'01.

[27] G. J. Simmons. The History of Subliminal Channels. IEEE Journal on Selected Areas in Communication, vol. 16, no. 4, May 1998.

[28] G. Tsudik and S. Xu. Accumulating Composites and Improved Group Signing. Asiacrypt'03.

[29] B. von Solms and D. Naccache, On Blind Signatures and Perfect Crimes, Computers & Security, 11(6), 1992, 581-583.

[30] A. Young and M. Yung. Kleptography: using Cryptography Against Cryptography. Crypto'97.

CHECKING AND SIGNING XML DOCUMENTS ON JAVA SMART CARDS
Challenges and Opportunities

Nils Gruschka, Florian Reuter and Norbert Luttenberger
Christian-Albrechts-University of Kiel

Abstract: One major challenge for digitally signing a document is the so called "what you see is what you sign" problem. XML as a meta language for encoding semistructured data offers new opportunities for a solution. The possibility for checking fundamental properties of XML-encoded documents (well-formedness, validity) can be used to improve the security of the signing process for such documents. In this paper we present an architecture for checking and signing XML documents on a smart card in order to enhance the control over the documents to be signed. The proposed architecture has successfully been used to implement a secure, smart card based electronic banking application for the financial transactions system FinTS.

Key words: Java smart cards, XML, digital signature, XML Schema, electronic banking

1.   INTRODUCTION

Smart card assistance for generating digital signatures [4] is current state of the art and best practice. This is mainly due to the fact that smart cards nowadays have enough processing power to produce digital signatures for documents by on-card resources (processor and memory) only. This way the owner's private signing key never has to leave the smart card: The signing key is and remains permanently stored in a tamper-proof environment.

A closer look at the signing process however reveals a still existing major security problem: the problem known as the "what you see is what you sign" problem. Before signing a document the signer usually wants to check the document's syntactic and semantic correctness.

When compared to the traditional process of signing a paper document with a handwritten signature, the difference can easily be identified: In the traditional case, it is relatively easy for the user to assert the correctness, because syntactic and semantic document checking and signature generation are in immediate context. Digitally signing an electronic document is completely different, because checking and signature generation are executed in two different fundamentally environments, exposing different characteristics—different with respect to security on the one hand and processor, memory, and display resources on the other hand.

Traditionally, the signing application computes the document's digest using a one-way hash function and sends the result to the smart card. The card encrypts the digest by an asymmetric cipher using the signing key stored on the card. The resulting value is the digital signature of the document. But, what is really signed is beyond the user's control. It might—for instance—be the digest for a manipulated document. Even if the smart card can be regarded as tamper-proof, the terminal (e.g. a PC) and the programs running on it are vulnerable to viruses and Trojan horses [3] [8] [9] [10]. Such evildoers might obviously also affect signing applications and let them produce valid signatures for—from the user's perspective—invalid documents. Such incidents invalidate the signing process in total. To solve this problem the whole or at least the major part of the signing process has be executed in a trusted environment. While trusted PCs are not yet available, one could use a signing application running on a trusted card reader (e.g. FINREAD [28]) or a smart card, which is our approach.

We propose an enhanced architecture which performs checking and signing of XML documents on Java smart cards, called JXCS architecture. The basic idea of JXCS is to shift the syntactic validation and hash value generation from the vulnerable PC to the trusted smart card. Syntactic validation imposes the following challenges and opportunities: Challenging is the need of processing XML documents on resource constraint Java smart cards. The opportunity of the approach is the possibility to perform syntactic and even semantic checks on the XML document in a tamper-proof environment which improves the security of the signing process.

We propose the need for three major checks on the XML documents to be signed: Well-formedness, validity and content acknowledgement using a class 3 card reader (i.e. a card reader including a display and a keypad). Taken together all three checks can defeat "what you see is what you sign" attacks.

The paper is organized as follows: Section 2 introduces relevant XML fundamentals. The checks are explained more detailed in chapter 3. Section 4 then proposes the JXCS architecture. Section 5 shows a sample implementation for this architecture. Finally section 6 gives the conclusion.

2.   XML PRELIMINARIES

2.1   XML well-formedness

An XML [11] document encodes semi structured data [13]. It is said to be well-formed if it contains at least one element and fulfills the well-formedness constraints given in the XML specification. Well-formedness is related to the syntactic format of markup, content, comments, document type definition, and so on, and it also ensures the usage of proper Unicode characters and the specification of their encoding. According to the XML specification every XML processing entity has to check and assert the well-formedness property.

The most important constraint is the logical structure of element tags. The tags must be properly braced, meaning that every start element has a corresponding end element, together forming a unique tree structure. For example `<a><b>Hello</b></a>` is well-formed, while `<a>Good</c>` and `<a><b>Bye</a></b>` are not well-formed.

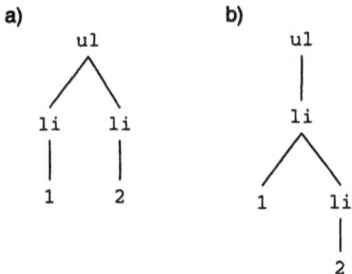

Figure 1: Possible interpretations for the not well-formed XML document: `<ul><li>1<li>2</ul>`

Well-formedness ensures the uniqueness of the XML documents' interpretation. Consider for example the following not well-formed XML fragment: `<ul><li>1<li>2</ul>`. The document allows the following two interpretations shown in figure 1: `<ul><li>1</li><li>2</li></ul>` for a) and `<ul><li>1<li>2</li></li></ul>` for b). In case of well-formedness the interpretation would be unique.

2.2   XML validity

A more restricting property compared to well-formedness of XML documents is its validity. An XML document is said to be *valid* if it is valid with respect to an associated document type declaration or schema [5] [6].

Document type definitions resp. schemata express tree grammars, against which the XML documents are checked.

A (regular) tree grammar is a 5-tuple $G = (\Sigma, D, N, P, n_S)$ where,

- $\Sigma = \Sigma_E \cup \Sigma_A$, with Σ_E is a finite set of element types and Σ_A is a finite set of attribute types,
- D is a finite set of data types,
- N is a finite set of non-terminals,
- P is a finite set of production rules of the form $n \rightarrow a(r)$, where $n \in N$ is a non-terminal, $a \in \Sigma$ is a symbol and r is either an element of D or a regular expression over the alphabet N with $L(r) \subset N^*$. Finally
- $n_s \in N$ is the starting non-terminal

Consider for example the following schema fragment

```
<xsd:element name="Amount">
  <xsd:complexType>
    <xsd:sequence>
      <xsd:element name="Value" type="xsd:decimal"/>
      <xsd:element name="Currency" type="xsd:string"/>
    </xsd:sequence>
  </xsd:complexType>
</xsd:element>
```

defining instances of the form:

```
<Amount>
  <Value>100</Value>
  <Currency>EUR</Currency>
</Amount>
```

The corresponding tree grammar for the above schema fragment has the following form:

$G = (\Sigma, D, N, P, n_S)$ with $\Sigma = \Sigma_E = \{\text{Amount, Value, Currency}\}$, $D = \{\text{xsd:string, xsd:decimal}\}$, $N = \{N_1, N_2, N_3\}$, $n_S = N_1$ and P containing the following production rules:

- $N_1 \rightarrow \text{Amount}(N_2\ N_3)$
- $N_2 \rightarrow \text{Value(xsd:decimal)}$
- $N_3 \rightarrow \text{Currency(xsd:string)}$

An import property of tree grammars is single-typeness, which means that for each production rule $n \rightarrow a(r)$, non-terminals in its content model *r* do not compete with each other. Single-typeness ensures unique parsing. Thus, in the following all tree grammars are assumed to be single-typed [15].

W3C XML Schema offers a rich variety of built-in data types which can be customized using `<restriction>`, `<union>` and `<list>`s. This allows to impose content-related constraints, e.g. the restriction of decimals to an upper bound:

```
<xsd:element name="Value">
    <xsd:restriction base="xsd:decimal">
        <xsd:maxInclusive value="100"/>
    </xsd:restriction>
</xsd:element>
```

2.3   XML Signature

XML Signature is a W3C recommendation [1] for digital signatures using an XML format. It specifies the required XML syntax and the processing rules for creating and representing XML signatures. In the context of the JXCS architecture we used this format and its related processing rules.

```
<Signature>
    <SignedInfo>
        <CanonicalizationMethod Algorithm=.../>
        <SignatureMethod Algorithm=.../>
        <Reference URI=...>
            <Transforms>...</Transforms>
            <DigestMethod Algorithm=.../>
            <DigestValue>...</DigestValue>
        </Reference>
    </SignedInfo>
    <SignatureValue>...</SignatureValue>
    <KeyInfo>...</KeyInfo>
</Signature>
```

Figure 2: XML Signature format (simplified)

An XML Signature is represented by a `Signature` element which has a structure as given in figure 2. A `Signature` element carries a sequence of three elements: a `SignedInfo` element, a `SignatureValue` element, and a `KeyInfo` element.

The `SignedInfo` element contains a sequence of a `CanonicalizationMethod` element, a `SignatureMethod` element and one or more `Reference` particles. A `Reference` particle is created for each signed object (XML document or arbitrary other content, the latter not regarded in this paper). The `Reference` particle includes a reference URI to the data object, which may be found externally or included in the same document. The `Reference` element further contains the digest value for the object (`DigestValue`), and the digest algorithm used (`DigestMethod`). For all signed data objects given in the `Reference` particles, a single signature algorithm element (`SignatureMethod`) and a canonicalization method element (`CanonicalizationMethod`) is available. (Canonicalization e.g. removes redundant whitespaces, sorts attributes, normalizes namespaces, etc.)

The signature value given in the `SignatureValue` element refers to the whole `SignedInfo` element using the signature algorithm and the user's signing key. The user's signing key may be included in the XML Signature in the `KeyInfo` element.

A signature according to the XML Signature recommendation is computed in five steps:
1. canonicalization of the document to be signed,
2. computing the digest of the canonicalized document,
3. generation of the `SignedInfo` element,
4. computation of the signature value,
5. building of the `Signature` element.

2.4   APIs for XML processing

There exist two different types of APIs for XML processing [14]: tree-based APIs like DOM or event-based APIs like SAX [18] or StAX [24]. Tree-based APIs in fact load the complete XML document into main memory which then can be accessed as an XML Infoset [12] like tree. Such kind of APIs put a great strain on system resources, especially if the document is large.

Event-based APIs report parsing events directly to the application through callbacks, and do not usually build internal trees. Applications implement handlers to deal with the different events, much like handling events in a graphical user interface.

With respect to the limited resources of a smart card the only useful choice for on card XML processing is an event-based API. We propose a SAX-like event-based API which produces the following events: begin-element(a), end-element(a), begin-attribute(a), end-attribute(a), char-content(c) where $a \in \Sigma$, $c \in$ Unicode*. The above XML document is represented by the following events:
- begin-element(Amount)
- begin-element(Value)
- char-content("100")
- end-element(Value)

- begin-element(Currency)
- char-content("EUR")
- end-element(Currency)
- end-element(Amount)

3. WHY SIGNING XML DOCUMENTS IS DIFFERENT

Why relying on XML for solving the "what you see is what you sign" problem? Our ideas can be summarized in two points:

1. If a document to be signed is either not well-formed in the sense of XML, or not valid in the sense of its accompanying schema, or both, than it must strictly be assumed that the document has been manipulated. In consequence, it has to be dropped, and the user has to be notified.

2. A smart card application can extract certain content items for display on the smart card reader from a structured and formally described document. The extraction and display operations are fully controlled by the tamper-proof smart card—which is the same environment that generates the digital signature.

The fundamental property of XML documents is well-formedness. According to the XML specification every XML processing entity has to check and assert this property. Regarding digital signing well-formedness is important, since it ensures the uniqueness of the XML documents' interpretation. Well-formedness also ensures the usage of proper Unicode characters and the specification of their encoding. This is also very important regarding digital signatures, since character set manipulation can be used to perform "what you see is what you sign" attacks [10].

Validity is a much more restrictive property of XML documents compared to well-formedness. A smart card which checks validity of XML documents with respect to a given schema before signing ensures due to the tamper resistance of the smart card that only certain types of XML documents are signed. Consider for example a smart card which contains your private key, but only signs XML documents which are valid with respect to a purchase order schema. You could give this card to your secretary being sure, that nothing else than purchase order is signed using your signature. Using additional constraints in the schema, e.g. the restriction of the maximum amount to 100 Euro eliminates the last chance of misusage.

When operated in a card reader containing a display the card can print selected parts of the content and request user confirmation. This brings a great approach to the solution of the "what you see is what you sign" problem.

Obviously, XML processing is not an easy task to perform on resource-constrained smart cards. The following table summarizes the challenging XML properties and the resulting opportunities for improving the signing process:

Table 1: Challenges and opportunities in signing XML documents

Challenge	Opportunity
Check well-formedness property on smart card	Confidence that only XML documents are signed
Check validity property on smart card	Confidence that only documents of a certain type are signed
Check validity property and additional content-related constraints (W3C XML Schema simple types) on smart card	Confidence that certain „lower bounds" on document content are observed (e.g. financial transaction not going over some critical amount)
Display selected content items on SmartCard reader	Confidence that even a manipulated document couldn't do any harm

4.   JXCS SMARTCARD SIGNING ARCHITECTURE

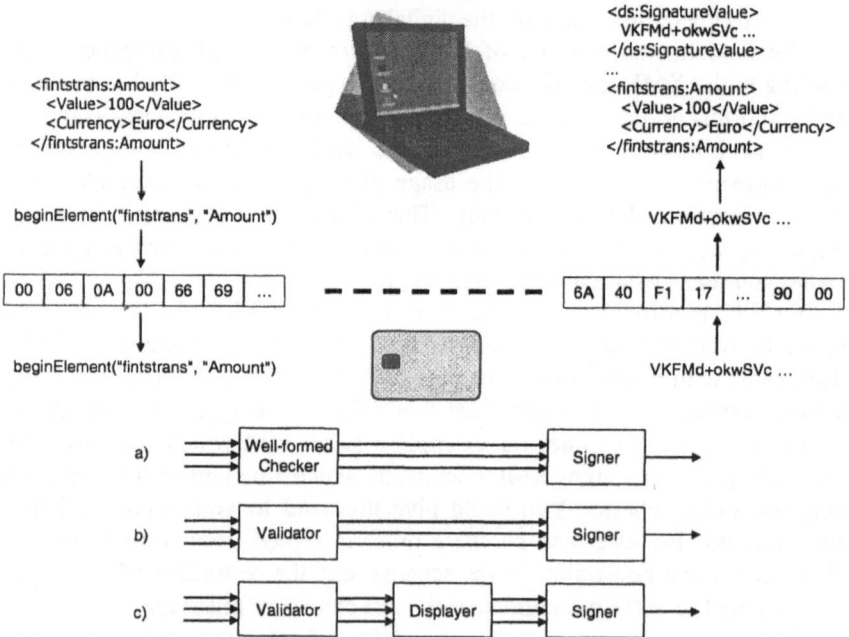

Figure 3: JXCS Architecture Overview

The JXCS architecture offloads all security-critical processing tasks to the tamper-proof smart card environment including:

- well-formedness and validity checking,
- document canonicalization and hashing, and
- signature value computation.

Figure 3 gives an overview over the JXCS architecture. The terminal runs an XML parser which analyzes the document to be signed. The parsing events are coded into APDUs and sent to the smart card. The events are forwarded to a chain of event handlers. These handlers process the total XML document sequentially event by event. Different handlers can be used in this chain according to the specific requirements (see for example figure 3, a) to c)). The chain normally contains event handlers for checking XML document properties like well-formedness or validity, and of course a handler for creating the signature. If an event causes one of the checks to fail, instantly an exception is thrown. If all checks on all events are successful, the signature value of the document is finally returned. From this value the XML Signature is created by the terminal computer. The following paragraphs describe the JXCS architecture handlers in detail.

4.1   Signing Handler

The signing handler's purpose is to digest the canonicalized document, to create the `SignedInfo` element and to compute the `SignatureValue`.

For the purpose of digesting the signing handler reconstructs the canonicalized document from the received parsing event; there is no need to send the original document to the smart card. This approach—based on a single event stream—enables the efficient combination of both checking and signing XML documents on resource constraint smart cards.

Table 2: Reconstructing the document from events

Event	Parameter 1	Parameter 2	Canon. Document
beginDocument			
beginElement	*name*		*<name*
beginElement	*prefix*	*name*	*<prefix:name*
addNamespace	*uri*		xmlns =*"uri"*
addNamespace	*prefix*	*uri*	xmlns:*prefix*="*uri*"
beginAttribute	*name*		*name*="
beginAttribute	*prefix*	*name*	*prefix:name*="
endAttribute			"
charContent	*data*		*data*
endElement	*name*		*</name>*
endElement	*prefix*	*name*	*</prefix:name>*
endDocument			

Table 2 shows how the canonicalized document is reconstructed from parsing events.

For each reconstructed document element the hash generator of the signing handler is triggered which that way digests the whole document piece by piece. Most hashing functions (like MD5 [26] or SHA-1 [27]) do not store the whole input data, but process the input stream block by block and only need a small additional digest buffer. Thus the hashing "piece by piece" is very space saving.

If the document has been completely parsed and transmitted to the card, the endDocument event is triggered and the signing object starts the actual signing process. The signer creates the `SignedInfo` (see above) element including the signature, canonicalization and digest algorithm. Furthermore the document's digest and the reference (created from the document's root element) is inserted.

The resulting SignedInfo fragment is signed using the user's private key, stored on the smart card. The resulting value is used to create the entire XML Signature.

As parsing events and the canonicalized document are semantically equivalent, the canonicalized document can be reconstructed from the parsing events, and the digital signature can be generated from the reconstructed document, and the signature is valid only for the parsed document and not for any other document.

4.2   Well-Formedness Checker

The following algorithm is used in the JXCS architecture to check well-formedness:

begin-element(a): Push a on the stack
end-element(a): Let a' be the stack's topmost element.
- If $a = a'$, pop a' from stack.
- Otherwise throw a not-wellformed exception.
begin-attribute(a): do nothing
end-attribute(a): do nothing
char-content(c): check if all characters in c are allowed characters

4.3   Validity Checker

The validity of XML documents with respect to a tree grammar (i.e. a document type definition or a schema) can easily be checked by the following algorithm using the proposed event-based API and a stack. The algorithm starts with an empty stack.

begin-element(a):

- If the stack is empty, find a production rule $n_s \to a(r)$ and push it onto the stack.
- Otherwise let $n' \to a'(r')$ be the stack's topmost production rule. Search the production rule $n \to a(r)$ where $n \in$ next(r').
  - o If none such exist, throw a not-valid exception.
  - o Otherwise change the actual production $n' \to a'(r')$ to $n' \to a'(n'\backslash r')$ and then push $n \to a(r)$ onto the stack.

end-element(a): Let $n' \to a'(r')$ be the stack's topmost production rule.

- If $a \neq a'$, throw a not-wellformed exception.
- Otherwise
  - o If $\varepsilon \notin L(r')$, throw a not-valid exception.
  - o Otherwise pop the production rule from the stack.

begin-attribute(a): same as for begin-element

end-attribute(a): same as for end-element

char-content(c): Let $n \to a(r)$ be the stack's topmost production rule.

- If $r \notin D$, throw a not-valid exception.
- If $c \notin r$, throw a not-valid exception.

The set next(r) = $\{a \in \Sigma \mid$ there exists a $\sigma \in \Sigma^*$ with $a\sigma \in L(r)\}$ describes the set of letters accepted next and $x\backslash a$ describes the derivative of x with respect to a, i.e. the expression which remains from a after parsing x. Both next(r) and $x\backslash a$ can be computed efficiently using a simple lookup table [25]. The above algorithm is correct, due to the fact that for a single type tree grammar the next non-terminal can be chosen uniquely [15, 17].

We use a compressed encoding of the production rules based on adjacency lists. Here the running time is $O(m\,n)$ with m is the length of the largest adjacency list. For a fixed tree grammar m is constant, so the running time is still $O(n)$ for a fixed schema or document type definition.

The space consumption is $O(h)$ where h is the height of the parsed tree. For non-recursive tree grammars the maximum height of an XML document is fixed, so the space consumption is limited for a fixed tree grammar. Therefore the memory consumption of the algorithm will not exceed the smart card's resources.

The adjacency list representation for an XML schema is generated off-card and installed on the card. A card can handle one or more schemata.

4.4　　Visual inspector for Class 3 Reader

A smart card reader with own display and numerical keyboard (often called "class 3 reader") offers further possibilities for checking the document before signing it. The data exchanged between card reader and smart card can neither be read nor changed by the terminal or any Trojan horse running on it. So unless the reader's firmware has not been modified, the card reader can be used as a secure display for the information send to the card.

A simple solution to the "what you see is what you sign" problem would be showing the complete document on the display prior to sending it to the smart card. This would indeed be useless for at most all practical purposes. These displays normally only have 1 to 3 rows with at most 20 characters each. Most users' acceptance for viewing a complete document on such a display would be very low.

Displaying selected parts of arbitrary XML documents otherwise is ineffective for checking the document to be signed. A single element may have totally different semantics in different contexts. If the user acknowledges e.g. the prompt

```
<Amount>
    <Value>1000</Value>
    <Currency>EUR</Currency>
</Amount>
```

he does not know the denotation of this element. He could buy a very cheap car or loose a lot of money by transferring it to the Nigeria connection. Even if all ancestors of the element are displayed additionally, the semantic is generally not unique.

This is solved by validating the XML document to a specific schema. In this case the context and semantic of an element is unambiguous. If the validator validates the documents e.g. against a schema for cash remittance, the user can be sure to sign a transfer of 1000 Euro, if he acknowledges the element shown above.

The functionality of the display component is the following. When the schema is transferred to the smart card, the most "critical" elements are marked for displaying. The display component reads the content of these elements from the event stream. These values have to be acknowledged by the user. If one is rejected an exception is thrown.

5.   FINTS SAMPLE IMPLEMENTATION

5.1   Sample Application: FinTS

FinTS [23] (formerly named HBCI) is a home banking computer interface developed by the German Central Banking Committee (ZKA). It defines a standard for home banking and specifies the relationship between customer products and bank systems. FinTS allows more flexible and convenient online banking than other systems. To ensure secure transactions over open networks, cryptographic functions and smart cards are used.

The actual version's communication is based completely on XML. All messages exchanged between the FinTS server and the FinTS clients are XML documents. Orders, like cash remittances, are signed using a personal smart card and coded as XML signature. The whole transaction message – containing one or more orders – is encrypted into an XML encryption document.

Banking transactions are obviously extreme secure relevant and a profitable target for attacks. The user wants assurance, that the document he creates using his online banking program is actually the one signed by the card and sent to his bank.

The assumed scenario for our sample application is an online-banking user who wants to sign cash remittances using a smart card and a class 3 reader. The signed remittances will then be sent to his bank's FinTS server.

5.2   Implementation

The proposed architecture was utilized to implement a client with a secure signing process for FinTS banking transactions. We implemented a client program on the terminal creating FinTS remittance transactions and sending it to the smart card. The smart card contains the user's private key and a signing component. It also contains a validator, validating the input document against a modified FinTS schema for a remittance. The schema has been altered in the way that the remittance value must be less or equal 100 and the currency must be EUR. Thus the smart card will sign only documents being a valid FinTS remittance with a maximum of 100 Euro. Thus the user has the assurance, that his signing card will never sign anything else, no matter what a Trojan horse would perform to the documents. Additionally the display component let the user acknowledge the remittance's most important content values like target account number. Thus even modifications on these values would be detected.

The sample client program is a GUI that creates a FinTS cash remittance document from the input values. This XML document is parsed into XML events, which are coded as TLV into request APDUs. Due to efficiency reasons not single events are sent to the card. Instead the amount of events fitting into an APDU is transferred, improving performance by up to 20%.

On the Java smart card a main applet and the following event handlers are implemented: validator, displayer and signer. The XML events are decoded from the APDUs by the applet, which calls the event handler's appropriate event methods. Once an exception is thrown by one of the handlers, an error code is returned to the client. In order to simplify the implementation the algorithms for hashing, signing, canonicalization are set statically. The final application needs approximately 13000 bytes plus the binary schema representation. The signing time increases linear with the document's size. For a typical cash remittance (ca. 1.2 KB) approximately 15 seconds are needed.

The validator is an implementation of the above described validity checking algorithm. In our sample application we provided the validator on the smart card with the FinTS schema. The original schema has 5537 bytes, the binary tree grammar only just 954 bytes. As shown above the stack's size is limited for a fixed schema. The stack can be implemented as list of fixed length. Thus dynamic object instantiating can be avoided, which is critical for a Java card without garbage collection. For the FinTS schema the maximum size needed for the validator's stack is 200 bytes.

The display component will be configured when uploading the schema. The simple type elements, that shall be displayed, are specified by an XPath [7] like expression. The displayer collects the content for these elements from the charContent events. He then waits for the acknowledgement for these values from the card reader. Due to the stringent master-slave-relationship between the terminal and the smart card, the card can not request the card reader for displaying these values. Instead, the PC sends the to-be-acknowledged values to the card reader contained in a special command. This command instructs the reader to display the value and send it to the card if it is acknowledged by the user. The displayer compares these values with the content extracted from the events. If this fails an exception is thrown. Figure 4 shows a sample output on the card reader's display.

Figure 4: Card reader waiting for user acknowledgement

As pointed out above the signing component creates the Signature fragment from the events. In order to minimize the communication between PC and card, not the total Signature fragment is returned to the client but only the documents digest and the signature of the SignedInfo fragment. As the algorithms are fixed, the client program can create the same Signature fragment from the digest and the signature values. And of course this is no lack of security.

The Java smart cards used are JCOP 21id from IBM. These cards are compliant to JavaCard 2.1.1, OpenPlatform 2.0.1' and FIPS 140-2 level 3. They have 30 Kbytes EEPROM as persistent Java heap, 590 bytes RAM as transient Java heap and 200 bytes RAM as Java stack. The card reader used is a class 3 reader (2 x 16 character display, numerical keypad) from ReinerSCT.

6.   CONCLUSION AND FUTURE WORK

In this article we have shown how processing XML documents on a smart card arises new opportunities for signing them. By checking properties like well-formedness and validity the user gains more control over the documents signed using his private key. We have also shown, that checking a document's validity according to a specific grammar, allows showing single elements from the document on a reader's display without losing the element's semantic. This way we can greatly improve the security of the signing process for XML document and even approach a solution for the "what you see is what you sign" problem.

Future research will focus on applying the XML processing smart card technology on XML communication protocols to improve e.g. Web Service security.

7.   ACKNOWLEDGEMENTS

We would like to thank Karsten Strunk and Jesper Zedlitz for implementing the secure FinTS signing client including an interface to a FinTS server system. We also like to thank Christian Friberg and Rainer Segebrecht of PPI Financial Services Kiel for supporting the implementation process.

8.   REFERENCES

[1] Mark Bartel et al. *XML-Signatur Syntax and Processing – W3C Recommendation 12 February 2002*. W3C (World Wide Web Consortium), 2002.

[2] John Boyer. *Canonical XML, Version 1.0 – W3C Recommendation 15 March 2001*. W3C (World Wide Web Consortium), 2001.

[3] Armin B. Cremers, Adrian Spalka, and Hanno Langweg. The Fairy Tale of 'What You See Is What You Sign' – Trojan Horse Attacks on Software for Digital Signatures. In

IFIP Working Conference on Security and Control of IT in Society-II (SCITS-II), Bratislava, Slovakia, June 2001.

[4] Whitfield Diffie and Martin E. Hellman. New directions in cryptography. *IEEE Transactions on Information Theory*, 22(6):644-654, November 1976.

[5] Henry S. Thompson et al. *XML Schema Part 1: Structures – W3C Recommendation 2 May 2001*. W3C (World Wide Web Consortium), 2001.

[6] Paul V. Biron and Ashok Malhotra. *XML Schema Part 2: Datatypes – W3C Recommendation 2 May 2001*. W3C (World Wide Web Consortium), 2001.

[7] James Clark, Steve DeRose. *XML Path Language (XPath) – W3C Recommendation 16 November 1999*. W3C (World Wide Web Consortium), 2001.

[8] Tim Redhead and Dean Povey. The Problem with Secure On-Line Banking. In *Proceedings of the XVIIth annual South East Asia Regional Conference (SEARCC'98)*, July 1998

[9] Arnd Weber. See What You Sign. Secure Implementation of Digital Signatures. In *Intelligence in Services and Networks: Technology for Ubiquitous Telecom Services* (IS&N'98), Springer-Verlag LNCS 1430, 509-520, Berlin, 1998.

[10] Audun Jøsang, Dean Povey, and Authony Ho. *What You See is Not Always What You Sign*. AUUG 2002 - Measure, Monitor, Control, September 2002

[11] Tim Bray et al. *Extensible Markup Language (XML) 1.0 (Third Edition) W3C Recommendation 04 February 2004*. W3C (World Wide Web Consortium), 2004.

[12] John Cowan, Richard Tobin. *XML Information Set (Second Edition) W3C Recommendation 4 February 2004*. W3C (World Wide Web Consortium), 2004.

[13] P. Buneman. *Semistructured data*. Tutorial in Proceedings of the 16th ACM Symposium on Principles of Database Systems, 1997

[14] Hiroshi Maruyama et al. *XML and Java: developing Web applications*. Pearson Education. 2nd ed. 2002.

[15] Makoto Murata, Dongwon Lee, and Murali Mani. *Taxonomy of XML Schema Languages using Formal Language Theory*. Extreme Markup Languages 2000, August 13-14, 2000. Montreal, Canada.

[16] Boris Chidlovskii. *Using Regular Tree Automata as XML Schemas*. IEEE Advances in Digital Libraries 2000 (ADL 2000). May 22 - 24, 2000. Washington, D.C.

[17] F. Neven. *Automata theory for XML researchers*. SIGMOD Record, 31(3), 2002.

[18] The SAX Project, URL: http://www.saxproject.org/

[19] IBM JCOP embedded security software. URL: http://www.zurich.ibm.com/jcop/

[20] Sun Microsystems: JavaCard 2.1.1 http://java.sun.com/products/javacard

[21] Global Platform Consortium: OpenPlatform 2.0.1'. URL: http://www.globalplatform.org/

[22] FIPS PUB 140-2: Security Requirements For Cryptographic Modules, May 2001. URL: http://csrc.nist.gov/publications/fips/fips140-2/fips1402.pdf

[23] FinTS Financial Transaction Services 3.0. URL: http://www.fints.org/

[24] JSR 173: Streaming API for XML. Java Community Process.

[25] Janusz A: Brzozowski. *Derivatives of regular expressions*. Journal of the ACM, 11(4), 1964.

[26] Ronald Rivest: The MD5 Message-Digest Algorithm, IETF RFC 1321, April 1992. URL: http://www.ietf.org/rfc/rfc1321.txt

[27] National Institute of Standards and Technology: Secure Hash Standard, April 1995. URL: http://www.itl.nist.gov/fipspubs/fip180-1.htm

[28] FINREAD. URL: http://www.finread.com/

XML AGENT ON SMART CARDS

Sayyaparaju Sunil
Kanwal Rekhi School of Information Technology,
IIT Bombay,
India.
sunil@it.iitb.ac.in

Dr. Deepak B. Phatak
Kanwal Rekhi School of Information Technology,
IIT Bombay,
India.
dbp@it.iitb.ac.in

Abstract   With the increase of resources and processing power on the Smart Cards, in recent days, their applicability has moved from a domain specific application to a more generic one. Multi-Application Frameworks, designed to host multiple applications on a card, also aim at increasing the interoperability between different vendors and their applications. This paper proposes a new model named *XML Agent*, which significantly enhances the inter-operability of the smart card applications. The XML Agent becomes an interface through which the external world can interact with the on-card applications. The XML Agent passes the commands and data to the appropriate application as per its requirements. This model will also facilitate the programmer in developing off-card applications, with minimal information about the on-card applications.

1.   Introduction

Smart Cards have a hardware specific firmware operating system, called *Card Operating System (COS)*, to manage all the resources on the card. Additionally, there may be a *Runtime Environment* which adds more functionality to the Card Operating System. The applications residing on the card are called *On-Card Applications*. The external applications, called the *Off-Card Applications*, communicate with the on-card applications by sending *Application Protocol Data Units (APDUs)*. The Card OS interprets these commands and performs the corresponding predefined operations.

Multi-Application Frameworks provides a platform to securely host more than one application on a card, making them multi-purpose. They define how the on-card applications should communicate and share data between each other. A firewall is deployed to protect and isolate applications residing on the same card. All the communication between on-card applications should pass through this firewall. Widely used on-card Multi-Application Frameworks are: Java Card and MULTOS.

Interoperability is an important aspect in Smart Card application development. Ideally, the off-card application on any terminal should be able to interact with an on-card application on any type of card. Even though the standards and on-card frameworks enforce a uniform interface, the problem is only partially solved. We will discuss this problem in more detail in the next section. Our proposed model significantly enhances the interoperability of the applications, however it doesn't solve the problem completely. It also eliminates the need to rebuild the off-card application to support a new type of card.

Abstraction for the developers is one of the most important principles in software engineering. Abstraction hides all the unnecessary details from the developer. It is difficult to define what is necessary and what is not. A vague definition can be: Whatever the developer is not using is unnecessary to him. He should not be bothered with the details of what he is not going to use. Then he can concentrate more on the programming logic.

Our model discussed in this paper provides powerful abstraction to the developers. In our model, the developers need to have very limited information about the on-card application, to develop an off-card application. As a result of this, there can be more number of market players who provide their own customized services and solutions to the consumers.

2.   Motivation

This section highlights some of the problems faced during the development of an application. Presently, all the problems may not appear to be very significant. But they will become more prominent once the smart card industry expands, and different players come up with their custom applications, terminals and cards.

Generally, the on-card and off-card applications are developed by the same developer. This is because the off-card application is tightly coupled with the on-card application. It will definitely be advantageous if this dependency is eliminated or at least reduced. Then all the application developers can easily develop different off-card applications based on a single on-card application. For example, based on an on-card application which maintains personal information of the card holder, numerous off-card applications like vehicle license, insurance policy, bank account etc. can be developed.

Applications on different cards may use different data formats and communication sequences to communicate with the off-card application on the terminal. By data format, we mean the data type and the type of encoding used to represent data. For example, the balance in an E-purse application may be an integer or a real number. In addition, if it is floating-point representation then some base(radix) is decided upon and a fixed number of bits is used to represent mantissa and exponent. By communication sequence, we mean the order in which the control commands and data are sent. This is decided upon during the design and implementation of the on-card application. The off-card applications cannot interact with the on-card application unless they know the exact details of the their data formats and communication sequence.

The card has to interact with a terminal whose architecture may be completely different from that of the card. Then the data formats used by the on-card and off-card applications may be different, but still the communication has to take place. For example, one system may be using 2 bytes to represent an integer while another system may be using 4 bytes. Same is the case with big-endian and little-endian representations. Similarly, for strings, different types of encoding like ASCII, UTF-8, UTF-16 can be used. If both the terminal and the card are following the same data formats, then there will not be any problem. If not, the off-card and/or on-card applications have to do proper transformations to get the exact data. This will incur extra overhead in processing.

There should be some mechanism to hide all these problems from the developers. These problems should be handled at a lower level without exposing them to the application developers. Then the developers can concentrate more on the application process logic rather than on the nitty-gritties of the underlying system.

3. Related Work

This section introduces some of the earlier work describing efforts made in the direction of integrating the XML with the Smart Card technology. Their objective was to increase the interoperability of the Smart Card applications. They also attempted to eliminate the need to rebuild the whole off-card application to support a new type of card.

3.1   Dynamic Card Service Factory

Open Card Framework(OCF) provides standardized high-level APIs for card programming [Ocfgim98]. Its **CardService** layer implements the logic of the off-card application for a specific Card Operating System. So there can be many possible CardService implementations for the same application. All knowledge about a family of CardServices is encapsulated in a **CardService-Factory**. For an application to support multiple cards, corresponding Card-Service implementations have to be bundled with it. This increases the size and loading time of the application. A bigger disadvantage is that, an off-card application cannot support card types which are not present during its development.

DynamicCardServiceFactory(DCSF) [DCSF] is a CardServiceFactory implementation. Initially, it downloads the up to date information about all known card types along with the location of CardServices for those cards and stores it in an XML file in the terminal. When a card is inserted into the terminal, using XML messages, it downloads the appropriate CardService implementation for the requested application and serves the card. This model allows the off-card application to support new cards without changing the application logic.

3.2   SmartX

This technology was introduced by Gemplus. It provides a solution for writing host-to-card protocols in XML. This allows the APDU protocol to be portable to any language and platform that has a SmartX implementation. It provides a framework where the application process logic is dissociated from the application protocol [Xavier99]. So it is required to implement the application logic only once. This methodology can support cards that are not present during the development of the off-card application.

The application protocol details are represented using an XML-based language called *SmartX Markup Language(SML)*. The SML documents, called *Dictionary* [SmartX00] contains card profiles which in turn will contain different card processes. Each card process is implemented for a given card profile using the card specific commands. The SmartX engine running on the terminal downloads the appropriate SML document from the dictionary and executes the application process logic. However, this technology is not in use. The development and support was withdrawn by its initiators.

4. XML Agent

In the previous section, we saw two models which enhance the interoperability of applications. Both the models run some sort of an engine on the terminals, which can download the up-to-date information about different types of cards. It will be beneficial if the need to dynamically download information can be avoided, without compromising the functionality provided by it. It is a costlier option to keep the terminals connected to internet round the clock. The above two models have serious limitations in situations where the terminal is not connected to internet.

With the increase of resources and processing power on the Smart Card, it is definitely not a bad idea to implement a small engine on the card itself. So our XML Agent is an engine which runs on the card. It doesn't demand any internet connectivity from the terminal. So the terminals can be deployed anywhere to perform all kinds of off-line transactions with the cards.

4.1 Basic Model

The XML Agent is a middleware between the off-card applications and the on-card applications. The off-card and the on-card applications communicate only through the XML Agent. In other words, XML Agent becomes an interface for the off-card applications to communicate with the on-card applications. The off-card applications communicate with the XML Agent in XML only. We chose XML because of the wide support available for it on different platforms.

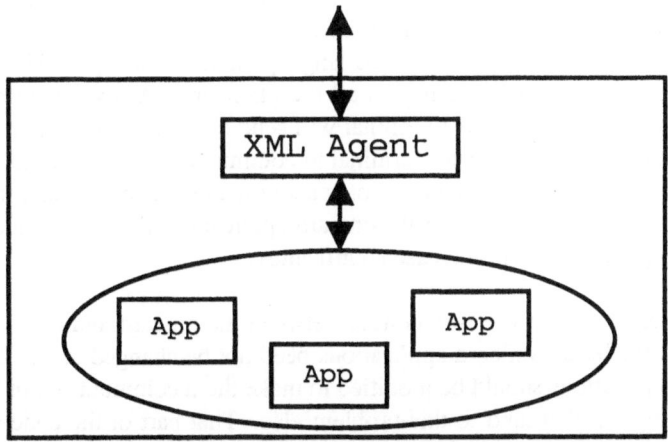

Figure 1. XML Agent

The XML Agent transforms the XML documents sent by the off-card applications into a byte stream which is understandable by the on-card applications. The on-card applications respond to the commands passed to them. This response reaches the XML Agent in a byte stream. It will construct an XML document out of the response and will send it to the calling off-card application. So the off-card applications communicate only in XML and the on-card applications communicate only in bytes. Eventually, the XML Agent is transforming the XML document to bytes and vice versa. It is important to note that the XML document can be sent as data in the standard APDU commands. So our model can fit in the existing standards and frameworks.

To transform the XML documents in this manner, the XML Agent should have a knowledge about the on-card applications. First, it should know which applications are installed on the card. Then, for each on-card application, the XML Agent should know the mapping between the fields in the XML document and the corresponding commands and its arguments which should be sent to the on-card application. It should also know the communication sequence for each on-card application. So the on-card application has to describe itself to the XML Agent, or the XML Agent has to be notified explicitly by the installing application. Similarly the information about the mapping rules should be specified by the on-card application or any external application. With this information, the XML Agent can parse the XML document, prepare a byte stream and pass it to the appropriate application.

4.2   Advantages

Ideally, all the data entering and leaving the card should be in XML only. In recent days, XML has become the defacto cross platform data representation standard. XML is a simple text-based language and is also extensible. There is wide support for XML parsing on all the platforms. As explained before, the off-card application on the terminal will send and receive only XML documents. It becomes easy for the terminal to communicate in simple characters and strings. The terminal need not worry about the details of the data formats and communication sequence of the on-card application. So the interoperability of the applications will increase significantly.

Since the XML Agent is a middleware between the on-card and off-card applications, the legacy on-card applications need not be changed at all. But the off-card applications should be modified to make them communicate in XML. But there is a workaround to this problem also. That part of the code which has to parse the XML document from the card and extract different fields can be written as wrapper functions to the original off-card application. So the migration from the existing model to our proposed model can be very smooth.

This model also allows the developers to develop an off-card application, with limited information about the on-card application. Consider a simple example of an on-card personal identification application. The XML document, which is sent out in response to a request may be as follows.

```
<Name: >Sunil< /Name: >
<Age>22< /Age>
<Homeplace>Vedureswaram< /Homeplace>
<State: Exp >Andhra Pradesh< /State: Exp >
<Country>India< /Country>
<Occupation>Student< /Occupation>
<SSN>123-45-6789< /SSN>
```

If suppose an external application wants to know just the name and social security number of a person, it can parse the whole XML document which will be transmitted by the XML Agent. Then it refers to the DTD specification of this data, understands the fields, and extracts the required information from it. In this example the application extracts the information between the tags <Name: >< /Name: > and <SSN>< /SSN>. The developer need not know about the data types of the values or the order in which they are arriving. He simply has to parse the XML document and extract the appropriate fields. This approach has an additional advantage. If the on-card application is changed by adding some more fields, or if some fields are removed (which no off-card application is using) then the already existing off-card applications need not be modified at all.

4.3 Disadvantages

It is obvious that the model demands more processing power on the card. It might be very costly to do the XML processing on currently available cards which have limited resources. So we decided not to validate the data against its DTD on the card. All the data which comes to the card is assumed to be valid data. The off-card application has to take care about this. Second disadvantage is that there is lot of movement of unnecessary data between OS, XML Agent and the applications. The verbose nature of XML makes this even more worse. So there should also be more memory on the card. But, we believe that these shortcomings can be overcomed with the increase in resources and processing power on the card.

5.   Implementation Alternatives

The basic model of XML Agent can be implemented in different ways. In this section we discuss the different approaches and their pros and cons. In the following discussion we will be using the terms 'Runtime Environment' and 'Card OS' interchangeably. The same discussion holds true in both the cases unless explicitly mentioned.

5.1   Along with Runtime Environment

The easiest and probably the most secure way is to integrate the XML Agent with the Runtime Environment and/or the OS. All the APDUs first come to the OS to be processed. So the OS (along with XML Agent) can do the job of translating the XML document to byte stream and send it to the appropriate on-card application. Since the OS has complete privileges over all the applications, the XML Agent can access any functions of the on-card application directly. It can trap the responses from the applications, prepare an XML document out of it and send it to the calling application. Since the XML Agent is integrated with the OS, the semantics of the on-card applications will not change. They will not be aware of the presence of the XML Agent. So there is no need to rebuild the existing on-card applications. The problem with this approach is that the developers are at the mercy of the card manufacturer to have an XML Agent. It also steals the opportunity from the developers to build their own Agents. The schematic representation is shown in the figure 2.

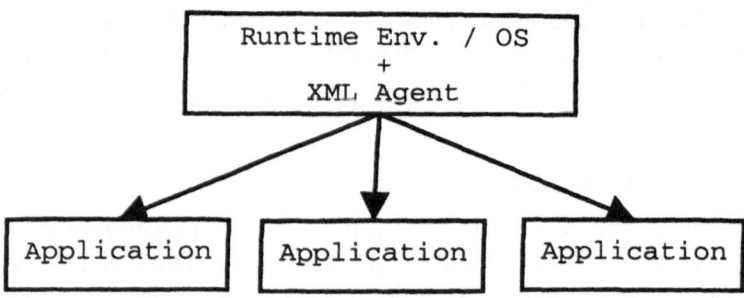

Figure 2.   XML Agent along with Runtime Environment

5.2   Between OS and Applications

In this method the OS passes the XML document to the XML Agent which will generate the byte stream and pass it to the appropriate application. There can be two possible approaches in this scheme. One in which the off-card

applications are aware of the XML agents, but no extra support from OS is needed to achieve this. Other method in which the off-card applications are ignorant of the presence of the XML agents, but some extra support is required from OS. The schematic representation is shown in the figure 3.

The first approach is where the off-card application is aware of the presence of the XML agent. The XML agent can be assigned a fixed/reserved Application Identifier (AID). The off-card applications can directly pass the XML document to XML agent along with the AID of the application to which the final data should be dispatched. The XML agent arranges the data accordingly and dispatches using delegation.

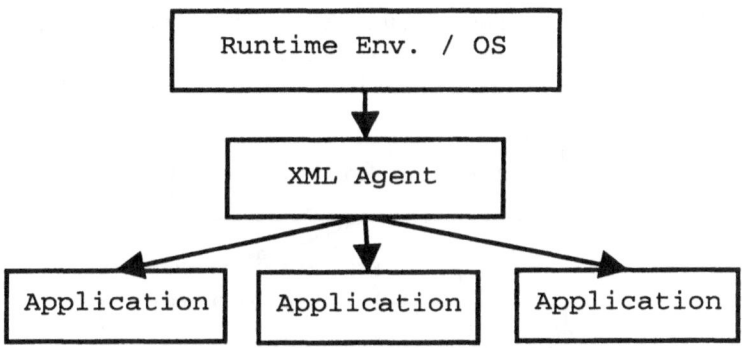

Figure 3.   XML Agent Between Runtime Environment and Applications

The second approach, where the off-card application is unaware of the presence of XML, requires some support from the OS. The OS should handover the XML document, sent by the off-card application, first to the XML agent. This should be done without the knowledge of the on-card and off-card applications. This is diversion from the ordinary flow of the data where the data is directly sent to the concerned application. The XML agent will then process the data and finally dispatch it to the destination application.

In both the methods the XML Agent is not privileged to access the on-card applications directly because it is just a standard application. So all the on-card application should be made shareable to it. Another alternative is to force that the XML agent should be given special privilege just to access the on-card applications directly. This is more sensible because other on-card applications should not be given a chance to misuse the fact that all applications are accessible by the XML Agent.

5.3   As a Standard Application

In this approach, the XML Agent has no special privileges. Here, instead of XML Agent invoking other applications, they will invoke the XML Agent and get their job done. The XML Agent behaves more or less like a library of functions. But the opportunity lies completely in the hands of the developers to implement their own specific XML Agent.

In this model, the applications will directly get the XML document passed by the OS. Instead of processing it, the applications should pass the data to the XML agent. The agent will arrange the data in required format and pass it back to the application. Then the application can proceed in the normal way of processing the commands and/or data. In this approach, there is lot of overhead in movement of data to and fro between the XML agents and the applications. So this is not a preferred way of implementing the XML Agent. The schematic representation is as shown in the figure 4.

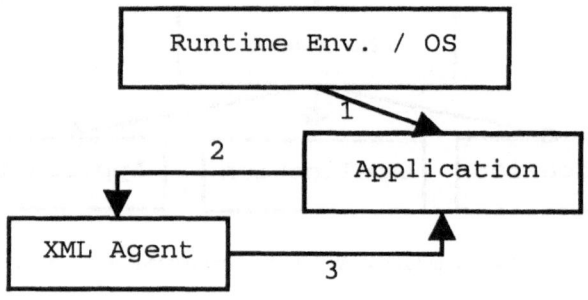

Figure 4.   XML Agent as a Standard Application

6.   Implementation Issues

In this section we will discuss some issues during implementation of XML Agent on two of the most popular Multi-Application Frameworks: Java Card [JavaCard] and MULTOS [MULTOS]. We will discuss the features which are supporting or hindering the implementation of XML Agent on them. We are not biased towards any of the frameworks. We view both of them just from a research perspective.

6.1   Java Card

Java Card technology adapts the Java platform for use on smart cards. It has a Java Card Runtime Environment(JCRE) which consists of Java Virtual

Machine(JVM), Java Card Framework APIs. As Java Card provides support for Java, we initially assumed that it will be most promising to implement our XML Agent on them. But later we found that the limited support (currently) provided by JVM on the card is not adequate to implement the XML Agent. The following features, which are important for our implementation are not supported [Chen00] on Java Card.

- Characters and Strings

- Dynamic Class Loading

- Object Serialization

XML is a text-based technology but Java Card currently has no support for characters and strings. It only support bytes, integer support is optional. So the characters should be encoded using some encoding scheme like Unicode or UTF. (UTF-8 has the advantage that it is fully compatible with the 7-bit ASCII encoding.) Processing the XML document will incur extra overhead of encoding and decoding. This overhead may be an overkill on the cards with limited processing power and resources.

Dynamic class loading is not supported in Java Cards. This feature allows the loading of classes, which share a common interface, even without any knowledge of their existence at the time of development of the main application. This is absolutely necessary in our model to support post issuance loading of card applets. There is no way in which the XML Agent can interact with the applets installed after it. So the XML Agent should be the last applet to be installed. And also it should hardcode the names of other classes which are already existing in the card. This is undesirable because there will be different applets on different cards. The size of the XML Agent will increase if we make it generic to handle the above situation.

Object Serialization will be helpful if present. In object serialization, rules can be defined to transform the object into a stream of bits. We can define our own rules to serialize an object. These serialization rules can be such that proper XML document is generated when the object is serialized. Then the on-card application has to send an object containing the response data to the XML Agent. The XML Agent can serialize the object, producing the XML document to be sent to the calling off-card application.

6.2 MULTOS

After encountering many limitations in Java Cards discussed in the previous section, we diverted our exploration towards MULTOS framework. We noticed that MULTOS has a very generic architecture to run programs. After analyzing the architecture we found that it can add more functionality to the XML Agent, than we had initially thought about. We have no doubt in saying that, currently MUTLOS does appear more appropriate than Java Card for implementing our idea. Two of the most promising features of MULTOS that help in the implementation are:

- Shell Application

- Delegation Mechanism

The shell application[MShell99] is implicitly selected when a MULTOS card is powered up. The shell application receives all the commands sent to the card, except the commands sent to the MULTOS Security Manager(MSM). The rationale behind this is to support non-ISO interface devices and commands which need to communicate with the cards (which are ISO standard). We can safely exploit this facility provided by MULTOS cards. We can load the XML Agent as a shell application. Since the commands first go to the shell application, we can send actual commands enclosed in XML document to the shell application directly. The XML Agent which is running as a shell application will unwrap the data and invoke the appropriate application and pass it the commands and data using delegation. It can do the processing in reverse direction as well. It can encapsulate the data returned by the application in an XML document and send it back to the off-card application.

The memory architecture and the delegation mechanism [MAPRM97] in MULTOS will be helpful in increasing the efficiency of the XML Agent. In MULTOS, some part of the memory is reserved as "public" area. All the applications can access this data space. During delegation, the delegator application creates a command APDU in public area and activates the delegate application. The delegate application will process the APDU in public area and then return the response in the public area itself. The delegator application will then examine the response set by the delegate application. Since there is no physical copy or movement of data between the applications, the delegation will be faster. In case of XML Agent, if there is no movement of XML documents between XML Agent and the applications, the performance of the system will definitely be better.

7.   Our Implementation

As we do not have the required resources to modify the Runtime Environment and OS of the card, we couldn't implement the first model where the XML Agent is integrated into the OS of the card. So, as a proof-of-concept we have implemented the second model where the XML Agent sits between

OS and the applications. The implemented model is one where the off-card application is aware of the presence of XML Agent on the card.

As the XML parser of the XML Agent is the heart of the problem, it should be as efficient as possible. We considered different parsing techniques[XML-Parsers] before choosing one. *Tree model* parsers, like DOM parser, are very expensive in this scenario because they will build a complete tree out of the whole XML Document. We did not find the *Push model* parsers, like SAX parsers, very useful because of the cumbersome callback mechanism. After exploring other alternatives, we found *Pull Model* parsers best suited for our requirements. We adopted the API[PullParser], which is actually developed for the Java language, for our implementation in C language. We implemented a simplified parser which can parse basic XML syntax. To decrease the parsing overhead, we are assuming that all strings are ascii encoded. In order to be able to load on currently available commonly used cards we didn't include the support for costly(in terms of processing) features like namespaces, references etc. Finally the size of the XML agent (ALU file for MULTOS) along with the parser came around 5KB. The XML parser alone came to a size of around 3KB.

We developed a very simple application which holds a value. It has only three functions *query(), incr()* and *decr()* to query, increment and decrement the value respectively. The APDU prefixes of these functions are stored in an array datastructure with the XML Agent. As the XML Agent is aware of this application, it can send APDUs accordingly. The XML document is sent as data in the APDU to the XML Agent. The XML command to increment the stored value by some amount is as follows:

<xml><appl><incr val="10"/></appl></xml>

The latest value is sent in response to this command. The XML document returned by the XML Agent is as follows.

<xml><appl><resp success="1" val="110"/></appl></xml>

8. Conclusion

The traditional approach of developing on-card and off-card portion of an application as a single cohesive bundle does not permit them to be made interoperable easily. There is a great need to have an interoperable framework which would be independent, not only of the differences in cards and terminals, but also of the data structure details of the on-card and the off-card application components.

In this paper, we proposed a novel framework to achieve this interoperability with the use of and XML Agent on the card. The XML Agent which runs as an engine on the card, transforms the incoming XML document to byte stream and send it to the on-card applications. Similarly, it will prepare an XML document encapsulating the response from the on-card application and send it back to the calling off-card application. This model also provides a powerful abstraction for the developers. It hides many details about the underlying architecture of the card and the terminal. Therefore the developers need to have only a limited information about the on-card application, to develop an off-card application.

The downside of the XML Agent is that it requires more processing power on the card. As the resources and processing power of the cards are increasing continuously, this may not be limitation for the cards in the future. XML Agent is a desirable thing because its advantages outweigh the disadvantages.

9.   Future Work

In this paper we have not explored security related concerns in depth. There may be some security compromises as a result of our on-card data sharing model. This is because the XML Agent is assumed to have the privilege of accessing any on-card applications directly. We will be exploring these security concerns in our future work.

References

[Chen00]   Zhiqun Chen. (2000). *Java Card Technology for Smart Cards: Architecture and Programmer's Guide*. The Java Series. Addison Wesley.

[DCSF]   Wangjammers. Dynamic Card Service Factory Website :
              http://www.wangjammers.com/smartcards/dcsf.html

[JavaCard]   Java Card Technology Homepage : http://java.sun.com/products/javacard/

[MAPRM97]   (1997). *MULTOS Application Programmers Reference Manual v1.0*. MULTOS. MAOSCO Limited.

[MDev00]   (1997). *MULTOS Developers Guide v1.3*. MULTOS. MAOSCO Limited.

[MShell99]   (1999). *Shell Applications, MULTOS Technical Bulletin* MULTOS. MAOSCO Limited.

[MULTOS]   MULTOS Homepage : http://www.multos.com

[Ocfgim98]   (1998). *General Information Web Document*. OpenCard Framework. OpenCard Consortium.

[PullParser] Common API for XML Pull Parsing :
              http://www.xmlpull.org/v1/doc/api/org/xmlpull/v1/package-summary.html

[SmartX00]   (2000). *SmartX User Guide v1.2*. Think Pulse.

[Xavier99]   Xavier Lorphelin. (1999). *Internet and Smart Card Application Deployment*. JSource, USA.

[XMLParsers] A Survey of APIs and Techniques for Processing XML :
              http://www.xml.com/pub/a/2003/07/09/xmlapis.html

Erratum to: Smart Card Research and Advanced Applications VI

Jean-Jacques Quisquater, Pierre Paradinas, Yves Deswarte and Anas Abou El Kalam (eds.)

This book was originally published with a copyright holder in the name of the publisher in error, whereas IFIP International Federation for Information Processing holds the copyright.

The updated original online version for this book can be found at
DOI 10.1007/978-1-4020-8147-7

J-J. Quisquater, et al. (eds.), *Smart Card Research and Advanced Applications VI,*
DOI 10.1007/978-1-4020-8147-7_21, © IFIP International Federation for Information Processing, 2017 E1

INDEX OF CONTRIBUTORS